Sensory Processing in the Mammalian Brain

SENSORY PROCESSING
IN THE MAMMALIAN BRAIN

Neural Substrates
and Experimental Strategies

Edited by
JENNIFER S. LUND
Center for Neuroscience
University of Pittsburgh

New York   Oxford
OXFORD UNIVERSITY PRESS
1989

Oxford University Press

Oxford   New York   Toronto
Delhi   Bombay   Calcutta   Madras   Karachi
Petaling   Jaya   Singapore   Hong King   Toyko
Nairobi   Dar es Salaam   Cape Town
Melbourne   Auckland

and associated companies in
Berlin   Ibadan

Library of Congress Cataloging-in-Publication Data
Sensory processing in the mammalian brain: neural substrates and
experimental strategies/edited by Jennifer S. Lund.
p.   cm.
Based on a conference held in the fall of 1986 at the University of
Pittsburgh in honor of Gerhard Werner and sponsored by the Center for Neuroscience.
Includes bibliographies and index.
ISBN 0-19-504554-8
1. Senses and sensation—Congresses.   2. Cerebral cortex—Congresses.
3. Afferent pathways—Congresses.   I. Lund, Jennifer S.
II. Werner, Gerhard, 1921–   . III. University of Pitttsburgh.
Center for Neuroscience.
[DNLM: 1. Brain—physiology—congresses.   2. Receptors, Sensory—
physiology—congresses.   3. Sensation—physiology—congresses.
WL 702 S478 1986]
QP431.S454 1989
612'.8—dc19
DNLM/DLC
for Library of Congress   88-5374   CIP

9 8 7 6 5 4 3 2 1

Printed in the United States of America
on acid-free paper

Foreword

This volume of essays, sponsored by the Center for Neuroscience at the University of Pittsburgh (CNUP), is compiled to honor Dr. Gerhard Werner, Professor of Psychiatry and Pharmacology at the University of Pittsburgh in recognition of his considerable research contributions to the field of sensory processing and his contribution to the development of neuroscience training and research at the University of Pittsburgh since 1965. The book had its origin in a conference held in the fall of 1986 at the University of Pittsburgh on the occasion of Dr. Werner's 65th birthday.

Gerhard Werner was born in 1921 in Vienna. He received his M.D. at the University of Vienna School of Medicine in 1945. His research was orginally in pharmacology, including the cardiovascular and renal systems. His early concentration on neuromuscular pharmacology, made notable contributions that included the identification of succinylcholine as a clinically significant innovation in anesthesiology. Later, with W. F. Riker, he demonstrated the cholinergic nature of the presynaptic nerve terminals at the neuromuscular junction. This work on the pharmacology of the central nervous system spurred his investigations into the effects of Rauwolfia alkaloids and the first neuropharmacological and behavioral studies using chlorpromazine, contributing to the demonstration of its clinical significance in psychopharmacology.

Dr. Werner's research in sensory processing began when he joined the Department of Physiology at John Hopkins in 1961. Here, initially under the tutelage of Vernon Mountcastle, he learned the techniques of CNS physiology and with Mountcastle and later Barry Whitsel conducted a series of outstanding studies on the representation of somesthetic sensation in thalamus and cortex. Dr. Werner's work has particularly emphasized the development of quantitative methods of data analysis with a view toward correlating psychophysics and neurophysiology. Most recently, he has been collaborating with Dr. Herbert Reitboeck at Philipps-Universität Marburg in quantitative work developing methods for recording and interpreting patterns of activity in clusters of cortical neurons.

Preface

This volume of essays on sensory processing in the mammalian nervous system came about as a result of a conference held in the fall of 1986* at the University of Pittsburgh, in honor of Dr. Gerhard Werner. The participants in this conference came together to discuss the views that neuroscientists interested in sensory processing are currently entertaining and the experimental approaches they are using and developing in their laboratories. The conference made clear the tremendous vitality and interest in this area of "systems" neuroscience research despite the emphasis today in neuroscience as a whole on molecular and cellular events. It was evident that over the years, sufficient data have accumulated to allow comparison of general organizational features between neural systems concerned with different senses. This comparison reveals considerable commonality of neural organization, so that the ideas being generated in research on any one sensory system are of considerable interest and applicability to the interpretation of events in the others.

The essays in this volume are not transcripts of the lectures delivered at the Pittsburgh conference; neither are they strictly research reports. Rather, they are intended to convey the essence of each author's particular field of research. The book's purpose is to provide examples of the types of experimental and conceptual approaches that are currently† being used in research on the neural substrates of sensation. By no means does it provide a complete picture of any one sensory system of the brain—for that a much wider reading is required. Of more importance to many readers, however, it gives illustrations of the themes common to different sensory systems, possible modes of experimental analysis, the importance of understanding the various components (from receptor organ to cerebral cortex) of any system, the difficulties of analyzing complex neuropils, and the experimental and conceptual ingenuity that researchers are currently applying to these studies.

The authors are distinguished investigators in their fields of sensory research, and each chapter offers readers an individual perspective. Their reference lists hold the clues to bodies of work, including their own, that should provide ample background to the history of the experimental investigation of each area. The essays may stimulate readers not familiar with sensory systems neuroscience to read the literature as it appears in the journals, if only to find out what the latest excitement is about. For molecular and cellular neuroscientists, the essays may act as a spur to relate their work to the development and function of these systems;

---

*The conference was entitled "Organizing Principles of Sensory Processing: The Brain's View of the World."
†These essays were completed in the Fall of 1987.

our increasing knowledge about the mature sensory systems certainly makes such endeavors promising. For sensory theorists the essays may encourage the use of real biological data and closer collaboration with neuroanatomists and neurophysiologists. Theoretical modelers have skills that vital to neuroscientists; the field urgently needs individuals who are trained in the biological and mathematical sciences and are prepared to combine these approaches. For psychologists, the essays may renew their perspective on investigations examining the substrates of brain function. It may spur those who have lost touch with current directions in research on the biology of sensory systems to renew their acquaintance with colleagues who battle with the anatomy and physiology of sensation and often need the input of those studying behaviors that reflect the activity and limitations of the sensory system. As we will see, those who study animal ecology can provide vital parts of the puzzle.

Most important of all, these essays provide a summary designed to intrigue those students still undecided on what direction their research in neuroscience will take. The wealth of different experimental approaches available today provide tools for analysis far richer than those of earlier years. The rewards of such research are considerable in terms of understanding brain function and the possibilities for rational treatment of neurological disorders.

Pittsburgh, Pa.   J.S.L.
March 1988

Contents

Contributors

Gary G. Blasdel   Department of Physiology, University of Calgary, Health Science Center, Calgary, Alberta, Canada T2N 4N1

James C. Boudreau   Sensory Science Center, University of Texas at Houston, Houston, Texas 77030

George E. Carvell   Department of Physical Therapy, School of Health Related Professions, University of Pittsburgh, Pittsburgh, Pennsylvania 15261

Lawrence J. Cauller   Neurobiology Department, Northeastern Ohio Universities College of Medicine, Rootstown, Ohio 44272

Mathew E. Diamond   Departments of Physiology and Mathematics, University of North Carolina School of Medicine, Chapel Hill, North Carolina 27514

Oleg V. Favorov   Departments of Physiology and Mathematics, University of North Carolina School of Medicine, Chapel Hill, North Carolina 27514

Esther P. Gardner   Departments of Physiology and Biophysics, New York University School of Medicine, New York, New York 10016

Heikki A. Hamalainen   Department of Psychology, University of Helsinki, Ritarikatu 5, SF-00170, Helsinki 17, Finland

Don H. Johnson   Department of Electrical and Computer Engineering, Rice University, Houston, Texas 77251-1892

Sharon L. Juliano   Department of Anatomy, Uniformed Services University of the Health Sciences, Bethesda, Maryland 20814

Douglas G. Kelly   Departments of Physiology and Mathematics, University of North Carolina School of Medicine, Chapel Hill, North Carolina 27514

Albert T. Kulics   Neurobiology Department, Northeastern Ohio Universities College of Medicine, Rootstown, Ohio 44272

Peter W. Land   Department of Neurobiology, Anatomy and Cell Science, University of Pittsburgh School of Medicine, Pittsburgh, Pennsylvania 15261

Sidney R. Lehky   Department of Biophysics, Johns Hopkins University, Baltimore, Maryland 21218

Michael Leon   Department of Psychobiology, University of California at Irvine, Irvine, California 92717

Jennifer S. Lund   Department of Psychiatry, University of Pittsburgh School of Medicine, Pittsburgh, Pennsylvania 15261

Raymond D. Lund   Department of Neurobiology, Anatomy and Cell Science, University of Pittsburgh School of Medicine, Pittsburgh, Pennsylvania 15261

Claude I. Palmer   Departments of Physiology and Biophysics, New York University School of Medicine, New York, New York 10016

Dennis P. Phillips Department of Psychology, Dalhousie University, Halifax, Nova Scotia, Canada B3H 4J1

Gian F. Poggio Bard Laboratories of Neurophysiology, Department of Neuroscience, Johns Hopkins University School of Medicine, Baltimore, Maryland 21205

Herbert J. Reitboeck Applied Physics and Biophysics, Philipps University Marburg, D-3550 Marburg, Federal Republic of Germany

Terrence J. Sejnowski Department of Biophysics, Johns Hopkins University, Baltimore, Maryland 21218

Daniel J. Simons Department of Physiology, University of Pittsburgh School of Medicine, Pittsburgh, Pennsylvania 15261

David L. Sparks Neurobiology Research Center, Department of Physiology and Biophysics, University of Alabama at Birmingham, Birmingham, Alabama 35294

David L. Tomko Vestibular Research Facility, Neuroscience Branch, Life Science Division, NASA Ames Research Center, Moffett Field, California 94035

Mark Tommerdahl Departments of Physiology and Mathematics, University of North Carolina School of Medicine, Chapel Hill, North Carolina 27514

Chiyeko Tsuchitani Sensory Science Center, Graduate School of Biomedical Sciences, The University of Texas Health Science Center at Houston, Houston, Texas 77030

Susan Warren Department of Anatomy, University of Mississippi Medical Center, Jackson, Mississippi 39216-4505

Gerhard Werner Department of Psychiatry, University of Pittsburgh School of Medicine, Pittsburgh, Pennsylvania 15213

Barry L. Whitsel Departments of Physiology and Mathematics, University of North Carolina School of Medicine, Chapel Hill, North Carolina 27514

William D. Willis, Jr. Marine Biomedical Institute and Department of Anatomy and Neurosciences, University of Texas Medical Branch, Galveston, Texas 77550:2772

Donald A. Wilson Department of Psychology, University of Oklahoma, Norman, Oklahoma 73019

Sensory Processing in the Mammalian Brain

I

SMELL AND TASTE

Smell and taste might be thought to represent the earliest sensory systems to evolve. Certainly those areas of the brain concerned with sensory input from the respiratory surfaces and taste organs are well developed in the most primitive of vertebrates; indeed, even single-celled organisms find these "chemical" senses invaluable! A system's early evolution, however, must not be equated with a primitive or simple nature. As we will see, both of these senses in the mammal show considerable sophistication. A most important theme from Chapters 1 and 2 continues throughout the volume: sensory receptors are specialized in each species, to match their life-style and normal food. Until the investigator understands the specialization and limitations of the particular sensory receptors, he or she can make little headway toward understanding the more central processing of the information these receptors provided in the brain.

Understanding the nature of the stimulus to which the receptor responds and the mode or coding of the response in the receptor is a significant problem. Knowledge of the animal's behavior and diet, its rearing and reproductive cycle, its sex and predators, can provide invaluable clues as the scientist struggles to find what alerts a sensory system. One issue these studies raise, which is beyond the scope of this book, is the question of how the receptors become specialized and matched to the animal's ecology. Is the match bred over generations into a perfectly assembled genetic code? Or is it determined for each individual by early experience and rearing condition? Probably, as in all things neural, it is determined by a mixture of both factors. The interested reader might enjoy the review by Pederson and colleagues (1985) of juvenile "imprinting" of the olfactory system, in which even in utero exposure to chemical substances determines later suckling preferences.

Both chapters in this section stress the difficulty experimenters have in controlling the stimulus presentation for smell and taste, compared, for instance, with the ease of preventing visual stimulation or controlling the quality of the visual image and where it falls on the retina. Students may be forgiven if they believe this issue is trivial compared with the difficulties of understanding the neural events, but control of stimulus presentation and quality is often the deciding factor when determining whether or not an experiment can be run.

The topographic distribution of different types of olfactory and gustatory receptors on the sensory surface, as described in these two chapters, is of considerable importance, because peripheral receptor topography is repeated in many structures of the brain receiving sensory relays. In the other sensory systems, in which the central relays have been more thoroughly studied, we see that the topography of sensory maps and their various transformations in geometry—particularly at the cortical level—can be useful in guiding analysis of the neuropil. It is not the literal map of the sensory surface that intrigues today's investigators, but rather the possibility that topography at a cortical level reflects or determines the means by which the brain interrelates and collates different types of information. Topographic mapping may assume new forms, going beyond sensory receptor layout. By means of highly geometric patterns of connectivity, these neural maps serve as a substrate for sophisticated analysis of the outside world, generating response properties that code stimulus features not distinguished by the receptors themselves.

Another feature of the olfactory and taste systems shared by other sensory modalities is the specialization of receptors so that different populations of receptors respond to different qualities in the sensory input. These receptors then feed their information into separate "parallel channels" to central neural targets. A considerable degree of neural processing, involving populations of local circuit neurons and information transfer between channels, occurs within each of the relay stations along the sensory pathways. This means that the investigator must carefully analyze each synaptic region along the relay in sequence, to compare the incoming messages with what is sent on to the next station in line.

Another clearly apparent feature of the olfactory system is the presence of centrifugal projections. The role of such feedback loops, which exist at almost every level of the sensory pathways, is debated continually, without much resolution as yet. Designing clear-cut experimental approaches to analyze these relays' function is a real problem because of the frequent difficulty in understanding the temporal relevance of the feedback process. The function of these pathways has been given every shade of significance from merely a means of conveying "your message has been received" to a mechanism essential to learning and memory.

Of wide interest is the use of 2-deoxyglucose as a metabolic marker for neural activity, a technique discussed in Chapter 1. As used to investigate neural systems (see Kennedy et al., 1976), the technique procedes as follows. An experimental animal, either alert or anesthetized, is injected with a radioactively labeled analog of glucose, 14C-2-deoxyglucose (2-DG). The animal is then immediately stimulated through one of its sensory systems, using particular stimulus paradigms, and the labeled sugar analog is rapidly taken up by metabolically active body tissues, including the brain. But, because the sugar analog cannot complete the normal metabolic cycle in the cells, it piles up to a degree that reflects the cell's rate of metabolic activity. When the brain from such animals is examined and compared with those of injected but nonstimulated control animals, elaborate patterns of high 2-DG uptake are often seen in regions of the brain served by the stimulated sensory system. The interpretation of these patterns of high 2-DG uptake is not simple, for they do not always relate clearly to known anatomic pathways or to the distribution of physiologically identified neural properties associated with the stimulus condition, although they can be extremely clear-cut and geometrically distributed. Readers are encouraged to form their own opinions as to the possible interpretation of labeling patterns including those described in the following pages. One real problem is that the animal must be sacrificed to see the pattern of label, which makes it difficult to determine whether 2-DG patterns shift their position in the neuropil with a change in stimulus. Determining whether a change in overall pattern configuration between stimulus conditions occurs is easier, but readers will find lengthy discussion in these chapters of what these patterns may mean. Chapter 12 describes an alternative method of mapping these shifting patterns in the live, anesthetized animal.

REFERENCES

Kennedy, C., DesRosiers, M. H., Sakwada, O., Shinohara, M., Reivich, M., Jehle, H. W., and Sokoloff, L. (1976). Metabolic mapping of the primary visual system of the monkey by means of the autoradiographic [14C] deoxyglucose technique. *Proc. Natl. Acad. Sci. (Wash.)* 73:4230–4234.

Pederson, P. E., Greer, C. A., and Shepherd, G. M. (1985). Early development of olfactory function. In E. M. Blass. (Ed.) *Handbook of Behavioral Neurobiology*, pp. 163–203. New York and London: Plenum.

1

Information Processing in the Olfactory System

DONALD A. WILSON AND MICHAEL LEON

The olfactory system has remained something of a mystery to neurobiologists, perhaps because we are a microsmatic species with a relatively low awareness of the sense. Understanding the means by which the olfactory system encodes and decodes information is not an easy task, given the lack of a clear physical dimension to characterize and control stimulus presentation (Engen, 1982; Lancet, 1986). Other aspects of the olfactory system, however, enhance its interest for the study of sensory processes and information processing. Phylogenetically, it is among the oldest sensory systems and encompasses the archicortex. Later-developing sensory systems may have retained some of the mechanisms originally evolved to deal with chemoreception, so that an understanding of olfactory system processes could shed light on processing in other systems. The olfactory bulb, the first central stage of olfactory processing, is a highly laminated structure that receives a single afferent input. The primary bulb output neurons lie in a tight band and project deep to the bulb surface, by means of the large lateral olfactory tract. The structure of the olfactory bulb thus allows easy identification of afferent and efferent fibers, as well as output and interneuron cell types for neurophysiologic or neuroanatomic analysis. Much research has been done in recent years on the coding of olfactory information and in this chapter we discuss the progress that has been made.

OVERVIEW OF OLFACTORY SYSTEM NEUROANATOMY

Olfactory stimuli are drawn across the olfactory receptor sheet during respiration. The olfactory receptor sheet, or olfactory epithelium, is located caudally in the airway and is highly convoluted in mammals. The receptive surface area of the epithelium varies greatly from species to species, from 100 cm^2 in dogs to only a few square centimeters in humans. The olfactory receptors themselves are bipolar neurons with short dendrites extending to the surface of the epithelium, ending in 10 to 20 modified cilia bathed in mucus. The cilia contain the olfactory receptor sites and are the site of olfactory transduction (Adamek et al., 1984;

Lancet, 1986). The nonmyelinated axons of receptor cells leave the epithelium and travel in bundles which eventually combine to form the olfactory nerve. The olfactory nerve passes through the cribiform plate, and then fans out over the surface of the ipsilateral olfactory bulb. In the bulb, olfactory nerve axons enter small clusters of neuropil called glomeruli, where the axons branch profusely and terminate on the dendrites of both primary output neurons and interneurons. The projection from the olfactory epithelium to the glomerular layer of the main olfactory bulb is topographic, with for example, lateral regions of the epithelium projecting to lateral regions of the bulb and medial regions projecting to its medial regions (Costanzo and O'Connell, 1978; Fujita et al., 1985; Jastreboff et al., 1984; Land, 1973; LeGros Clark, 1951; Schwob and Gottlieb, 1986).

The mammalian olfactory bulb is organized in discrete layers, with receptor input located superficially and centrifugal inputs and bulb output located deep to the surface of the bulb. For detailed descriptions of olfactory bulb neuroanatomy, see Shepherd (1972) and Macrides and Davis (1983) (Figure 1-1). Immediately deep to the olfactory nerve layer is the glomerular layer, which contains glomeruli and glomerular-layer neuron cell bodies. The glomeruli are spherical regions of dense synaptic contact between olfactory receptor axons and olfactory bulb neurons. In the glomeruli, the olfactory nerve forms excitatory synapses on both the dendrites of glomerular layer neurons and the distal dendrites of the primary output neurons of the olfactory bulb, the mitral and tufted cells. In the rabbit and rat, each mitral cell receives input from a single glomerulus. About 1000 receptor cells converge on a single mitral cell, with approximately 25 mitral cells per glomerulus (Shepherd, 1972). Mitral cell bodies are arranged in a narrow layer deep to the surface of the bulb.

Excitability of the bulbar output neurons is controlled primarily through the inhibitory interneurons (Jahr and Nicoll, 1982; Mori and Takagi, 1978), periglomerular and granule cells, which form dendrodendritic contacts with both mitral and tufted cells (Price and Powell, 1970b). In addition, the dense centrifugal and commissural input the bulb receives acts primarily through these inhibitory interneurons (Nakashima et al., 1978; Price and Powell, 1970a). The olfactory bulb receives massive centrifugal inputs from the locus coeruleus (noradrenergic), median raphe nucleus (serotonergic), and medial septum/diagonal band complex (cholinergic) as well as direct or indirect inputs from all brain regions to which the olfactory bulb projects (reviewed in Macrides and Davis, 1983).

Both mitral and tufted cells project out of the bulb; however, their target sites differ, leading to the concept of parallel processing of olfactory information (Macrides et al., 1985). Mitral cells project primarily to the ipsilateral olfactory and lateral entorhinal cortices, but also to the anterior olfactory nucleus, the nucleus of the lateral olfactory tract, the hypothalamus, and the septum/diagonal band areas (Haberly and Price, 1977; Macrides et al., 1985). Figure 1-2 displays some of the major connections of the olfactory system. Tufted cells can be divided into several subcategories, based on the location of their somata in the olfactory bulb and their axonal projection pattern. Internal tufted cells, with somata just superficial to the mitral cell body layer, have projection patterns very similar to those of mitral cells. Both middle tufted cells and the majority of external tufted cells have somata located in the superficial external plexiform layer as well as in glo-

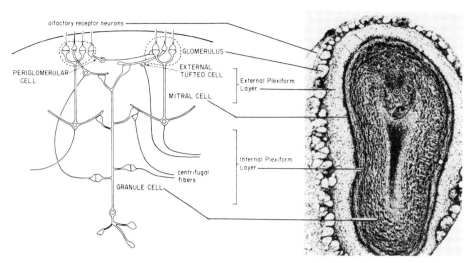

Figure 1-1 Diagram of mammalian olfactory bulb circuitry and coronal section through the bulb at the level of the anterior accessory olfactory bulb (AOB), stained for cell bodies. The olfactory bulb is the first central structure for olfactory information processing. See the text for a description of cell types and synaptic interactions.

merular layer, and they project to the olfactory tubercle, anterior olfactory nucleus, and anterior olfactory cortex (Haberly and Price, 1977; Scott, 1981). Additional external tufted cells form intra- and interbulbar connections (Schoenfeld et al., 1985). These associational connections are topographic both within and between the bulbs (Schoenfeld et al., 1985). External tufted cells forming the commissural pathway terminate on relay neurons in the ipsilateral pars externa of the anterior olfactory nucleus, which then project to the contralateral bulb. The differences in cell body location and projection patterns of bulb output neurons

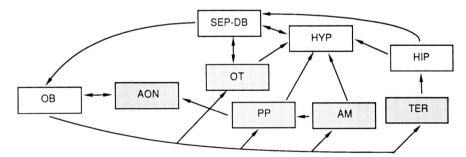

Figure 1-2 Primary pathways of olfactory information processing in the central nervous system. Shaded boxes receive monosynaptic input from the olfactory bulb. Note that all the structures that receive input from the olfactory bulb (OB) directly or indirectly send centrifugal input back to the bulb. Other structures that receive bulb output or send input to the bulb are discussed in the text. Abbreviations: OB, olfactory bulb; AON, anterior olfactory nucleus; OT, olfactory tubercle; PP, pyriform cortex; AM, amygdaloid complex; TER, transitional entorhinal cortex; SEP-DB, septum and diagonal band; HYP, hypothalamus; HIP, hippocampus.

permits accurate electrophysiological identification of cell types during in vivo recording by electrode location and response to antidromic stimulation. Comparing the response patterns of different cell types helps us understand the olfactory coding processes (see following).

The rich synaptic network, abundance of interneurons and dense centrifugal inputs to the olfactory bulb suggest that extensive processing of olfactory information occurs within the bulb itself. Because of the extensive projections of the olfactory bulb, this information has direct access to higher order processing centers, such as the hippocampus (Wilson and Stewart, 1978) and frontal cortex (Tanabe et al., 1975; Yarita et al., 1980).

At least two additional pathways for olfactory stimuli access to the central nervous system (CNS) exist: the trigeminal nerve and the vomeronasal system. These systems will not be dealt with in this review (for a recent review see Keverne et al., 1986).

NEURAL CODING OF OLFACTORY STIMULI

Problems of Analysis

Olfactory information processing involves three major problems of interpreting neurophysiological data: (1) stimulus selection (2) stimulus control, and (3) response definition. The first problem involves selecting odorants themselves. Most odors encountered in the natural environment are mixtures of several odors, and defining primary odors has proven to be difficult (Amoore, 1970). In nonmammalian olfactory systems, however, responses to odor mixtures have been found not to be simple linear summations of responses to the components of those mixtures (Derby and Ache, 1984). Thus, analysis of responses to odor mixtures can lead to a different set of response characteristics from those obtained by use of a simpler set of pure odors. The number, type, and concentration of odor stimuli used in descriptive olfactory physiology vary a great deal from laboratory to laboratory. Single-molecule stimuli such as amyl acetate, butanol, and camphor, however, are commonly used in experimental protocols, thus allowing some direct comparisons to be made between studies.

The second major problem in olfactory neurobiology is controlling the odor stimulus. Odor molecules diffuse away from the odorant with time, forming a concentration gradient over the distance from the source. These odor molecules are then drawn across the receptor sheet at varying rates and concentrations, in the respiring animal. Thus, in much the same way that saccadic eye movements change the intensity of light that hits a particular photoreceptor, the continual cycle of inhalation and exhalation constantly varies the stimulus concentration at any one point in the olfactory epithelium. Controlling the exact onset and termination of the stimulus, and the concentration during that stimulus, therefore, is much more difficult than simply turning a valve on and off (Meredith, 1986; Kauer, 1974). This problem is critical to interpreting responses to odors. The flow of air through the convoluted olfactory epithelium and the time required for odor

molecules to travel through the mucus layer overlying the receptors also compli-
cate the control of stimulus duration and concentration.

One method of dealing with this problem is to draw air across the epithelium
artificially with a vacuum, thus allowing a constant stimulus to be presented over
a long period of time (Meredith, 1986). Alternatively, the vacuum can be applied
in short, controlled pulses, producing an "artificial sniff" (Harrison and Scott,
1986; Macrides and Chorover, 1972; Mair, 1982). Another method is to open the
nasal passage and apply the stimulus directly to a small, defined region of the
receptor sheet (Kauer and Moulton, 1974; Mair and Gesteland, 1982). A fourth
method is to monitor respiration and analyze olfactory responses in relation to the
respiratory phase (Chaput and Holley, 1985; Pager, 1985).

The third major problem in the study of olfactory system responses is defining
the response to the stimulus. Has a response occurred and what are its features?
For example, olfactory receptor cells primarily demonstrate simple excitatory responses monotonically related to stimulus intensity (Lancet, 1986; Trotier and
MacLeod, 1983; Sicard, 1986), but responses of olfactory bulb mitral and tufted
cells can be complex mixtures of excitatory and suppressive components. These
responses are further influenced by the duration of the stimulus, the respiration
cycle, and the odor concentration, which complicates the analysis of how the
response to a stimulus relates to the information in that stimulus. As we discuss,
in the following, however, much progress has been made in interpreting these
responses and in understanding how they may code olfactory information.

Olfactory System Response Characteristics

Neurophysiological analysis of olfactory coding has dealt primarily with the response properties of olfactory receptor cells, as well as of mitral and tufted cells.
In contrast, very little work has been done on the response properties of neurons
at higher levels of the olfactory system.

Olfactory Receptors

The response of olfactory receptors to odors has been examined by recordings of
the compound action potential in the olfactory nerve (Adrian 1950; Mozell, 1970),
recordings of the summated olfactory receptor generator potentials through elec-
troolfactography (EOG) (Doving, 1964; Mozell, 1964), and extra- and intracellular recordings from individual olfactory receptor cells (Getchell, 1977; Getchell
and Sheperd, 1978; Sicard, 1986; Mathews, 1972a; Trotier and MacLeod, 1983).
Single-unit and intracellular recordings demonstrate that the vast majority of receptors are excited by odor stimulation (Getchell, 1977; Getchell and Shepherd,
1978; Sicard and Holley, 1984; Trotier and MacLeod, 1983; Mathews, 1972a).
The magnitude and latency of excitatory responses are monotonically related to
stimulus intensity—the greater the odor concentration, the greater the increase
in firing rate and the shorter the latency to respond (Sicard and Holley, 1984;
Mathews, 1972a). However, some olfactory receptor cells are spontaneously active, and occasionally suppressive responses are observed (Sicard and Holley,
1984).

Individual receptor cells respond to many odors, and a single odor stimulates many receptors (Gesteland et al., 1982; Mathews, 1972a; Sicard and Holley, 1984). The response specificity of olfactory receptor cells therefore appears to be low (although see Sicard, 1986). The collection of odors to which a given receptor cell is sensitive, however, distinguishes it from other cells. It is not known whether the broad sensitivity of receptor cells is due to multiple receptor molecule or binding site types on each individual cell or to the broad sensitivity of the odor receptor molecules themselves. Odor sensitivity is also topographically distributed across the epithelium, and not all regions of the receptor sheet are equally sensitive to all odors (MacKay-Sim et al., 1982; Moulton, 1976; Thommesen and Doving, 1977). Another factor that may influence receptor responsiveness is the location of the receptor cell in relation to the direction of airflow over the mucosa (Mozell, 1964; 1970). Different odor molecules are adsorbed through the mucosa at different rates; thus, the mucus layer overlying the receptor sheet can act as a chromatograph (Mozell, 1964; 1970). Molecules become separated along the receptor sheet according to their solubility in the mucus, and thus the concentration of odor molecules reaching the receptors varies across the receptor sheet. The multiple receptor and the chromatographic mechanisms are both probably involved in coding olfactory quality at the olfactory receptor sheet.

Olfactory Bulb Neurons

Most electrophysiological analyses of the olfactory bulb response to odors have concentrated on the mitral/internal tufted cells that project from the bulb. Mitral/tufted cells are easily isolated and identified for extra- or intracellular recording, and they exhibit robust responses to odors. However, these second-order neurons respond to odor stimulation in much more complex patterns than olfactory receptor cells. Although olfactory receptor neuron input to mitral/tufted cells is excitatory (Nicoll, 1972), spontaneously active mitral/tufted cells can respond to odors with excited or suppressed phases, or a mixture of excitatory and suppressive phases (Chaput and Holley, 1985; Kauer, 1974; Mair, 1982; Mathews, 1972b; Meredith, 1986). Furthermore, a mitral/tufted cell that demonstrates an excitatory response to an odor at one concentration may be suppressed with the same odor at another concentration (Chaput and Lankheet, 1987; Kauer, 1974; Mair, 1982; Meredith, 1986). Most evidence suggests that these suppressive responses result from lateral, feedback, or feedfoward inhibition. Periods of suppressed spontaneous activity recorded extracellularly in response to odor stimulation correspond to intracellularly recorded hyperpolarization (Hamilton and Kauer, 1985).

The complex relationship between mitral/tufted cell response patterns and stimulus intensity may tell us a great deal about the circuitry underlying olfactory coding in the olfactory bulb. It has been hypothesized (Kauer, 1974; Kauer and Moulton, 1974; Kauer and Shepherd 1977; Meredith, 1986) that few mitral/tufted cells are excited at low odor concentrations. Cells with the lowest threshold for a particular odor may be determined in part by their location in relation to the odor-specific pattern of glomerular activation induced by the odor (Jourdan et al., 1980; Stewart el al., 1979; see the following for more detail). These low-threshold cells then induce a surrounding region of inhibition, thus suppressing activity of neighboring mitral/tufted cells. As the odor concentration increases more cells become excited, including some that were suppressed at the lower concentration. This increased excitation, however, can evoke more lateral and feedback inhibition, so that a single neuron may exhibit a brief burst of excitation and subsequent period of suppression during a single prolonged stimulus presentation (Kauer, 1974; Meredith, 1986). Thus, the complex, nonmonotonic relationship between response patterns and odor concentration (although see Harrison and Scott, 1986) may reflect simple underlying changes in the number of neurons activated by the odor and in the magnitude of lateral and feedback inhibition (Wilson and Leon, 1987). Furthermore, this relationship may be critical for coding odor intensity, given that the system is designed to respond to a stimulus constantly changing in intensity as it is inhaled and exhaled across the olfactory epithelium.

Response specificity of mitral/tufted cells is low, as it is in olfactory receptors (e.g., Mathews, 1972a). Individual mitral/tufted cells may respond to many odors, although they can vary their response pattern to different odors (Kauer, 1974; Mair, 1982; Mathews, 1972a; Mathews, 1972b, Meredith, 1986). For example, a mitral/tufted cell can be excited by a low concentration of one odor and inhibited by the same concentration of another odor (Kauer, 1974; Mair, 1982; Meredith, 1986). This pattern of responsiveness and response specificity may depend on the particular receptors that provide input to the mitral/tufted cell (MacKay-Sim et al., 1982; Thommesen and Doving, 1977). In fact, some evidence supports the presence of mitral/tufted cell receptive fields in the olfactory epithelium, with individual mitral/tufted cells receiving input from a limited region of the olfactory receptor sheet (Adrian, 1953; Kauer and Moulton, 1974; Costanzo and Mozell, 1976; Costanzo and O'Connell, 1980). Thus, given the spatial distribution of responsiveness to different odors in the olfactory epithelium, these receptive fields could provide the basis for response specificity of second-order neurons in the olfactory bulb.

Mitral/tufted cell receptive fields in the olfactory receptor sheet are characterized by a topographical relationship, with mitral/tufted cells on the lateral side of the olfactory bulb receiving input from the lateral side of the olfactory epithelium, and cells on the medial side receiving input from the medial olfactory epithelium (Costanzo and Mozell, 1976; Costanzo and O'Connell, 1978; Land, 1973; LeGros Clark, 1951). Similar topographic relationships can be seen in the dorsal–ventral plane (Land, 1973; Schwob and Gottlieb, 1986). It should be emphasized, though, that not all mitral/tufted cells have defined receptive fields, and generally those that do, have broad fields overlapping extensively with those of other mitral/tufted cells (Costanzo and Mozell, 1976; Costanzo and O'Connell, 1980; Kauer and Moulton, 1974). It has been suggested (Costanzo and O'Connell, 1980) that the mitral/tufted cells that respond to large regions of the olfactory epithelium (without defined receptive fields) code for odor intensity, whereas cells with receptive fields are more important for coding odor quality. Lateral inhibitory interactions between mitral/tufted cells may serve to enhance the signal:noise ratio and further define the stimulus (Wilson and Leon, 1987).

Two other lines of work suggest some degree of differential spatial responsiveness to odors in the olfactory bulb. First, extensive use or disuse of receptor input to mitral/tufted cells through prolonged exposure to a single odor can result in death or shrinkage of those cells. This death/shrinkage occurs in odor-specific patterns across the mitral cell layer (Laing and Panhuber, 1978; Pinching and Doving, 1974); it is not clear whether it occurs in regions activated by the odor or in regions that are not responsive to the odor and thus experience prolonged disuse. However, there is some evidence that behavioral detection thresholds for the exposure odor are unchanged following prolonged exposure to odor, while prolonged odor deprivation can elevate detection thresholds (Laing and Panhuber, 1980; although see Cunzeman and Slotnick, 1984).

Another line of evidence suggesting a spatial distribution of odor sensitivity in the olfactory bulb comes from autoradiographic analysis of metabolic activity during odor stimulation using the radiolabeled 2-deoxyglucose (2-DG) technique. During exposure to odor, 2-DG is most heavily taken up in the glomerular layer of the olfactory bulb (Sharp et al., 1977). Exposure to a particular odor does not induce uniform 2-DG uptake across the glomerular layer; rather each odor induces an odor-specific spatial pattern (Jourdan et al., 1980; Stewart et al., 1979). Although patterns for different odors may overlap, they are relatively constant in different animals (Stewart et al., 1979). Furthermore, use of a high-resolution 2-DG technique has shown that individual glomeruli activated by a particular odor are uniformly active (Lancet et al., 1982). Together, these data provide further evidence that odor quality is at least partially coded spatially in the olfactory bulb and suggest that the glomerulus may act as functional unit in coding olfactory information.

Although the evidence suggests a spatial component in coding of odor quality in both the olfactory epithelium and olfactory bulb (Costanzo and O'Connell, 1980; Kauer and Moulton, 1974; Wilson and Leon, 1988), no spatial coding appears to occur at higher levels of processing. For example, no odor-specific 2-DG uptake patterns exist in the olfactory cortex (Sharp et al., 1977). Furthermore, it should be emphasized that the cellular elements underlying 2-DG uptake in the olfactory bulb are unknown; that is, it is unclear whether 2-DG uptake primarily reflects presynaptic activity in the olfactory nerve, postsynaptic activity in dendrites, or even glial processes. Thus, a spatial distribution of focal 2-DG uptake in response to an odor does not necessarily suggest a spatial distribution of odor sensitivity within the bulb or spatial coding of odor quality at that level. In addition, lesions in large regions of the olfactory bulb produce no apparent selective loss of odor sensitivity or induction of specific anosmias (McBride et al., 1985). Finally, the extreme amount of overlap among single-unit receptive fields (Kauer and Moulton, 1974; Costanzo and O'Connell, 1980), receptor projection patterns (Costanzo and O'Connell, 1978; Land, 1973), and focal 2-DG uptake patterns (Jourdan et al., 1980; Stewart et al., 1979) strongly suggests that spatial coding is not the only mechanism of odor quality coding.

The response of olfactory bulb neurons to odor stimuli is influenced not only by the quality and intensity of an odor, but also by activity in centrifugal inputs to the bulb from the rest of the brain (Chaput, 1983; Pager, 1978; Potter and Chorover, 1976; Shepherd, 1972). Centrifugal inputs provide a primarily inhibitory or suppressive control over the olfactory bulb; thus, eliminating centrifugal input (olfactory peduncle sectioning) results in hyperresponsive single units (Chaput, 1983; Potter and Chorover, 1976; Wilson and Leon, in preparation). Centrifugal inputs are also critical in olfactory bulb responses to biological or learned odors (Cattarelli, 1982; Gray et al., 1986; Pager, 1978; Sullivan et al., 1987).

In addition to mitral/tufted cells, other identified neurons in the olfactory bulb respond to odors (Harrison and Scott, 1986; Onoda and Mori, 1980; Wellis and Scott, 1987). In fact, external plexiform and glomerular layer interneurons and tufted cells have lower thresholds for olfactory stimulation than mitral cells (Onoda and Mori, 1980; Schneider and Scott, 1983). Furthermore, responses of external plexiform and glomerular layer neurons are much simpler than those of mitral/tufted cells. For example, the former are more likely to demonstrate simple excitatory responses in phase with the respiratory cycle, whereas mitral/tufted cells have the more complex temporal responses described earlier (Onoda and Mori, 1980). The lower threshold and simpler responses of non-mitral/tufted cells suggests they may be subject to less synaptic interactions and modulation than the mitral/tufted cells.

Olfactory Bulb Projection Sites and Higher Order Processing

Few detailed studies of responses to olfactory stimulation have been done in structures that receive direct or indirect olfactory inputs beyond the olfactory bulb. Areas that have been examined include the olfactory cortex (Haberly, 1969; Giachetti and MacLeod, 1975), the amygdala (Cain and Bindra, 1972), the hippocampus (Yokota et al., 1967), the frontal cortex (Onoda et al., 1984; Tanabe et al., 1975; Yarita et al., 1980), the dorsal medial nucleus of the thalamus (Kogure and Onoda, 1983; Yarita et al., 1980), and the lateral hypothalamus (Komisaruk and Beyer, 1972; Scott and Pfaff, 1970; Scott and Pfaffman, 1972). The following summarizes the results of these studies:

1. Single units in all of these areas respond (excited or suppressed) to odor stimuli.
2. Some evidence suggests that the selectiveness of single-unit responses may increase with synaptic distance from the olfactory bulb (Tanabe et al., 1975; Yarita et al., 1980); that is, given a battery of odor stimuli, olfactory bulb neurons tend to respond to many of the odors whereas thalamic and neocortical neurons respond to increasingly fewer of the stimuli. Neurons in the lateral hypothalamus, however, show almost no response specificity (Scott and Pfaffman, 1972).
3. Thalamic, hypothalamic, and cortical olfactory responsive regions may be more sensitive to biologically significant odors (such as urine and feces) than the olfactory bulb (Onoda et al., 1984; Scott and Pfaff, 1970).

The role of these higher structures in olfactory processing is suggested by results from the behavioral literature. Surprisingly, extensive lesions of olfactory bulb primary and secondary projection pathways and target structures do not impair behavioral olfactory discriminations. Lesions of the posterior lateral olfactory tract (Slotnick, 1985; Slotnick and Berman, 1980; Staubli et al., 1986), the olfactory cortex (Staubli et al., 1987), the anterior amygdala (Eichenbaum et al. 1986; Slotnick, 1985), or the dorsal medial nucleus of the thalamus (Eichenbaum et al., 1980; Staubli et al., 1987) have little or no influence on normal olfactory discriminations. However, lesions of several of these structures impair performance on olfactory memory tasks, particularly the posterior lateral olfactory tract (Staubli et al., 1986), the anterior limb of the anterior commissure (Bennett, 1968), the amygdala (Eichenbaum et al. 1986), and the dorsal medial nucleus (Eichenbaum et al., 1980; Slotnick and Kaneko, 1981; Staubli et al., 1987).

FUTURE PROSPECTS

Although much progress has been made toward understanding olfactory coding, there is much room for advancement. Several areas of research currently show great promise for advancing our understanding of olfactory processes and processing.

Receptor Transduction Processes

Great strides have recently been made toward understanding the mechanisms of sensory transduction at the olfactory receptor cilia (Adamek et al., 1984 Pace et al., 1985; Vodyanoy and Murphy, 1983). This work was recently reviewed by Lancet (1986). Evidence now exists that binding an odor molecule to a transmembrane protein receptor activates a GTP-binding protein that, in turn, modulates adenylate cyclase to produce cyclic AMP. The cyclic AMP subsequently activates a protein kinase, which phosphorylates ion-channel proteins, opening the channel and depolarizing the receptor cell. The transmembrane olfactory receptor protein may have a variable extracellular binding region, which would allow for binding of various odor molecules. The enzymatic cascade induced by this binding would result in amplification at each step (e.g., a single GTP-binding protein could modulate many adenylate cyclase molecules). A large number of ion channels could therefore be opened by binding at a single olfactory receptor site. A detailed understanding of the mechanisms of olfactory receptor cell specificity (or lack of it) has yet to be achieved.

Olfactory Bulb Processing

One line of work that should provide some insight into how information is coded in the olfactory bulb is an examination of how olfactory bulb responses to odors are modified by experience. The effects of hunger (Pager et al., 1972) and emotional state (Cattarelli et al., 1977) as well as olfactory learning in young (Coopersmith and Leon, 1984; Sullivan and Leon, 1986; Wilson et al., 1987) and mature animals (Viana Di Prisco and Freeman, 1985) on olfactory bulb responses to specific odors have been examined. This work has demonstrated the following:

1. Olfactory bulb neural responses can be modified by an animal's current state and prior experience (Cattarelli et al., 1977; Pager et al., 1972; Wilson et al., 1987; Viana di Prisco and Freeman, 1985).

2. These modifications often demonstrate spatial patterns in the bulb (Wilson and Leon, 1988; Viana Di Prisco and Freeman, 1985).
3. Centrifugal inputs are involved in this bulbar information processing (Cattarelli, 1982; Gray et al., 1986; Pager, 1978; Sullivan et al., 1987).
4. Olfactory bulb output provides information about odor quality, quantity, and behavioral significance.

Together, these results demonstrate that extensive processing of information occurs in the olfactory bulb itself, and that the prior history of an animal can modify the response patterns of neurons one synapse from the olfactory receptors.

Role of Higher Structures in Olfactory Information Processing

As described previously, little is known about olfactory coding in structures other than the olfactory bulb. The behavioral and lesion work discussed earlier however, has begun to suggest that many primary and secondary target sites of the olfactory bulb may be more involved in forming and storing olfactory associations and memories than in olfactory discrimination per se. The exception to this may be the coding and discrimination of complex odor mixtures. It has been suggested that discrimination of complex olfactory stimuli involves sequential processing through multiple synaptic connnections, whereas discrimination of simpler stimuli may require substantially less processing (Lynch, 1986). In support of this, recent evidence suggests that following olfactory cortex lesions, rats can discriminate simple odors much more easily than mixtures of those odors (complex odors) (Staubli et al., 1987). Continued research into the role of higher structures in olfactory coding should prove fruitful.

CONCLUSIONS

Our understanding of the nature of olfactory information processing has improved in recent years. Detailed investigations of the neuroanatomy and neurophysiology of the olfactory bulb have led to testable hypotheses about the mechanisms of stimulus–response relationships in this structure. Knowledge of the diverse centripetal connections of the bulb has prompted physiologic and behavioral investigations of the role of these primary and secondary target sites in olfactory function. Although few definitive statements about the nature of olfactory coding can be made at this time, the application of new techniques and the exploration of responses throughout the olfactory system under varying conditions should provide valuable insights into this process.

ACKNOWLEDGMENTS

This research was supported by grants BNS-8606786 from The National Science Foundation to DAW and ML and MH00371 from National Institute of Mental Health to ML.

REFERENCES

Adamek, G. D., Gesteland, R. C., Mair R. G., and Oakley B. (1984). Transduction physiology of olfactory receptor cilia. *Brain Res.* 310:87–97.

Adrian, E. D. (1950). The electrical activity of the mammalian olfactory bulb. *EEG Clin. Neuophysiol.* 2:377–388.

Adrian, E. D. (1953). Sensory messages and sensation. The response of the olfactory organ to different smells. *Acta Physiol. Scand.* 29:5–14.

Amoore, J. E. (1970). *Molecular Basis of Odor.* Springfield, IL: Charles C. Thomas.

Bennett, M. H. (1968). The role of the anterior limb of the anterior commissure in olfaction. *Physiol. Behav.* 3:507–515.

Cain, D. P., and Bindra, D. (1972). Responses of amygdala single units to odors in the rat. *Exp. Neurol.* 35:98–110.

Cattarelli, M. (1982). The role of the medial olfactory pathways in olfaction: behavioral and electrophysiological data. *Behav. Brain Res.* 6:339–364.

Cattarelli, M., Vernet-Maury, E., and Chanel, J. (1977). Modulation de l'activite du bulbe olfactif en fonction de la signification des odeurs chex le rat. *Physiol. Behav.* 19:381–387.

Chaput, M. (1983). Effects of olfactory peduncle sectioning on the single unit responses of olfactory bulb neurons to odor presentation in awake rabbits. *Chem. Sens.* 8:161–177.

Chaput, M., and Holley, A. (1985). Responses of olfactory bulb neurons to repeated odor stimulations in awake freely-breathing rabbits. *Physiol. Behav.* 34:249–258.

Chaput, M., and Lankheet, M. J. (1987). Influence of stimulus intensity on the categories of single-unit responses recorded from olfactory bulb neurons in awake freely-breathing rabbits. *Physiol. Behav.* 40:453–462.

Coppersmith, R., and Leon, M. (1984). Enhanced neural response to familiar olfactory cues. *Science* 225:849–851.

Costanzo, R. M., and Mozell, M. M. (1976). Electrophysiological evidence for a topographical projection of the nasal mucosa into the olfactory bulb of the frog. *J. Gen. Physiol.* 68:297–312.

Costanzo, R. M., and O'Connell, R. J. (1978). Spatially organized projections of hamster olfactory nerves. *Brain Res.* 139:327–332.

Costanzo, R. M., and O'Connell, R. J. (1980). Receptive fields of second-order neurons in the olfactory bulb of the hamster. *J. Gen. Physiol.* 76:53–68.

Cunzeman, P. J., and Slotnick, B. M., (1984). Prolonged expsoure to odors in the rat: effects on odor detection and on mitral cells. *Chem. Sens.* 9:229–239.

Derby, C. D., and Ache, B. W. (1984). Quality coding of a complex odorant in an invertebrate. *J. Neurophysiol.* 51:906–924.

Doving, K. J. (1964). Studies of the relation between the frog's electro-olfactogram (EOG) and single unit activity in the olfactory bulb. *Acta Physiol. Scand.* 60:150–163.

Eichenbaum, H., Fagan, A., and Cohen, N. J. (1986). Normal olfactory discrimination learning set and facilitation of reversal learning after medial-temporal damage in rats: implications for an account of preserved learning abilities in amnesia. *J. Neurosci.* 6:1876–1884.

Eichenbaum, H., Shedlack, K. J., and Eckmann, K. W. (1980). Thalamocortical mechanisms in odor-guided behavior. I. Effects of lesions of the mediodorsal thalamic nucleus and frontal cortex on olfactory discrimination in the rat. *Brain Behav. Evol.* 17:255–275.

Engen, T. (1982). *Perception of Odour.* New York: Academic Press.

Fujita, S. C., Mori, K., Imamura, K. and Obata, k. 1985). Subclasses of olfactory receptor cells and their segmented central projections demonstrated by a monoclonal antibody. *Brain Res.* 326:192–196.

Gesteland, R. C., Yancey, R. A., and Farbman, A. I. (1982). Development of olfactory receptor neuron selectivity in the rat fetus. *Neurosci.*, 7:3127–3136.

Getchell, T.V. (1977). Analysis of intracellular recordings from salamander olfactory epithelium. *Brain Res.* 123:275–286.

Getchell, T. V., and Shepherd, G. M. (1978). Responses of olfactory receptor cells to step pulses of odour at different concentrations in the salamander. *J. Physiol.* (Lond.) 282:521–540.

Giachetti, I., and MacLeon, P. (1975). Cortical neuron responses to odours in the rat. In: *Olfaction and Taste*, D.A. Denton and J.P. Coghlan (eds.), New York Academic Press.

Gray, C. M., Freeman, W. J., and Skinner, J. E. (1986). Chemical dependencies of learning in the rabbit olfactory bulb: acquisition of the transient spatial pattern change depends on norepinephrine. *Behav. Neurosci.*, 100:585–596.

Haberly, L. B. (1969). Single unit responses to odor in the prepyriform cortex of the rat. *Brain Res.* 12:481–484.

Haberly, L. B., and Price, J. L. (1977). The axonal projection patterns of the mitral and tufted cells of the olfactory bulb in the rat. *Brain Res.* 129:152–157.

Hamilton, K. A., and Kauer, J. S. (1985). Intracellular potentials of salamander mitral/tufted neurons in response to odor stimulation. *Brain Res.* 338:181–185.

Harrison, T. A., and Scott, J. W. (1986). Olfactory bulb responses to odor stimulation: analysis of response pattern and intensity relationships. *J. Neurophysiol.* 56:1571–1589.

Jahr, C. E., and Nicoll, R. A. (1982). An intracellular analysis of dendro-dendritic inhibition in the turtle *in vitro* olfactory bulb. *J. Physiol.* (Lond.) 326:213–234.

Jastreboff, P. J., Pedersen, P. E., Greer. C. A., Stewart, W. B., Kauer, J. S., Benson, T. E., and Shepherd, G. M. (1984). Specific olfactory receptor populations projecting to identified glomeruli in the rat olfactory bulb. *Proc. Natl. Acad. Sci. USA* 81:5250–5254.

Jourdan, F., Duveau, A., Astic, L., and Holley, A. (1980). Spatial distribution of [14C]2-deoxyglucose uptake in the olfactory bulbs of rats stimulated with two different odours. *Brain Res.* 188:139–154.

Kauer, J. S. (1974). Response patterns of amphibian olfactory bulb neurons to odour stimulation. *J. Physiol.* (Lond.) 243:695–715.

Kauer, J. S., and Moulton, D. G. (1974). Responses of olfactory bulb neurons to odour stimulation of small nasal areas in the salamander. *J. Physiol.* (Lond.) 243:717–737.

Kauer, J. S., and Shepherd, G. M. (1977). Analysis of the onset phase of olfactory bulb unit responses to odour pulses in the salamander. *J. Physiol.* (Lond.) 272:495–516.

Keverne, E. B., Murphy, C. L., Silver, W. L., Wysocki, C. J., and Meredith, M. (1986). Non-olfactory chemoreceptors of the nose: the vomeronasal and trigeminal systems. *Chem. Sens.* 11:119–133.

Kogure, S., and Onoda, N. (1983). Response characteristics of lateral hypothalamic neurons to odors in unanesthetized rabbits. *J. Neurophysiol.* 50:609–617.

Komisaruk, B. R., and Beyer, C. (1972). Responses of diencephalic neurons to olfactory bulb stimulation, odor, and arousal. *Brain Res.* 36:153–170.

Laing, D. G., and Panhuber, H. (1978). Neural and behavioral changes in rats following continuous exposure to an odor. *J. Comp. Physiol.* 124:259–265.

Laing, D. G., and Panhuber, H. (1980). Olfactory sensitivity of rats reared in an odorous or deodorized environment. *Physiol. Behav.* 25:555–558.

Lancet, D. (1986). Vertebrate olfactory reception. *Annu. Rev. Neurosci.* 9:329–355.

Lancet, D., Greer, C. A., Kauer, J. S., and Shepherd, G. M. (1982). Mapping of odor-related neuronal activity in the olfactory bulb by high-resolution 2-deoxyglucose autoradiography. *Proc. Natl. Acad. Sci. USA* 79:670–674.

Land, L. J. (1973). Localized projection of olfactory nerves to rabbit olfactory bulb. *Brain Res.* 63:153–166.

LeGros Clark, W. E. (1951). The projection of the olfactory epithelium on the olfactory bulb in the rabbit. *J. Neurol Neurosurg. Psychiat.*, 14:1–10.

Lynch, G. S. (1986). *Synapses, Circuits and the Beginnings of Memory*. Cambridge: MIT Press.

MacKay-Sim, A., Shaman, P., and Moulton, D. G. (1982). Topographic coding of olfactory quality: odorant-specific patterns of epithelial responsivity in the salamander. *J. Neurophysiol.* 48:584–596.

Macrides, F., and Chorover, S. L. (1972). Olfactory bulb units: activity correlated with inhalation cycles and odor quality. *Science* 175:84–87.

Macrides, F., and Davis, B. J. (1983). The olfactory bulb. In: P. C. Emson (ed.) *Chemical Neuroanatomy*, New York: Raven Press.

Macrides, F., Schoenfeld, T. A., Marchand, J. E., and Clancy, A. N. (1985). Evidence for morphologically, neurochemically and functionally heterogeneous classes of mitral and tufted cells in the olfactory bulb. *Chem. Sens.* 10:175–202.

Mair, R. G. (1982). Response properties of rat olfactory bulb neurons. *J. Physiol.* (Lond.) 326:341–359.

Mair, R. G., and Gesteland, R. C. (1982). Response properties of mitral cells in the olfactory bulb of the neonatal rat. *Neurosci.* 7:3117–3125.

Mathews. D. F. (1972a). Response patterns of single neurons in the tortoise olfactory epithelium and olfactory bulb. *J. Gen. Physiol.* 60:166–180.

Mathews, D. F. (1972b). Response patterns of single units in the olfactory bulb of the rat to odor. *Brain Res.* 47:389–400.

McBride, S. A., Slotnick, B. M., Graham, S. J., and Graziadei, P. P. C. (1985). Failure to find specific anosmias in rats with olfactory bulb lesions. *Chem. Sens.* 10:410.

Meredith, M. (1986). Patterned response to odor in mammalian olfactory bulb: the influence of intensity. *J. Neurophysiol.* 56:572–597.

Mori, K., and Takagi, S. F. (1978). An intracellular study of dendrodendritic inhibitory synapses on mitral cells in the rabbit olfactory bulb. *J. Physiol.* (Lond.) 279:569–588.

Moulton, D. G. (1976). Spatial patterning of response to odors in the peripheral olfactory system. *Physiol. Rev.* 56:578–593.

Mozell, M. M. (1964). Evidence for sorption as a mechanism of the olfactory analysis of vapors. *Nature* 203:1181–1183.

Mozell, M. M. (1970). Evidence for a chromatographic model of olfaction. *J. Gen. Physiol.* 56:46–63.

Nakashima, M., Mori, K., and Takagi, S. F. 81978). Centrifugal influence on olfactory bulb activity in the rabbit. *Brain Res.* 154:301–316.

Nicoll, R. A. (1972). Olfactory nerves and their excitatory action in the olfactory bulb. *Exp. Brain Res.* 14:185–197.

Onoda, N., Imamuta, K., Obata, E., and Iino, M. (1984). Response selectivity of neocortical neurons to specific odors in the rabbit. *J. Neurophysiol.* 52:638–651.

Onoda, N., and Mori, K. (1980). Depth distribution of temporal firing patterns in olfactory bulb related to air-intake cycles. *J. Neurophysiol.* 44:29–39.

Pace, U., Hanski, E., Salomon, Y., and Lancet, D. (1985). Odorant-sensitive adenylate cyclase may mediate olfactory reception. *Nature* 316:255–258.

Pager, J. (1978). Ascending olfactory information and centrifugal influxes contributing to a nutritional modulation of the rat mitral cell responses. *Brain Res.* 140:251–269.

Pager, J. (1985). Respiration and olfactory bulb unit activity in the unrestrained rat: statements and reappraisals. *Behav. Brain Res.* 16:81–94.

Pager, J., Giachetti, I., Holley, A., and LeMagnen, J. (1972). A selective control of olfactory bulb electrical activity in relation to food deprivation and satiety in rats. *Physiol. Behav.* 9:573–579.

Pinching, A. J., and Doving, K. B. (1974). Selective degeneration in the rat olfactory bulb following exposure to different odours. *Brain Res.* 82:195–204.

Potter, H., and Chorover, S. L. (1976). Response plasticity in hamster olfactory bulb: peripheral and central processes. *Brain Res.* 116:417–429.

Price, J. L., and Powell, T. P. S. (1970a). An electron-microscopic study of the termination of the afferent fibres to the olfactory bulb from the cerebral hemispheres. *J. Cell. Sci.* 7:157–187.

Price, J. L., and Powell, T. P. S. (1970b). The mitral and short axon cells of the olfactory bulb. *J. Cell. Sci.* 7:631–651.

Schneider, S. P., and Scott, J. W. (1983). Orthodromic response properties of rat olfactory bulb mitral and tufted cells correlate with their projection patterns. *J. Neurophysiol.* 50:358–378.

Schoenfeld, T. A., Marchand, J. E., and Macrides, F. (1985). Topographic organization of tufted cell axonal projections in the hamster main olfactory bulb: an intrabulbar associational system. *J. Comp. Neurol.* 235:503–518.

Schwob, J. E., and Gottlieb, D. I. (1986). The primary olfactory projection has two chemically distinct zones. *J. Neurosci.* 6:3393–3404.

Scott, J. W. (1981). Electrophysiological identification of mitral and tufted cells and distributions of their axons in olfactory system of the rat. *J. Neurophysiol.* 46:918–931.

Scott, J. W., and Pfaff, D. W. (1970). Behavioral and electrophysiological responses of female mice to male urine odors. *Physiol. Behav.* 5:407–411.

Scott, J. W., and Pfaffmann, C. (1972). Characteristics of responses of lateral hypothalamic neurons to stimulation of the olfactory system. *Brain Res.* 48:251–264.

Sharp, F. R., Kauer, J. S., and Shepherd, G. M. (1977). Laminar analysis of 2-deoxyglucose uptake in olfactory bulb and olfactory cortex of rabbit and rat. *J. Neurophysiol.* 40:800–813.

Shepherd, G. M. (1972). Synaptic organization of the mammalian olfactory bulb. *Physiol. Rev.* 52:864–917.

Sicard, G. (1986). Electrophysiological recordings from olfactory receptor cells in adult mice. *Brain Res.* 397:405–408.

Sicard, G., and Holley, A. (1984). Receptor cell responses to odorants: similarities and differences among odorants. *Brain Res.* 292:283–296.

Slotnick, B. M. (1985). Olfactory discrimination in rats with anterior amygdala lesions. *Behav. Neurosci.* 99:956–963.

Slotnick, B. M., and Berman, E. J. (1980). Transection of the lateral olfactory tract does not produce anosmia. *Brain Res. Bull.* 5:141–145.

Slotnick, B. M., and Kaneko, N. (1981). Role of dorsomedial thalamic nucleus in olfactory discrimination learning in rats. *Science* 214:91–92.

Staubli, U., Fraser, D., Kessler, M., and Lynch, G. (1986). Studies on retrograde and anterograde amnesia of olfactory memory after denervation of the hippocampus by entorhinal cortex lesions. *Behav. Neural Biol.* 46:432–444.

Staubli, U., Schottler, F., and Nejat-Bina, D. (1987). Role of dorsomedial thalamic nucleus and piriform cortex in processing olfactory information. *Behav. Brain Res.* 25:117–129.

Stewart, W. B., Kauer, J. S., and Shepherd, G. M. (1979). Functional organization of rat olfactory bulb analysed by the 2-deoxyglucose method. *J. Comp. Neurol.* 185:715–734.

Sullivan, R. M., and Leon, M. (1986). Early olfactory learning induces an enhanced olfactory bulb response in young rats. *Devel. Brain Res.* 27:278–282.

Sullivan, R. M., Wilson, D. A., Do, J., and Leon, M. (1987). Noradrenergic control of neural and behavioral correlates of early olfactory learning. *Soc. Neurosci. Abstr.* 13:1402.

Tanabe, I., Iino, M., and Takagi, S. F. (1975). Discrimination of odors in olfactory bulb, pyriform-amygdaloid areas, and orbitofrontal cortex of the monkey. *J. Neurophysiol.* 38:1269–1283.

Thommesen, G., and Doving, K. B. (1977). Spatial distribution of the EOG in the rat: a variation with odour quality. *Acta Physiol. Scand.* 99:270–280.

Trotier, D., and MacLeod, P. (1983). Intracellular recordings from salamander olfactory receptor cells. *Brain Res.* 268:225–237.

Viana Di Prisco, G., and Freeman, W. J. (1985). Odor-related bulbar EEG spatial pattern analysis during appetitive conditioning in rabbits. *Behav. Neurosci.* 99:964–978.

Vodyanoy, V., and Murphy, R. B. (1983). Single-channel fluctuations in bimolecular lipid membranes induced by rat olfactory epithelial homogenates. *Science* 220:717–719.

Wellis, D. P., and Scott, J. W. (1987). Odor induced responses of morphologically identified rat olfactory bulb neurons. *Soc. Neurosci. Abstr.* 13:1411.

Wilson, D. A., and Leon, M. (1987). Evidence of lateral synaptic interactions in olfactory bulb output cell responses to odors. *Brain Res.* 417:175–180.

Wilson, D. A., and Leon, M. (1988). Spatial patterns of olfactory bulb single unit responses to learned olfactory cues in young rats. *J. Neurophysiol.* 59:1770–1782.

Wilson, D. A., Sullivan, R. M., and Leon, M. (1987). Single-unit analysis of postnatal olfactory learning: modified olfactory bulb output response patterns to learned attractive odors. *J. Neurosci.* 7:3154–3162.

Wilson, R. C., and Stewart, O. (1978). Polysynaptic activation of the dentate gyrus of the hippocampal formation: an olfactory input via the lateral entorhinal cortex. *Exp. Brain Res.* 33:523–4534.

Yarita, H., Iino, M., Tanabe, T., Kogure, S., and Takagi, S. F. (1980). A trans-thalamic olfactory pathway to orbitofrontal cortex in the monkey. *J. Neurophysiol.* 43:69–85.

Yokota, T., Reeves, A. G., and MacLean, P. D. (1967). Intracellular olfactory response of hippocampal neurons in awake, sitting squirrel monkeys. *Science* 157:1072–1074.

2

Analysis of Mammalian Peripheral Taste Systems

JAMES C. BOUDREAU

The main problem facing any investigator in chemosensory research is to determine the nature of the natural chemical stimuli optimally affecting the system under study. Only when the natural taste stimuli are known can we be sure how chemical information is encoded neurally, how receptors are organized, or how an animal's taste systems relate to human systems.

Taste physiologists commonly assume that the experimental animal has four basic tastes, which are adequately represented both psychophysically and physiologically by four compounds (salt, sugar, HC1, and quinine). Human beings have far more than four taste sensations, however, and experimental animals are widely divergent in their sensitivity to chemical compounds. A more cautious approach to the problem is to examine a variety of compounds naturally present in an animal's diet and to evaluate these compounds on the basis of peripheral neural responses subjected to systems analysis. In a systems approach, the component studied may be black-boxed and its function considered solely in terms of the transformation of the input signals. The main problem in applying this method to the study of a system is that the investigator must introduce into the system an array of naturally occurring relevant signals, and the output signal must be relevant to the organism itself.

Sensory systems are readily amenable to systems analysis because the input signals often can be exactly specified in physical or chemical terms and readily manipulated. Output signals may represent the entire informational output of the cell: the transmitted spike train, since the spike is used by the nervous system in transmitting information over long distances. From peripheral sensory ganglion cells, all information about the stimulus must be contained in the spike discharge. By varying the stimulus in a controlled manner and quantitatively measuring spike output, one can completely describe the information-processing function of the neuron under study.

It is customary in systems analysis to block-diagram the system under study, specifying only those conditions relevant to signal processing. Thus, the entire animal may be black-boxed and treated in terms of input and output signals (Figure 2-1). Such an approach applies particularly well to taste physiology, because

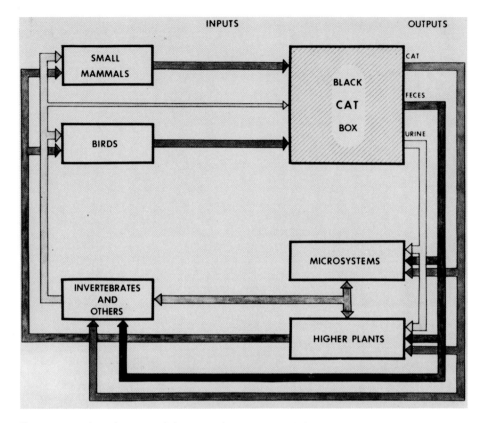

Figure 2-1 Flow diagram of the natural ecosystem of the cat, with the main chemical input and output signals indicated. This ecological orientation is necessary for studying taste systems.

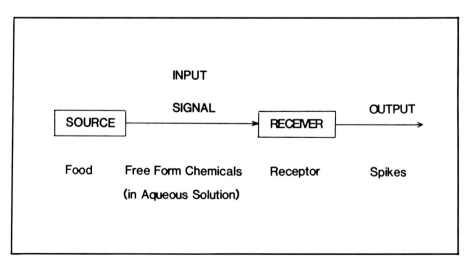

Figure 2-2 Flow diagram of a communication system, with the main components indicated. The model is applied to the taste system as indicated.

an animal tastes what it naturally eats. The peripheral sensory system can be treated as a communications system (Figure 2-2). The system has a source, a signal, and a receiver with an output signal. In this model, food is the source; the free-form, largely nonvolatile compounds are the signal; and the receptors are the receivers, and the spike are the output signal. The medium is water based: saliva and food juices. In practice, the main variables are chemical solutions and spike trains. The chemical solutions are described as precisely as possible with modern techniques, although solution chemistry has more unknowns than knowns. The spike trains are described as precisely as possible with computer-assisted measurements, although simple spike counts have supplied most of the data in our own laboratory. The input signals are varied and the resulting spike changes monitored. The primary chemical variable that my colleagues and I have varied is quality. Our goal has been to use the spike output to specify optimal chemical stimuli, in terms of excitation and inhibition.

APPLICATION OF SYSTEMS ANALYSIS TO TASTE

Our neurophysiological approach has been to make metal electrode recordings from the sensory ganglion cell bodies in the geniculate (facial nerve) and petrosal (glossopharyngeal) ganglia of anesthetized animals(Figure 2-3). This was a for-

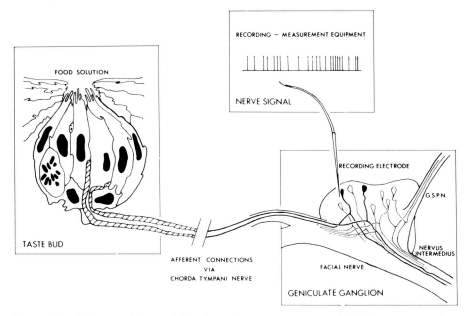

Figure 2-3 Diagram of the peripheral vertebrate taste system with the sensory ganglion cells of the geniculate that innervate the taste buds. Pulse signals sent to the brain from the periphery are the output measures used to represent the chemical input signals. Spontaneous activity was common, and both inhibition and excitation served as output measures. The geniculate ganglion innervates taste buds of the fungiform papillae on the front of the tongue and on the palate. The petrosal ganglion of the glossopharyngeal nerve was also studied. GSPN = greater superficial petrosal nerve.

tuitous choice, because it permitted long-term extracellular recordings from sensory neurons with their peripheral and central extremities intact. Units, especially in the geniculate ganglion, could be held for the extended times (often 3 to 4 hours) necessary to specify completely their neurophysiological and chemical stimulus variables. Neurophysiological measures included spontaneous and evoked activity and receptive field papillae system mapping with latency measures. Neurophysiological data were recorded on magnetic tape, together with voice commentary, stimulus markers, and standardized triggered pulses (using an amplitude discriminator coupled to an oscilloscope with a level marker). The triggered pulses with stimulus markers were also stored on magnetic disk in 2100S HP on-line computer system. Special computer programs were available to process the spike trains.

The spike trains recorded from first-order taste neurons have some unusual characteristics. All taste neurons have a certain level of spontaneous activity, which is often highly complex. Bursting—that is, spikes appearing at short, relatively fixed intervals—is common, and "grouping," or the appearance of a pseudodischarge, is also not unusual. With optimal stimuli, two types of discharge can occur. In one, the spikes occur tonically, usually with a fairly rapid decline in the first few seconds. In the other type, the spikes may appear in groups, often after a long latency. The first type of discharge is common to most geniculate neurons, the second to some geniculate ganglion units and most petrosal ganglion units. Examples of spontaneous and evoked discharges recorded from peripheral sensory ganglion cells are presented in Figure 2-4.

The first preparation examined was the geniculate ganglion of the cat. The experiment started with some of the standard solutions used by taste physiologists, but in terms of the four basic tastants, the cat was a recalcitrant responder: there was virtually no sugar response, little response to quinine, and the threshold to NaCl virtually was above 50 mM. Only acids elicited large responses from a fair number of units. The responses to the four compounds seemed to have little relevance to a cat's diet, unless it consumed seawater and vinaigrette. As an exclusive carnivore, the cat does not encounter alkaloids or sugar, so the general lack of response to these two compounds made sense. Virtually all the cat's natural foods contain high levels of sodium, but pH levels below 5.0 are not encountered in meat.

The course of action was first indicated by the strong responses of cat taste units to meat solutions. Chicken, pork liver, pork kidney, heart, beef, and tuna were chopped up and steeped in distilled water. These solutions proved to be highly reactive to cat taste units, especially pork liver, pork kidney, and tuna. But what were the stimulating compounds in these solutions? At the time of this experiment (the late 1960s), virtually all food flavor was considered to come from smell. An examination of biochemistry textbooks indicated a bewildering variety of compounds present in animal tissues. Many of those present in free form or in fair concentration were tested on the cat, but the big breakthrough came with the discovery of articles by Solms (1968; 1971) on the main compounds present in a water extract of meat, with an indication of which are known to be taste-active in the human. Virtually all of the compounds present on Solms' list were tested for cat taste activity. Whenever one compound was found to be active, the

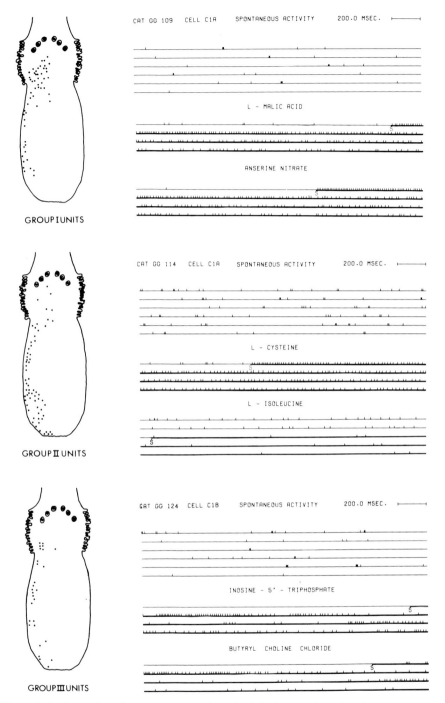

Figure 2-4 Examples of spontaneous and evoked discharges from the three major types of units recorded from the cat geniculate ganglion. From the top to bottom are the acid unit (group I), amino acid unit (group II), and an X (nucleotide) unit (group III). In the figures of tongues, to the left, dots indicate where the different units of the three neural groups projected. Note that the amino acid unit is inhibited when stimulated with L-isoleucine. [From Boudreau et al. (1977)].

biochemistry textbooks were consulted and similar compounds were tested. Thus, all the common amino acids were tested, as well as most nucleotides, most of the acids in the Krebs cycle, and others. Then modifications of the most active molecules, peptides of active amino acids, and so forth were tested. Eventually more than 600 compounds were examined on the cat's taste system, with single-unit discharges as the basic measure for evaluating the different compounds.

Before a detailed analysis of the chemical stimulus question could be made, the problem of unit heterogeneity had to be resolved. Our studies of cat taste units had indicated that the units were a heterogeneous collection, but only when the new taste chemistry was combined with the physiological measures was this problem solved satisfactorily. With only a few effective stimuli, often active only at high concentrations, it proved difficult to separate units from one another on a purely chemical stimulus basis. With an improved understanding of stimulus chemistry, however, subsets of units could be maximally activated with compounds that were inactive or even inhibited the other units that, in turn, could be affected by other types of compounds. The most effective excitatory stimulus for one subset of neurons, for example, was L-proline, a compound completely inactive with all other units.

The new chemistry, together with the standard neurophysiological measures, permitted separation of the cat units into three major groups (Figure 2-4), formerly labeled groups I, II, and III and now labeled acid, amino acid, and X units; the X units are further divisible into two subgroups. Some of the measures used to separate the units are shown in Figure 2-5. In the upper graph, the spontaneous activity rate of the unit is plotted against its latency to electrical stimulation. The high-rate, medium-latency amino acid units are interposed between the low-rate, short-latency acid units and the low-rate, long-latency X units. The X units are also separated from the others because they display the unusual grouping of spikes when activated (See Figure 2-4). The three different unit groups could also be separated by purely chemical response measures, such as the response to malic acid and L-proline. The real clincher for the separation of unit groups occurs when the physiological and chemical stimulus response measures are combined. Thus, if all units are arranged with respect to latency, and their discharges to selected compounds are plotted above them, their separation into three distinct populations is clear (Figure 2-5B), even though some compounds activate two or even three unit groups. All told, the case for distinct unit groups within the cat taste system is as compelling as that for any other peripheral sensory system.

Once the cat geniculate ganglion taste units were subdivided into three relatively homogeneous groups, we could vigorously attack the problem of the chemistry of the stimulus. The power of systems analysis with single-unit recordings is most manifest in this description of the chemistry.

The amino acid units of the cat responded to certain amino acids and to di– and triphosphate nucleotides. The most excitatory compounds were the imino acid L-proline and the amino acid L-cysteine, followed by L-orithine, L-histidine, and L-alanine. A structure function study (Boudreau et al., 1975) found that pyrrolidine, the heterocyclic moiety of L-proline, was even more excitatory than L-proline and imidazole was more excitatory than L-histidine—evidence that the carboxyl groups did not participate in stimulus activation. These heterocyclic nuclei

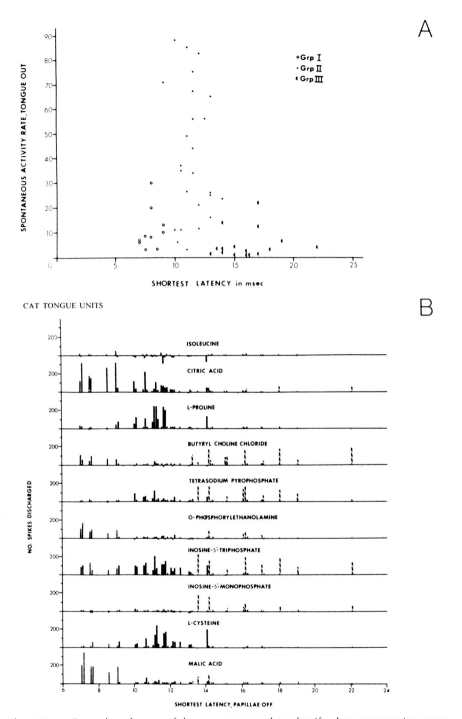

Figure 2-5 Examples of some of the measures used to classify chemoresponsive cat geniculate ganglion units. (A) A plot of spontaneous activity rates against latency to electical stimulation. Note that a unit group with high rates of spontaneous activity (amino acid units) is interposed between a short-latency, low-rate unit group (acid units) and a long-latency, low-rate unit group (X or nucleotide units). (B) Units arranged according to latency to electrical stimulation. The spike counts indicate the discharges elicited by different compounds (50 mM concentration). [From Boudreau and Alev (1973).]

form a rigid simple basis for structure function series. Examining a group of related heterocyclic compounds made it possible to describe the excitatory and inhibitory properties of the heterocycles with respect to two physical chemical parameters: ring size and pKa. A pKa discharge plot (Figure 2-6), reveals that pyrrolidine and l-pyrroline,—nonaromatic five-member nitrogen heterocycles,— were the most stimulating compounds tested, and the aromatic five-member nitrogen heterocycle pyrrol and eight-member nonaromatic heterocycle azaoctane were the most inhibitory. When first tested, pyrrol inhibited the neuron for minutes, the strongest inhibitory effect encountered. Considering that amino acid unit inhibitors are avoided (White and Boudreau, 1975), pyrrol must be the worst-tasting compound that exists for the cat.

The acid unit group is the second largest major group of taste units in the cat's geniculate ganglion. By examining the compounds that stimulate this unit group, it was possible to describe all active and inactive compounds with respect to a common chemical property—for example, Brønsted acidity. Active compounds were Brønsted acids, which donated protons, and inactive compounds were Brønsted bases, which accepted protons (Boudreau and Nelson, 1977). The acids were excitatory and the bases inhibitory. The discovery of some nitrogenous Bronsted acids that functioned near neutral pH answered the vexing question of how acid units could function in meat systems, because compounds with imidazole (a nitrogen heterocycle with a pKa of 6.3) are abundant and in free form in vertebrate tissues. As with the amino acid units, heterocyclic compounds functioning as Brønsted acids were the most effective stimuli between pH 5.0 and 7.0. Especially

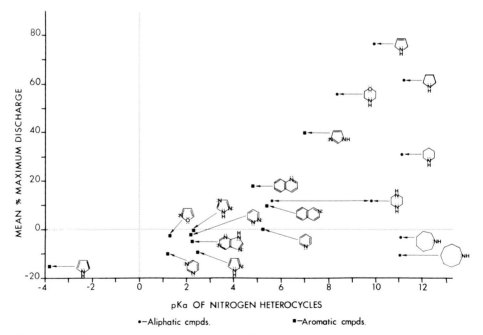

Figure 2-6 Relationship of neural response of cat geniculate ganglion amino acid units to the pKa of certain heterocyclic compounds. The most effective stimulus for the cat amino acid units is L-proline, which contains pyrrolidine. [From Boudreau et al. (1975).]

stimulating were the heterocycles thiazolidine, pyridine, and imidazole. Heterocycles are primary stimuli in the cat, as flavor chemists have been discovering
elsewhere.

Of course, the systems analysis approach, which yields the relationship between chemicals and spikes in the neurophysiological preparation, does not guar-
antee behavioral relevance of these chemicals to the organism. To test whether
the chemicals were significant to the cat, and to examine the possible significance
of the inhibition of spontaneous activity, the cat was offered a choice between
water (actually 50 mM saline) and either an exciter or inhibitor in saline. A diluent
of 50 mM saline was chosen because the acid units in the cat were discharged by
distilled water and the objective of the behavioral experiment was to restrict activity to amino acid units. The outcome of the experiment was clear; the cat avidly
drank solutions of amino acid unit exciters and avoided inhibitors (Figure 2-7).
Since the inhibitory solutions excited no neural group in the geniculate ganglion,
the experiment indicated a neural avoidance function for peripheral inhibition (White
and Boudreau, 1975).

Once the problem of unit groups and their stimulus chemistries had been worked
out satisfactorily in the cat, we began a series of comparative studies involving
different mammals. The emphasis of these studies was on stimulus chemistry, an
area of mammalian taste still little understood. The cat chemistry studies had led
us far from NaCl and HCl, and the prospect of new food chemical signals functioning in different species was tantalizing. The situation was like a vision physiologist discovering new aspects of light or an auditory physiologist finding different nuances of sound. The taste signals control the eater and determine the fate
of the eaten. They link diverse organisms and constitute the cement in ecosystems.

The success of cat studies led us to examine other species and to relate taste
to diet, through stimulus chemistry. The next animal chosen for neurophysiological analysis was the dog, a closely related carnivore. As in the cat, neurons
in the dog could be parceled into three neural groups: Brønsted acid, amino acid,
and X groups. Only the Brønsted acid group seemed identical to the corresponding
group found in the cat. The amino acid group of the dog responded well to the
amino acids that were stimulatory in the cat, but it also responded to some amino
acids that were not excitatory in the cat and to sugars, which were completely
inactive in the cat. There were few X units, and unlike those of the cat, some
discharged well to furaneol, ethyl maltol, and methyl maltol (Figure 2-8). Furaneol is a compound that occurs naturally in many fruits (Pickenhagen et al.,
1981), thus providing an oral sensory compliment for the canine's ingestion of
fruit, which is totally absent from the felid's natural diet. This taste system functions in the seed-dispersing properties of the dog.

After examining the taste systems of the fissiped carnivores, the next animal
investigated was the goat, an herbivore. Although the goat geniculate ganglion
(GG) taste systems could be subdivided into three neural groups, none was identical to a carnivore group (Boudreau et al., 1982). The acid units were partly
distinct from those seen in the carnivore, and the X units could be discharged
only by alkaloids. Most astonishing, however, the amino acid system, which was
prominent in the carnivore, was missing in the goat. In addition, the goat had a
taste system we had not seen before, which responded only to solutions with

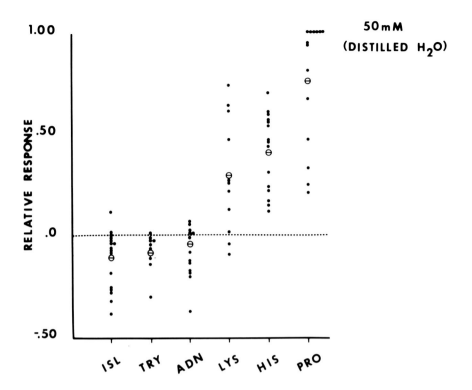

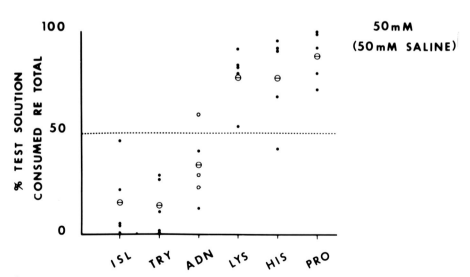

Figure 2-7 (*Upper plot*) Neural response of cat amino acid units to different solutions. (*Lower plot*) Behavioral response of the cat to different solutions. Note that neural inhibition is associated with avoidance and excitation with consumption. [From White and Boudreau (1975)].

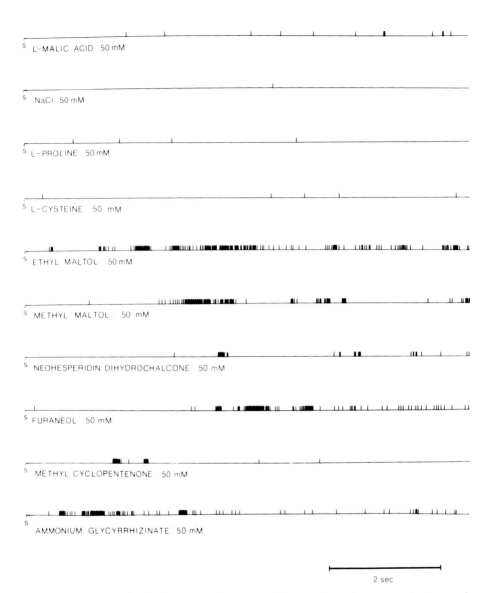

Figure 2-8 Response of a dog X unit (furaneol unit) to a variety of compounds. Furaneol, ethyl maltol, and methyl maltol are important flavor constituents for many fruits. The natural diet of the canid includes fruit.

sodium (Na) and lithium (Li) ions. It turns out that Na is in short supply in most plants and is often a population-limiting compound for herbivores.

Subsequent studies on the rat uncovered taste systems similar to those in the goat, plus an amino acid–responsive system quite distinct from those in the cat or dog. Also studied in the rat, for the first time, were the taste systems of the petrosal ganglion (PG), the cranial sensory ganglion that innervates the posterior tongue and oral cavity. The systems in this ganglion have not yet been completely studied. Nevertheless, at least four neural subsystems can be recognized in this ganglion—two distinct from those in the rat geniculate ganglion and two similar to them (Table 2-1).

The taste systems of the rat geniculate ganglion seem quite similar to those of the goat geniculate ganglion except for the rat amino acid units (found in both the rat geniculate ganglion (GG) and the rat petrosal ganglion). The acid, salt, and alkaloid units all had their counterparts in the goat geniculate ganglion. The GG salt units of the rat showed the same exclusivity of response to Na and Li salts (Figures 2-9 and 2-10), and the other two groups also had similar response properties (Boudreau et al., 1985). The two groups of units found exclusively in the rat petrosal ganglion (PG salt and PG acid units) differed from rat GG salt and GG acid units (Boudreau et al., 1987). The PG salt units, for instance, fired to salts other than Na and Li salts.

The studies on the four species are summarized in Table 2-2. At least nine neural groups can be identified in these four species, even though the petrosal ganglion was examined in only the rat. At the present time we are studying the geniculate ganglion of the pig, an animal that uses food systems largely distinct from those used by the other four species. At least four neural groups can be identified in the pig, two of which seem different from those encountered in other species.

The main criteria used to classify the units in Table 2-2 were stimulus response measures; that is, the units discharged or were inhibited by different chemical compounds. In addition, other criteria were used to supplement the chemical stimulus–response differentiation. Thus, the two main groups in the cat (acid and amino acid units) are further differentiated by spontaneous activity measures, latency to electrical stimulation, the area of tongue innervated, and differential responses to solution temperature (Boudreau and Alev, 1973; Ishiko and Sato, 1973; Nagaki et al., 1964). These two unit groups have been found by four independent groups of investigators in three countries. The two main groups in the rat (acid

Table 2-1 Rat Taste Systems in the Geniculate (GG)and Petrosal (PG) Ganglia

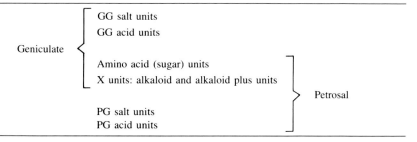

	GG salt units
Geniculate	GG acid units
	Amino acid (sugar) units
	X units: alkaloid and alkaloid plus units	Petrosal
	PG salt units
	PG acid units

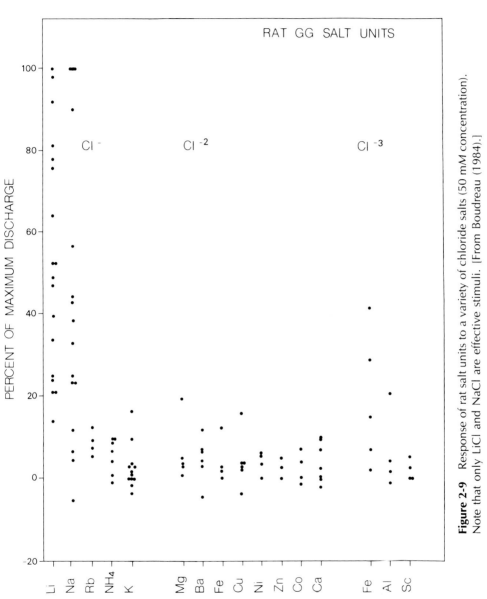

Figure 2-9 Response of rat salt units to a variety of chloride salts (50 mM concentration). Note that only LiCl and NaCl are effective stimuli. [From Boudreau (1984).]

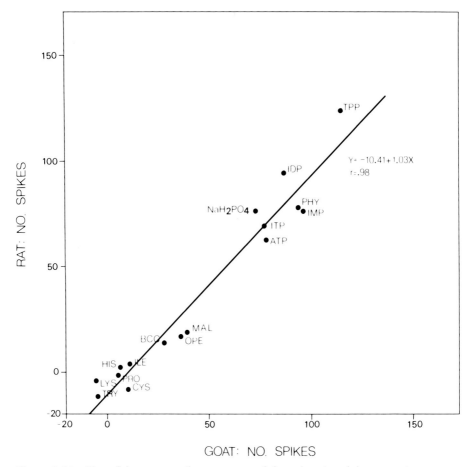

Figure 2-10 Plot of the mean spike responses of the salt units of the rat and goat to a variety of compounds. All compounds eliciting discharges greater than 50 spikes are sodium salts. Note that the salt units in the two species are quantitatively identical. Nonstandard abbreviations: BCC, butyrl choline chloride; MAL, malic acid; OPE, O-phosphoroethanolamine; TPP, tetrasodium pyrophosphate; PHY, phytic acid. [From Boudreau et al. (1985).]

units and salt units) have a similar research history. Different researchers using different techniques and different stimuli tend to see the same unit groups. Peripheral unit taste groups have also been described in fish (Caprio, 1975) and invertebrates (Bauer et al., 1981; Johnson et al., 1984).

   This comparative work has led to a modular view of peripheral taste systems, which is schematically represented in Figure 2-11. Here the different neural groups are seen to have distinct receptors that respond to distinct types of chemical stimuli. Both positive and negative aspects of the chemical signals are indicated (e.g., Brønsted acids and Brønsted bases), because inhibition is common.

   The implications of the power of accurate determination of natural stimuli extend beyond the initial experiments. Quantitative relationships between chemicals and spikes are the basis for concepts concerning receptor mechanisms. The

Table 2-2 Mamalian Peripheral Neural Groups

Neural Group	Animal	Stimuli
GG[a] salt system	Rat and goat only	Na^+ and Li^+
GG acid system	Rat and goat different from cat and dog	Brønsted acids
GG amino acid system	Cat and dog	Proline, cysteine, hydroxyproline, lysine, alanine
GG nucleotide system (Xa units)	Cat only	ITP, ATP, etc.
GG furaneol system (probably mainly PG) (X units)	Dog only	Furaneol, ethyl, maltol, methyl maltol
PG[b] amino acid (also in GG)	Rat	Sugar, saccharin, amino acids
PG alkaloid system (also in GG)	Rat and goat	Atropine
PG acid system	Rat	Restricted set of carboxylic acids
PG salt system	Rat	KCl, $CaCl_2$, $MgCl_2$, NaCl

[a] GG = geniculate ganglion (facial nerve).
[b] PG = petrosal ganglion (glossopharyngeal nerve).

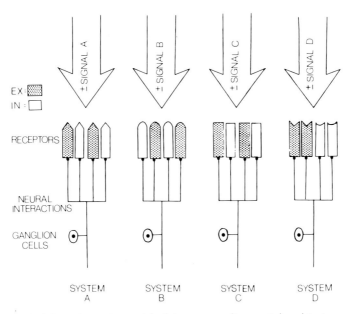

Figure 2-11 Modular schematic model of the mammalian peripheral taste system based on input–output analysis of peripheral sensory ganglion cells. A mammal's facial nerve taste system consists of three or four basic modules. Each module is attuned to different nutritional chemical signals, which have both excitatory and inhibitory properties. Within each module are at least two types of receptors, one excitatory and the other inhibitory. The spike output from the neuron is a result of the interaction of excitatory and inhibitory signals in the peripheral branches of the sensory ganglion cell dendrites. [From Boudreau et al. (1985).]

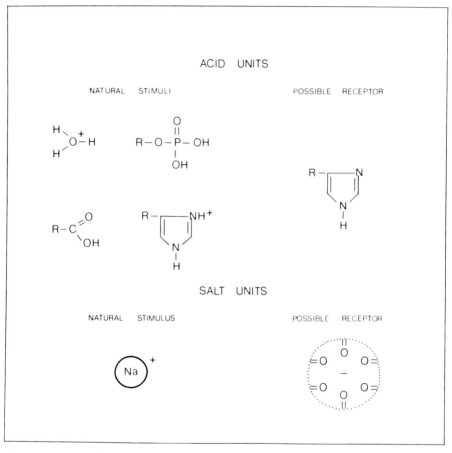

Figure 2-12 Specification of the active stimuli and possible receptors for the GG acid units (especially carnivore) and the GG salt units. Transduction is assumed to involve conduction of either protons (acid system) or sodium (salt system). [From Boudreau et al. (1983).]

specification of the stimulus has formed the basis for two theoretical formulations of taste receptor activation. In both cases, transport of charged particles is postulated as the active force for receptor activation (Figure 2-12). For the acid units, the presumed mechanism is proton transport; for the GG salt units it is sodium transport (initially proposed by DeSimone et al., 1981).

COMPARISON OF NEUROPHYSIOLOGY AND HUMAN TASTE SENSATIONS

Food chemists have provided us with at least 13 human sensations that have a certain amount of psychophysics and chemistry (Table 2-3). The sensations in Table 2-3 are arranged according to the cranial nerve believed to carry most of the information. Sensations carried by the trigeminal nerve (N. V) include a burning sensation (Grew, 1682; Sinden et al., 1976), cool and warm sensations, two

Table 2-3 Nerves and Chemical Sensations

Free nerve ending systems

Trigeminal nerve (N. V)
    Astringent—dry
    Astringent—tangy
    Pungent$_1$
    Pungent$_2$
    Warm
    Cool
    Burning

Taste bud systems

Facial nerve (N. VII)	*Glossopharyngeal nerve (N. IX)*
Sweet-1	Sweet-2
Bitter-1	Bitter-2
Salty	Metallic
Insipid	Umami-1
Sour	Umami-2
	Vagus nerve (N. X)
	?

---

astringent sensations (Sanderson et al., 1976), and two pungent sensations, since capsaicin and piperidine apparently activate different receptor systems (Stevens and Lawless, 1986).

Four sensations attributable to the facial nerve are the salty, sour, sweet-1, and bitter-1 sensations. The salty sensation is associated with relatively high concentrations of inorganic ions, particularly Na and Li. The sour sensation is elicited by various Brønsted acids, and there are indications that proton-donating nitrogen groups may be active at neutral pH. Thus, the sour sensation to histidine compounds can be attributed to the proton-donating capabilities of the imidazole ring (Boudreau and Nelson, 1977; Boudreau, 1978). The "-1" postscripts of sweet-1 and bitter-1 distinguish them from similar sensations elicitable from the back of the mouth. Sweet-1 is evoked by low-concentration solutions of inorganic salts, sugars, and various nitrogen compounds, especially amino acids such as L-hydroxyproline and L-alanine. Bitter-1 is associated with hydrophobic amino acids and alkaloids (Boudreau, 1978).

The sensations of sweet-2 and bitter-2 are elicited from posterior oral loci innervated by the glossopharyngeal nerve by chemical stimuli distinct from those that act on the front of the mouth. Dihydrochalcones are active stimuli for sweet-2 (Dubois et al., 1981), and the bitter-2 sensation is elicited by certain salts such as $MgSO_4$, and probably by various polyphenols. Additional sweet and bitter sensations could probably be distinguished. The sweet-tasting proteins thaumatin and monellin have been found to stimulate fungiform papillae maximally on the lateral edge of the tongue as opposed to sucrose that stimulates the tip (Van der Well and Arvidson, 1978). Certain foods seem to elicit a bitter sensation localized to the foliate papillae.

The umami sensation is of primary importance in many foods and its measurement is obligatory in food flavor research. Yamaguchi (1979) has reviewed the chemistry and psychophysics of the umami sensation. Two distinct types of food compounds elicit this sensation: a group of L-amino acids and a group of 5'-ribonucleotides and their derivatives (especially inosine monophosphate, [IMP], adenosine monophosphate [AMP], and guanosine monophosphate [GMP]. This sensation is most effectively elicited from the back of the tongue by monosodium glutamate (MSG). Since the glossopharyngeal nerve innervates the posterior tongue and the back of the mouth, this nerve may be associated with the umami sensation. Although the umami sensation is elicited by MSG, a mixture of IMP and MSG produces more sensation than the two tasted separately. Umami substances, when added to food, increase the total taste intensity and enhance certain flavor characteristics such as palatability, mouthfulness, and impact.

The list of sensations could be extended considerably because many taste sensations, such as grassy, acid, seafood, nutty, burnt, moldy, and papery, have been studied inadequately. Once the door is opened to allow consideration of more than four tastes, the whole concept of taste and food flavor is altered. (Flavor is usually described as the odor produced when a food is put in the mouth!) Food scientists attempting to synthesize the flavor of foods such as potatoes, crab, and abalone have discovered that all the essential compounds are nonvolatile (Solms, 1971, Konosu, 1979). Food flavor consists mostly of taste.

With the new view of taste physiology provided by single-unit systems analysis, it is possible to take an objective, unbiased view of the relationship between human sensations and mammalian neural taste systems. The first discovery is evidence that human sensations is associated with activity within a specific neural group. This "law of specific nerve energy" was predicted by early German psychophysicists. The only modification to be added to this concept is that inhibition within a neural group may signal a sensation distinct from excitation; thus, excitation of human GG salt units seems to signal salty and inhibition signals insipidity (Boudreau, 1984).

In attempting to match mammalian neural groups with human taste sensations (Boudreau, 1986), one discovers that the human taste system is far more complex than that of any of the animals studied. Furthermore, no one animal approximates the human; rather a composite of systems from different species is required to make even a rough match. Also, for some human sensations no adequate animal analogs exist, nor does there seem to be a human equivalent for some animal systems. Some suggested analogs of the neural systems presumed to underlie various human sensations are presented in Table 2-4. In this table, neural group was matched to sensation on the basis of stimulus chemistry.

Probably the only human neural taste system that is identical to one of those studied neurophysiologically is that believed to underlie the salty sensation (Boudreau, 1984). The stimuli eliciting the human sour sensation are quite similar to those that excite the cat and dog GG acid systems, although nitrogen acids in the human have been inadequately studied (Boudreau and Nelson, 1977). The human sweet-1 sensation is elicited in part by stimuli that are active on the cat amino acid system (Boudreau 1978, 1986). The human sweet-2 sensation is evoked by compounds somewhat similar to those that excite the dog's furaneol system (Dubois et al., 1981). No excitatory analog of a general bitter sensation is present in

Table 2-4 Relationships of Human Sensations to Animal Taste Systems

Taste Sensation	Possible Analog	Food System (Suggested)
Salty—Insipid	Like rat and goat GG salt system	Sodium compounds
Sour	Like cat and dog GG acid system	Many foods, both positive and negative
Sweet-1—bitter-1 (?)	Like cat GG amino acid system, in part	Animal tissues
Sweet-2	Like dog GG furaneol system	Fruit
Umami-1	Unlike any mammal system described, but like lobster glutamate system (Johnson et al., 1984)	Roots and tubers, tomatoes,shellfish, mushrooms
Umami-2	(Like cat GG nucleotide system, in part)?	Fish, mushrooms, carrion
Bitter-2	(PG alkaloid system)?	Alkaloids
Metallic	(rat PG salt)?	Mushrooms, seeds (?)

any experimental animal, because most bitter stimuli are neurophysiologically inactive or inhibitory; the few alkaloids that stimulate the rat and goat alkaloid system are exceptions. No analog exists in mammalian neurophysiology for umami-1 (Boudreau, 1986). The link between many human taste sensations and neurophysiologically delineated neural groups is tenuous and speculative. Apparently, the similarity between animal taste systems and those of the human is predicated more on the types of food systems used than on the animal's evolutionary standing, since the lobster seems to have as many taste systems in common with the human as the goat does.

ACKNOWLEDGEMENTS

I thank Hanh Le for assistance on this manuscript. This work was supported in part by NSF research grants.

REFERENCES

Bauer, U., Dudel J., and Hatt H. (1981). Characteristics of single chemoreceptive units sensitive to amino acids and related substances in the crayfish leg. *J. Comp. Physiol.* 144:67–74.

Boudreau, J. C. (1978). Cat and human taste responses to L-α-amino acid solutions. In: G. Charalambous and G. Inglett (eds.). *Flavor of Foods and Beverages*, pp. 232–246, New York: Academic Press.

Boudreau, J. C. (1984). A peripheral neural correlate of the human fungiform papillae salty, insipid sensations. *Chem. Senses*: 9:341–353.

Boudreau, J. C. (1986). Neurophysiology and human taste sensations. *J. Sensory Studies* 1:185–202.

Boudreau, J. C., and Alev, N. (1973). Classification of chemoresponsive tongue units of the cat geniculate ganglion. *Brain Res.* 54:157–175.

Boureau, J. C., Anderson, W., and Oravec, J. (1975). Chemical stimulus determinants of cat geniculate ganglion chemoresponsive group II unit discharge. *Chem. Senses Flavour* 1:495–517.

Boudreau, J. C., Do, L. T., Sivakumar, L., Oravec, J., and Rodriquez, C. A. (1987). Taste systems of the petrosal ganglion of the rat glossopharyngeal nerve. *Chem. Senses* 12:437–458.

Boudreau, J. C., Hoang, N. K., Oravec, J., and Do, L. T. (1983). Rat neurophysiological taste responses to salt solutions. *Chem. Senses* 8:131–150.

Boudreau, J. C., and Nelson, T. E. (1977). Chemical stimulus determinants of cat geniculate ganglion chemoresponsive group I discharge. *Chem. Senses and Flavour* 2:353–374.

Boudreau, J. C., Oravec, J., and Hoang, N. K. (1982). Taste systems of goat geniculate ganglion. *J. Neurophysiol.* 48:1226–1241.

Boudreau, J. C., Oravec, J., White, T. D., Madigan, C., and Chu, S. P. (1977). Geniculate neuralgia and facial nerve sensory systems. *Arch. Otolaryngol.* 103:473–481.

Boudreau, J. C., Shivakumar, L., Do, L. T., White, T. D., Oravec, J., and Hoang, N. K. (1985). Neurophysiology and geniculate ganglion (facial nerve) taste systems, species comparison. *Chem. Senses* 10:89–127.

Caprio, J. (1975). High sensitivity of catfish taste receptors to amino acids. *Comp. Biochem. Physiol.* 52:247–251.

DeSimone, J. A., Heck, G. L., and DeSimone, S. K. (1981). Active ion transport in dog tongue, a possible role in taste. *Science* 214:1039–1041.

Dubois, G. E., Crosby, G. A., Lee, J. F., Stephenson, R. A., and Wang, P. C. (1981). Dihydrochalcone sweeteners. Synthesis and sensory evaluation of a homoserine-dihydrochalcone conjugate with low aftertaste, sucrose-like organoleptic properties. *J. Agric. Food Chem.* 29:1269–1276.

Grew, N. (1682). *The Anatomy of Plants with an Idea of a Phylosophical History of Plants, and Several Other Lectures*, W. Rawlins (ed.) London, England.

Ishiko, N., and Sato, Y. (1973). Gustatory coding in the cat chorda tympani fibers sensitive and insensitive to water. *Jap. J. Physiol.* 23:275–290.

Johnson, B. R., Voight, R., Borroni, R., and Atema, J. (1984). Response properties of lobster chemoreceptors, tuning of primary taste neurons in walking legs. *J. Comp. Physiol. A* 155:593–604.

Konosu, S. (1979). The taste of fish and shellfish. In: J. C. Boudreau (ed.). *Food Taste Chemistry*, pp. 185–203. Washington, DC: American Chemical Society.

Nagaki, J., Yamashita, S., and Sato, M. (1964). Neural response of cat to taste stimuli of varying temperatures. *Jap. J. Physiol.* 14:67–89.

Pickenhagen, W., Velluz, A., Passerat, J. P., and Ohloff, G. (1981). Estimation of 2,5 dimethyl-4-hyroxy-3 (2H)-furanone (Furaneol) in cultivated and wild strawberries. *J. Sci. Food Agric.* 32:1132.

Sanderson, G. W., Ranadive, A. S., Eisenberg, L. S., Farrell, F. J., Simons, R., Manley, C. H., and Coggon, P. (1976). Contribution of polyphenolic compounds to the taste of tea. In: G. Charalambous and I. Katz (eds.). *Phenolic, Sulfur, and Nitrogen Compounds in Food Flavors*, p. 14–46. Washington, DC: American Chemical Society.

Sinden, S. L., Deah, K. L., and Aulenback, B. B. (1976). Effect of glycoalkaloids and phenolics on potato flavor. *J. Food Sci.* 41:520.

Solms, J. (1968). Geschmackstoffe und Aromastoffe des Fleisches. *Die Fleischwirtschaft* 48:287–290.

Solms, J. (1971). Nonvolatile compounds and the flavor of food. In: G. Ohloff and

A. F. Thomas, (eds.). *Gustation and Olfaction, an International Symposium,* pp. 92–110. New York: Academic Press.

Stevens, D. A., and Lawless. H. T. (1986). Sequential interactions of oral chemical irritants. Abstracts of International Symposium on Olfaction and Taste, p. 89. Snowmass Village, CO: Association for Chemoreception Sciences.

Van der Well, H., and Arvidson, K. (1978). Quantitative psychophysical studies on the gustatory effects of the sweet tasting proteins thaumatin and monellin. *Chem. Senses and Flavour* 3:291–305.

White, T. D., and Boudreau, J. C. (1975). Taste preferences of the cat for neurophysiologically active compounds. *Physiol. Psychol.* 3:405–410.

Yamaguchi, S. (1979). The umami taste. In: J. C. Boudreau (ed.). *Food Taste Chemistry,* pp. 33–51. Washington, DC: American Chemical Society.

II

TOUCH AND PAIN

The history of experimental investigations of the somatosensory system is long and interesting. Chapters 3 through 5 explore in detail one aspect of this field—the activity of the cortical neuron ensembles concerned with tactile sensation. Significant advances in the experimental study of this aspect of somatosensory input have been made in recent years, perhaps because, conceptually, the physical framework of body surface and stimulus quality is not difficult to appreciate. Conceptually and experimentally more difficult to work with, the sense of joint and limb position involves a complex interrelation of sensory input with motor control. Uncertainties over the nature of the central mapping features of joint position present difficult puzzles for researchers in this field (Mountcastle and Powell, 1959; McCloskey, 1978; and Iwamura et al., 1985).

Chapters 3, 4, and 5 examine the way information from single tactile receptors in the skin is collated and used by ensembles of neurons at the cortical level. The nature of the cortical representation is discussed at length, and the authors' views fairly represent the currently held concepts. Gardner and her coauthors show us the remarkable transformation that occurs in the cortex, where information derived from many single skin receptors is collated by single cortical neurons that "survey" a whole array of receptors. These cortical neurons can respond to a pattern of activation across the array of receptors. Some cells respond selectively to a sequential activation that travels in only one direction across the array, in a temporally coded fashion, thus coding for speed and direction of movement which are qualities not appreciated at the receptor level. This generation of new analytical properties at a cortical level is a typical feature of sensory cortex organization, which we will see described in many of the following chapters on other sensory systems. The analytic property of the sensory cortex neuropil interests neuronal modelers, and the nature of the function and basis for the generation of new properties interests sensory cortex investigators.

Chapter 4 describes a most intriguing cortical area concerned with somato-sensation; this is the region of cortex that contains a representation of the long sensory whiskers that form a tactile fan around the snout of rodents and other mammals. Many rodents can draw this fan of stout whiskers rapidly and repeatedly over an object encountered in a stroking or "whisking" motion, apparently to recognize the contours of the surface much as we would use our hands to feel our way around in the dark, but probably far more efficiently. The cortical representation of this snout region has a clearly defined anatomy that lends itself to somatosensory studies. Chapter 4 shows how studies in widely different animal species can help us understand features of organization common to all sensory cortical areas. Particularly interesting are the excitatory and inhibitory interactions across the representation of the entire whisker array, and the correlation of anatomical and physiological information. The 2-DG technique is a component of these studies.

The same 2-DG technique is discussed in Chapter 5, which stresses the dynamic quality of cortical neuron responses and explores the recruitment of neuron ensembles and gating of neuron activity between different cortical depths. The original concept of the cortical column, with similar response properties from pia to white matter, has evolved considerably over the years; it will be apparent from Chapters 4 and 5 that response properties differ distinctly at different points between pia and white matter in the cortical neuropil. Chapter 5 describes the complex and puzzling patterning of activity in stripelike arrays, and the relationship between the pattern of metabolic activity seen in 2-DG and unit recording data remains open to interesting debate.

Stripelike patterns of activity and anatomical connections are common features of sensory cortex. Their function is unclear, even when their patterns of anatomical connection have been defined. It may be that developmental constraints on nerve fiber populations that compete with each other for terminal territory in the same region, but both needing to construct a topographic map, lead to alternating stripelike arrays of terminations of the competing pathways. Much has been published on this phenomenon (see, e.g., Hubel et al., 1977; Von der Marlsburg and Willshaw, 1976; Constantine-Paton and Law, 1978; Constantine-Paton and Ferrari-Eastman, 1987), but as yet no role for the segregation patterns is evident. Nonetheless, the development of these topographic projections and segregation patterns may determine how sensory information is processed in the neuropil; these patterns of nerve fiber termination may indeed be an important constraint or necessary element in the development of analytic capability by the neuropil. This question of the importancce, in analytic terms, of the geometric patterns of sensory cortex is currently of great interest to sensory system investigators.

An experimental approach that attempts to overcome the problems of recording from single cells is presented in Chapter 6. Kulics and Cauller use an array of recording points distributed in depth to gather information simultaneously from different laminae in the cortical neuropil of behaving monkeys. The multiple-point electrode can gather information on the local field potential and on multiunit activity and provides data for comparison of activities over time, permitting analysis of intrinsic processing in cortical depth. Chapter 6 introduces current source density analysis, and use of this method in the alert animal raises the possibility of determining interesting relationships among physiological activity, intracortical circuitry, and behavior.

The somatosensory inputs that allow discrimination of the quality of objects and surfaces must be closely linked with those systems that recognize and control limb and body position. A third sensory system, intimately associated with these two, signals pain and potentially damaging stimuli. Experimental work on the pain system is of the utmost importance in devising appropriate clinical treatments, yet it understandably raises concern that the experimental subjects may suffer. In Chapter 7 Willis makes clear what limitations these concerns must place on experimental paradigms in this field of research. Studies carried out on anesthetized animals and on alert animals able to avoid the stimulus if distressed by it have produced a wealth of information. But, of course, anesthetics are effective blockers of neural pain pathways, particularly in the more central neuropil of thalamus and cortex. Thus, for the investigator concerned with the many interesting changes in appreciation of pain and its emotional connotations—factors governed by central neural pathways—there are real obstacles. Chapter 7 shows how far researchers have progressed while operating within these constraints.

REFERENCES

Constantine-Paton, M., and Law, M. I. (1978). Eye-specific termination bands in tecta of three-eyed frogs. *Science* 202:639–641.

Constantine-Paton, M., and Ferrari-Eastman, P. (1987). Pre- and postsynaptic correlates of interocular competition and segregation in the frog. *J. Comp. Neurol.* 255:178–195.

Hubel, D. H., Wiesel, T. N., and LeVay, S. (1977). Plasticity of ocular dominance columns in monkey striate cortex. *Phil. Trans. R. Soc. Lond.* (Biol.) 278:377–409.

Iwamura, Y., Tanaka, M., Sakamoto, M., and Hikosaka, O. (1985) Vertical neuronal arrays in the postcentral gyrus signaling active touch: a receptive field study in the conscious monkey. *Exp. Brain Res.* 58:412–420.

McCloskey, D. I. (1978). Kinesthetic sensibility. *Physiol. Rev.* 58:763–820.

Mountcastle, V. B., and Powell, T. P. S. (1959). Central nervous mechanisms subserving position sense and kinesthesis. *Bull. Johns Hopkins Hosp.* 105:173–200.

Von der Malsburg, C., and Willshaw, D. J. (1976). Mechanism for producing continuous neural mappings. Ocular dominance stripes and ordinal retinotectal projections. *Exp. Brain Res.* (Suppl.) 1:463–469.

3

Touching the Outside World: Representation of Motion and Direction Within Primary Somatosensory Cortex

ESTHER P. GARDNER, HEIKKI A. HAMALAINEN,
CLAUDE I. PALMER, AND SUSAN WARREN

One of the most challenging problems in modern neurophysiology is to understand how the brain processes sensory information and derives specific attributes or "features" from complex stimuli. Sensory systems are particularly sensitive to changes in the spatial location of stimuli and have developed mechanisms for detecting movement. This has been well studied in the visual system, in which motion-sensitive and direction-selective neurons were first described in the cerebral cortex by Hubel and Wiesel (1968). In 1972, Gerhard Werner and colleagues published a report of direction selectivity of neurons in the somatosensory cortex of monkeys (Whitsel et al., 1972). This was the first demonstration of feature-detecting properties in the somatosensory system, and it suggested that similar information-processing mechanisms might operate in a variety of sensory systems.

The basic hypothesis we propose is that feature detection results from integration of extensive information from a population of receptors. Individual receptors in almost all sensory systems have very small, circumscribed receptive fields. They respond to a limited set of stimuli in a restricted portion of the sensory sheet. For example, receptive fields of mechanoreceptors on the hands of humans and monkeys cover only a few square millimeters. When objects are grasped in the hand and manipulated, information is conveyed to the central nervous system from a large population of tactile sensors. To sense the global properties of objects, such as form, motion, and texture, information from individual receptors must be integrated in some coherent manner. We have proposed that the mechanism of this integration involves the specific organization of synaptic inputs on feature-detection neurons (Warren et al. 1986a). Such neurons respond only when a specific conjunction of receptors is activated.

In the somatosensory system, feature-detection neurons require inputs to be presented in a specific spatial or temporal configuration on a large area of skin. Certain neurons in the cerebral cortex respond most vigorously when an object moves across the skin in a particular direction, thus activating mechanoreceptors

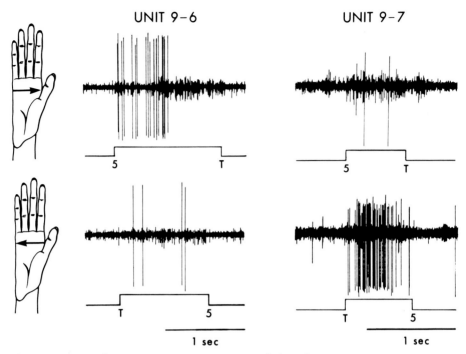

Figure 3-1 Two direction-sensitive neurons recorded on the same penetration in area 2. Cell 9-6 responds strongly to stroking of the palm in the radial direction, whereas cell 9-7 responds to motion in the ulnar direction. [Reproduced with permission from Costanzo and Gardner (1980).]

in a specific temporal sequence (Whitsel et al., 1972, 1979; Hyvarinen and Por-anen, 1978; Costanzo and Gardner, 1980). Examples of this type of feature-detection neuron are presented in Figure 3-1, which shows two direction-sensitive neurons recorded on a single electrode penetration in area 2. Both cells had receptive fields on the palm, but opposite direction preferences. Cell 9-6 was strongly excited by stroking the palm from beneath digit 5 toward the thumb, and it gave only a weak response to stroking of the skin in the opposite direction. Cell 9-7 responded vigorously to motion toward the ulnar margin of the palm, and was insensitive to motion in the radial direction. Responses to motion in other directions was intermediate between the preferred and null directions. Thus each cell had a preferred sequence of receptor activation.

Other cortical neurons respond only when a large number of receptors are turned on simultaneously, such as when an object is grasped in the hand (Hyvarinen and Poranen, 1978; Iwamura and Tanaka, 1978). These cells sometimes require that the edges be oriented in a particular direction, responding to a highly specific combination of simultaneously activated receptors. This integration of specific sensory information is necessary to sense such global properties of objects as form, motion, and texture.

Investigations of the mechanisms underlying motion sensitivity and direction selectivity in different sensory systems are useful for elucidating the basic mechanisms for feature detection by the nervous system. One would like to know which mechanisms are amodal, and generalizable to different sensory pathways, and which are specific to individual sensory modalities. Direction selectivity in both the visual and somatosensory systems involves two distinct, yet complementary mechanisms: suppression of activity in the least preferred direction and facilitation in the most preferred direction (Barlow and Levick, 1965; Hyvarinen and Poranen, 1978; Costanzo and Gardner, 1980; Newsome et al., 1986; Warren et al., 1986b). In this chapter, we present evidence that similar facilitatory and inhibitory mechanisms underlie motion sensitivity of cerebral cortical cells.

CORTICAL RESPONSES TO MOVING GRATINGS

In previous studies, we demonstrated that moving stimuli crossing the skin provide an effective means of exciting somatosensory cortical neurons (Costanzo and Gardner, 1980; Gardner, 1984; Warren et al., 1986a,b). Most cells with tactile receptive fields respond more vigorously to tangential movement across the skin than to simple pressure. Furthermore, the path of movement does not have to be continuous to be sensed as motion. When we rolled grating wheels with variably spaced teeth across the skin (Figure 3-2), we found that this semipunctate stimulus was as effective as brushing the skin in exciting movement-sensitive cells. During the rolling movement, only the grating ridges contacted the skin; the gaps between ridges were not stimulated.

Three types of movement-sensitive responses were observed in the primary somatosensory (SI) cortex. About 37 percent of the neurons responded equally well to motion in both longitudinal and transverse directions (*motion-sensitive neurons*). They were prevalent in cortical areas 3b and 1, and their receptive fields tended to be located on glabrous skin. Almost 60 percent of the neurons responding to motion displayed some direction sensitivity, showing preferences for movement in one or more directions. *Direction-sensitive neurons* were more common on hairy skin than on glabrous skin, and were usually found in areas 1 and 2. A small number of cortical neurons showed clear orientation sensitivity to moving stimuli. *Orientation-sensitive neurons* responded more vigorously to transverse moving stimuli than to motion along the longitudinal axis. Unlike directive-sensitive neurons, these cells did not differentiate motion in opposing directions, but responded equally well to motion in ulnar and radial directions or proximal and distal directions.

Although the gratings ranged in spatial period from 0.8 to 9.6 mm, we found little differentiation of textures in the firing patterns of cortical neurons (Warren et al., 1986b). For example, the direction-sensitive neuron shown in Figure 3-2 responded vigorously to rolling the grating from the elbow to the wrist, but showed almost no activity when the direction of motion was reversed. The direction preference for distal motion was maintained even when the spatial period of the grating was increased fivefold. Thus, cortical neurons can apparently integrate information from widely spaced points in their receptive fields, as long as the stimuli are presented in the appropriate temporal sequence. Signaling of motion, and its direction, apparently does not require continuous trajectories across the skin, nor do immediately adjacent points have to be stimulated to provide information about

MONK 308   UNIT 35-09

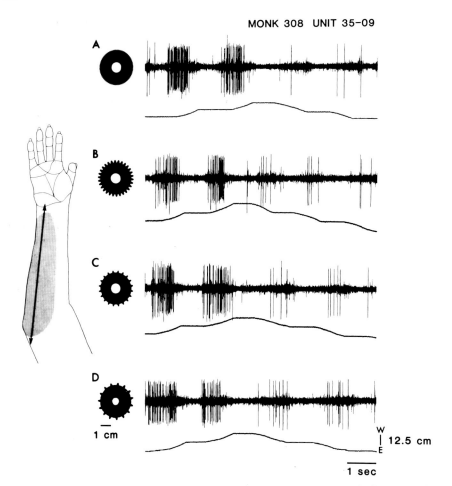

Figure 3-2   Direction-sensitive neuron insensitive to the spatial period of grating wheels rolled up and down the forearm. The cell fires at high rates to distal movements from elbow to wrist, and is weakly excited by motion in the opposite direction. The trace below the unit records indicates output of a potentiometer monitoring rotation of the grating wheel. Upward or downward deflections indicate a complete sweep of the grating across the receptive field in the indicated direction; during flat portions of the trace, the grating was lifted from the skin and repositioned for a new stimulus. [Reproduced with permission from Warren et al., (1986a).]

the direction of motion. The essential feature of motion required to activate direction-sensitive neurons appears to be the order in which peripheral receptive fields are activated.

OPTACON STIMULATION

We have developed a new method for studying motion sensitivity on the skin using a series of closely spaced probes on the skin surface. By strobing the individual probes at appropriate intervals, one can produce the *illusion* of motion on the skin, much as is done in the visual system by rapidly flashing lights at neighboring locations in the visual field (Newsome et al., 1986). Simulation of motion across the skin using a dynamic display has the advantage that one can precisely locate the specific points activated at a particular moment. Furthermore, one can independently and rapidly vary the spatial and temporal characteristics of the moving pattern by computer control. We have adapted the OPTACON (Telesensory Systems, Mt. View, CA) to simulate movement of bars and stripes on the hands of humans and monkeys.

The OPTACON is normally used as a sensory substitution aid for the blind and deaf, which translates printed text or speech into vibrating patterns that are swept across the skin (Bliss et al., 1970). It consists of an array of 144 miniature probes placed over an 11- by 26-mm rectangle of skin on the finger. The probes are arranged in a matrix of 6 columns and 24 rows. Adjacent rows are separated by 1.1 mm, and the columns are 2 mm apart. Blind individuals use this device with an optical sensor, which scans printed text character by character. The patterns of light and dark images viewed by the sensor are translated into a corresponding pattern of vibrations on the tactile array. A user can "view" these patterns by placing a finger over the vibrating probes, and can thus "read" normal texts such as the daily newspaper. The OPTACON has also been widely used in psychophysical studies of tactile discrimination in humans (reviewed in Craig and Sherrick, 1982; Craig, 1985). Despite use of this device by more than 10,000 blind individuals, no electrophysiological studies of the neural activity generated by OPTACON spatial patterns have been published.

In the experiments described in the following sections, single-cell responses to the OPTACON were recorded from cutaneous mechanoreceptors innervating the glabrous skin of the monkey's hand, and from cerebral cortical neurons in areas 3b, 1, and 2. Single pulses (each of four millisecond duration) were applied simultaneously to the central four columns of one or two rows on the OPTACON, under computer control. When pulsed simultaneously, the individual probes in a row are sufficiently close together to evoke the sensation of an edge or bar pressed against the skin. The active row was shifted proximally or distally to the immediately adjacent row (1.1 mm away), or to rows 2.2 or 4.4 mm away, at intervals of 10, 20, 40, or 80 ms. High-frequency stimulus presentation is perceived as a continuous sweep of an edge across the skin, whereas low-frequency stimuli feel pulsatile or punctate. We will demonstrate that the difference between apparent motion and punctate stimuli is reflected in the properties of the cortical spike train.

RESPONSES OF CUTANEOUS MECHANORECEPTORS TO OPTACON STIMULATION

The fingertips of humans and monkeys have the highest density of mechanoreceptors, and are the tactile equivalent of the fovea in the retina. Glabrous skin on the fingers is characterized by a regular array of ridges arranged in circular groups forming the fingerprints. These papillary ridges contain four types of mechanoreceptors: Meissner's corpuscles, Merkel cells, Ruffini endings, and Pacinian corpuscles. Meissner's corpuscles and Merkel cells are the most numerous mechanoreceptors in glabrous skin; they lie close to the surface of the epidermis and have small receptive fields covering three or four papillary ridges. Pacinian corpuscles and Ruffini endings lie in the dermis or subcutaneous tissue; they have large receptive fields with rather indistinct borders.

Physiologically, one can distinguish these different types of mechanoreceptors by their receptive field contours and their responses to steady pressure on the skin (Darian-Smith and Kenins, 1980; Vallbo and Johansson, 1978). Meissner's corpuscles and Pacinian corpuscles fire on–off responses to pressure and are *rapidly adapting (RA) receptors;* they are sensitive to the movement of stimuli into the skin and to changes in pressure. Merkel cells and Ruffini endings fire throughout the period of stimulation, and are therefore *slowly adapting (SA)* mechanoreceptors; Merkel cells are called SAI mechanoreceptors, and Ruffini endings are designated SAII receptors. We used these criteria to classify mechanoreceptor axons dissected from the medial and ulnar nerves in barbiturate-anesthesized monkeys.

Meissner's Afferents (RAs)

RAs were extremely responsive to OPTACON stimulation, firing one or two impulses to stimulation of each row within their receptive field (Figure 3-3). What is striking in the responses of RAs is the *great fidelity* and *reproducibility* of the responses to the rows within the receptive field. Responses at each field position occurred with identical latencies on each trial and were either present on all trials or completely absent. RAs followed both high and low rates of stimulation; interspike intervals were approximately 10, 20, or 40 ms, corresponding to the interpulse intervals on the OPTACON. Average firing rates were highest to 100 Hz stimuli, and decreased exponentially (Figure 3-4).

Responses were relatively independent of location within the receptive field. Most RAs fired one spike/pulse, yielding uniform sensitivity throughout the receptive field (Figure 3-3, *top*). The remainder fired two spikes, to probe indentation and retraction, when stimulated in the middle of their fields (see Figure 3-3, *bottom*). Receptive fields spanned two to five OPTACON rows, depending on their location on the hand.

Only minor changes in *total spike output* were observed as a function of tem-

Figure 3-3 Responses of two RA mechanoreceptors to sequential activation of rows 1 to 24 of the OPTACON stimulator. Diamonds above the records indicate the time of activation of specific rows on the OPTACON. (A) Each row was turned on once with a 4-ms pulse, and then the immediately adjacent row was activated 10, 20, or 40 ms later (100, 50, and 25 Hz). Impulse replicas show responses to five sequential stimulus presentations; PSTHs below rasters show averages of 15 (100 Hz), 10 (50 Hz), and 5 (25 Hz) sweeps across the OPTACON. The ordinate plots the probability of firing in each 1-ms bin. Unit 2-10 (Monkey 83), located on the radial margin of the distal phalanx of digit 3, responded to two rows. (B) Pairs of rows, spaced six rows apart, were turned on simultaneously, and shifted in tandem across the OPTACON in the proximal direction. Spike bursts separate into two groups when spacing of active rows exceeds the field diameter. The same format is used as in Part A. Unit 202-7 (Monkey 301), located on the ulnar margin of the palm, responded to activation of rows 14 through 16.

RA mechanoreceptors

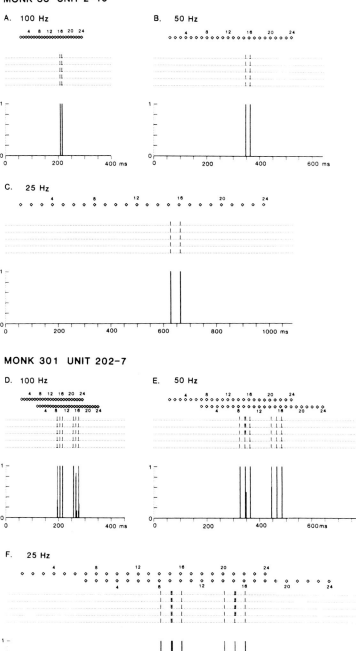

MONK 83 UNIT 2-10

A. 100 Hz

B. 50 Hz

C. 25 Hz

MONK 301 UNIT 202-7

D. 100 Hz

E. 50 Hz

F. 25 Hz

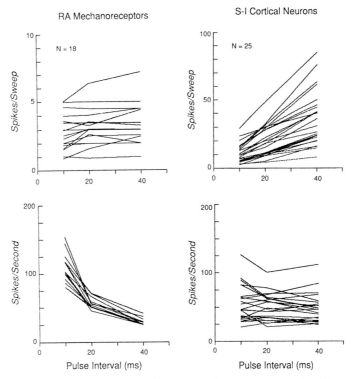

Figure 3-4   Comparison of responses of 18 RA mechanoreceptors and 25 SI cortical neurons to OPTACON stimulation. (*Top panel*) Total spikes evoked per sweep as a function of interpulse interval (*Bottom panel*) Average firing rates as a function of pulse interval.

poral frequency of stimulation in more than 80% of the cells studied. The range of responses observed is illustrated in Figure 3-4. Eleven of 18 RAs showed flat input-output functions, indicating no effect of pulse frequency on spike output. Five cells showed increases of 1 or 2 spikes per sweep as the interpulse interval was lengthened from 10 to 40 ms, and two cells showed strongest activity at 20 ms intervals.

Responses to pairs of rows separated by 1 to 12 mm were similar to those evoked by single rows. The paired rows correspond to stripes of different spatial frequencies. Individual stripes were resolved only when their spacing exceeded the diameter of the receptive field. For example, the cell illustrated in the lower panels of Figure 3-3 was activated by pulsing rows 14, 15, and 16 on the tactile array. The center row was most effective, producing activity at both indentation and retraction from the skin. We observed a pause in activity with six row separations, as the active rows straddled the edges of the field. Responses to each stripe were distinct, and nearly identical. RA mechanoreceptors can thus follow repetitive stimulation of their receptive fields, showing little habituation of responses.

RAs did not sum inputs from two rows within their receptive fields, responding only to the shorter latency row. If two active rows were separated by a blank row, their neural representations were merged into a single, broad bar, and could not be resolved. There was no gap in the spike discharge corresponding to the blank row. Thus, resolution of stripes seems to be limited by the receptive field diameter.

Pacinian Corpuscles (PCs)

PCs also responded to OPTACON stimulation, but fired more irregular responses than did RAs. Intertrial variability was greater, and multiple spike bursts were more common. PCs fired up to four spikes per pulse, yielding greater spikes per sweep values than the RAs. PC receptive fields ranged from 5 to 24 rows, reflecting a greater sensitivity to high-frequency vibration (Verrillo, 1966; Talbot et al., 1968). Total spike output decreased slightly at high stimulus frequencies, because of occlusion of multiple-strike responses from subsequent stimuli. RAs had better spatial resolution than PCs, because of their smaller receptive fields and more regular responses.

Slowly Adapting (SA) Mechanoreceptors

SAI and SAII mechanoreceptors appear to be totally insensitive to activation of the OPTACON. They responded only to contact of the receptive field with the tactile array, but no modulation of the spike discharge occurred when single bars or multiple stripe bars were displayed on the array. All 53 SA receptors tested were unresponsive to OPTACON activation. A similar insensitivity of SA afferents to OPTACON patterns was found by Johnson and colleagues and reported in an unpublished study using the standard OPTACON (personal communication). The indentation and retraction cycle of the probes appears to be too rapid for excitation of SA mechanoreceptors. Thus, SAs contribute only noise to the CNS from skin contact with the OPTACON, but no useful information concerning spatial patterns.

The insensitivity of SA mechanoreceptors to spatial patterns on the OPTACON is particularly unfortunate, because Johnson and Lamb (1981) and Phillips and Johnson (1981) have shown that SA mechanoreceptors have the best resolution of spatial detail when tested with Braille dot patterns or gratings pressed against the skin. This may explain why the reading rates of the blind are much poorer with the OPTACON than with Braille.

In summary, dynamic patterns on the OPTACON excite only RA and PC mechanoreceptors. Their firing rates are tightly linked to the frequency of stimulation over the range of 25 to 100 Hz; average rates are highest to 100-Hz stimuli, and decrease with decreasing frequency. Most RAs showed no change in total spikes per sweep when the frequency was raised from 25 to 100 Hz, but PCs displayed slight decreases at high frequencies.

CORTICAL RESPONSES TO OPTACON STIMULATION

A remarkable transformation of sensory information occurs between the peripheral nerves and primary somatosensory cortex, involving enlargement of receptive fields, and filtering of high-frequency responses. Timing and spacing between stimuli play key roles in determining cortical firing patterns. The closer the active rows are in space and in time, the more mutual attenuation of activity occurs in somatosensory cortex. We propose that this filtering regularizes firing, yielding a sensation of continuous motion across the skin.

Receptive Fields

In SI cortex of alert monkeys, OPTACON stimulation activates motion-sensitive neurons whose preferred stimulus is light brushing or stroking of the skin. These cortical neurons integrate information from mechanoreceptors innervating wide areas of skin; as a consequence, they can be activated from almost any part of the OPTACON array. Figure 3-5 shows typical responses of an area 1 neuron to sweeping the active row across the entire OPTACON. This neuron had its recep-

S-I Cortical Neuron

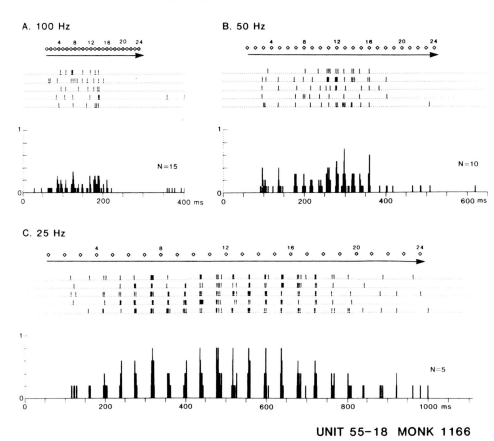

UNIT 55-18 MONK 1166

Figure 3-5 Impulse replica rasters and PSTHs for a cortical neuron responding with phasic bursts of impulses to OPTACON pulses. Unit 55-18, Monkey 1166, recorded in area 3b; receptive field on the medial margin of digit 2. (Same format as Figure 3-2.)

tive field on the distal and middle phalanges of digit 2. Its receptive field is greatly enlarged compared with RA mechanoreceptor receptive fields; the cell can be activated from almost the entire tactile array, rather than from the two to five rows observed in peripheral nerves. Total spikes per sweep in the cortex were approximately an order of magnitude larger than those for mechanoreceptive afferents (see Figure 3-4 *top*). Cortical neurons thus are able to view the entire tactile array, rather than little pieces of it.

Temporal Frequencies

Temporal frequency of stimulation seems to be the major determinant of cortical firing patterns. Low-frequency stimuli evoke short, highly regular bursts of impulses, separated by silent periods (Figure 3-5C). Responses are phase locked to the OPTACON pulses, and typically occur at latencies of 18 to 20 ms. Activity is strongest at the onset of stimulation, or in the middle of the receptive field.

   Whereas in the peripheral nerve responsiveness altered little as the frequency of stimulation was changed, in the cortex, frequency following is most apparent at low rates of stimulation. The number of spikes evoked per pulse diminishes as the pulse frequency rises to 50 Hz (Figure 3-5B), but phase locking to OPTACON pulses is still apparent in the spike train. At higher stimulus frequencies (100 Hz), frequency following is further degraded, and many of the rows that are effective at low frequencies evoke no response at all. Greater variance is also observed in the spike trains elicited by 100-Hz stimuli than at lower rates. A similar falloff in frequency following has been observed by Mountcastle and co-workers (1969) and by Ferrington and Rowe (1980) in SI neurons tested with vibratory stimuli at a fixed position in their receptive fields.

   The receptive field size of cortical neurons depends on frequency of stimulation. Apparent receptive field dimensions expand when tested with slower rates of stimulation. In Figure 3-5, responses were evoked from rows 2 to 14 at 100 Hz, with only sporadic activity from more proximal rows. Decreasing the temporal frequency to 25 Hz expanded the excitable territory to cover rows 2 through 22.

   A more quantitative evaluation of firing patterns is illustrated in Figure 3-4. The average number of spikes discharged per OPTACON pulse increased linearly as the interpulse interval was raised from 10 to 40 ms, but it increased only slightly more at pulse intervals of 80 ms. The increased responsiveness was large, ranging from 20 to 50 spikes per sweep in most cells studied; this change in firing was more than an order of magnitude greater than that observed for the most sensitive RA mechanoreceptors.

   The decreased responses and receptive field shrinkage observed at brief interpulse intervals suggest a *time-dependent inhibitory process*. Poststimulus inhibition appears strongest at 10-ms pulse intervals, decreasing to prestimulus levels by 40 ms. We demonstrated previously this kind of in-field inhibition, using stationary stimuli in the receptive field (Gardner and Costanzo, 1980a; 1980b), and paired moving stimuli (Gardner and Costanzo, 1980c). The peripheral nerve experiments clearly demonstrate that the reduced responsiveness to high-frequency stimulation is not a result of receptor adaptation, but rather of central neuronal processing. The total spikes per sweep value of mechanoreceptors is independent of frequency, whereas SI cortical neurons show clear response attenuation at high frequencies (Figure 3-4, *top*).

Temporal frequency of stimulation is the major parameter determining the firing characteristics of cortical cells during simulated motion across the skin. These firing characteristics are shown in greater detail in peristimulus time histograms, (PSTH), which average the responses for a period 10 ms before to 40 ms following each OPTACON pulse (Figure 3-6). Each PSTH displays an averaged response to a series of OPTACON pulses, independent of their field location. Activity is clearly phase locked to the OPTACON pulses, as indicated by the dotted lines below the abscissa marking the timing of responses to the preceding and subsequent stimuli. Peaks in the activity profiles correspond exactly to the interpulse interval. Cortical activity is clearly attenuated at 100 Hz compared to activity at 50 and 25 Hz. The activity pattern changes from continuous, frequency-modulated activity at high frequencies to strong bursts separated by long silent periods at low stimulus rates.

The magnitude of the neural transformation is most apparent if one compares cortical PSTHs with those from RA mechanoreceptors (Figure 3-6, *left*). RA responses occur at 4 to 5 ms latencies, or 8 to 9 ms latencies, depending on the field location in the trajectory. Little or no attenuation of activity occurs as the interpulse interval is shortened. Therefore, mechanoreceptors fire at much higher rates to the more rapidly applied stimuli, and average firing rates decrease in parallel with the drop in stimulus temporal frequency (see Figure 3-4). Clearly there is a correlation between the firing rates of peripheral nerves and the degree of cortical attenuation of activity.

The activity patterns displayed in the PSTHs parallel the sensations evoked by the different stimulus frequencies when four column bars are swept across the human index finger. At 100 Hz, it produces the sensation of continuous, smooth motion across the skin; sensations from the individual probes blur, merging into each other. A 25-Hz rate produces a distinct series of punctate taps that stutter their way across the skin.

While the spike train is dramatically transformed by different rates of stimulus presentation, *average firing rates* in the cortex are relatively uniform over this range of frequencies. Although the number of spikes evoked per OPTACON pulse increases as a function of temporal period, the long pauses between bursts at low frequencies often lead to a lower average number of spikes per second. The average firing rates of the 16 neurons illustrated in Figure 3-4 dropped or remained constant as interpulse intervals were raised from 10 to 40 ms.

Velocity of Apparent Motion

To determine whether the attenuation of cortical responses at high stimulus frequencies is due to the timing between stimuli or to the apparent velocity at which the moving stimulus crosses the skin, we used different combinations of spacing and timing on the OPTACON. For example, sequential activation of rows spaced either 1 mm apart at 100-Hz rates, 2 mm apart at 50 Hz, or 4 mm apart at 25 Hz yields the same apparent velocity of motion across the skin (110 mm/s). Stim-

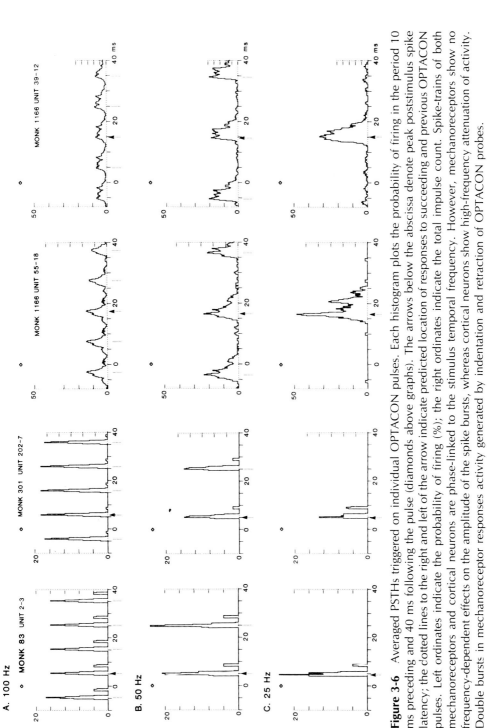

Figure 3-6 Averaged PSTHs triggered on individual OPTACON pulses. Each histogram plots the probability of firing in the period 10 ms preceding and 40 ms following the pulse (diamonds above graphs). The arrows below the abscissa denote peak poststimulus spike latency; the dotted lines to the right and left of the arrow indicate predicted location of responses to succeeding and previous OPTACON pulses. Left ordinates indicate the probability of firing (%); the right ordinates indicate the total impulse count. Spike-trains of both mechanoreceptors and cortical neurons are phase-linked to the stimulus temporal frequency. However, mechanoreceptors show no frequency-dependent effects on the amplitude of the spike bursts, whereas cortical neurons show high-frequency attenuation of activity. Double bursts in mechanoreceptor responses activity generated by indentation and retraction of OPTACON probes.

uli of the same duration seem to elicit approximately the same total number of impulses, regardless of the combination of spatial and temporal frequencies used to stimulate the skin. However, details of the spike train change dramatically, depending on the temporal frequency of stimulation. The number of *spikes per pulse increased,* and firing characteristics changed from *continuous low frequency* activity to *pulsatile bursts,* as intermediate rows were eliminated from the spatial pattern. The constant number of spikes per sweep observed for stimuli of the same apparent velocities seems to result from a balance between a lower number of spikes per stimulus pulse and higher rates of stimulation for tightly spaced, rapidly presented patterns. Firing patterns of cortical neurons thus seem to be determined by the *temporal frequency of stimulation* rather than by the apparent velocity of motion.

In a parallel series of psychophysical studies (Gardner and Sklar, 1986), we found that the ability of humans to discriminate the direction of motion across the skin at a given velocity is significantly better with rapidly presented dense patterns, than with the widely spaced, lower frequency patterns. This improved direction selectivity may result from the greater amount of information provided by the more numerous mechanoreceptors activated by dense arrays.

CONCLUSION

We have seen that the pathways leading up to the cortex and intracortical networks filter information so that low-frequency stimuli are less severely attenuated than those of high frequencies. High-frequency pulses are perceived as smoothly moving stimuli, and they elicit continuous, regular firing by cortical neurons. Low-frequency stimuli are perceived as pulsatile and punctate, and their cortical responses are burstlike. Thus, the difference between moving and punctate stimuli may be encoded by the properties of the cortical spike train.

What have we learned from simulated motion that we did not already know from more traditional stroking and brushing? We view two results as particularly significant. When one speaks of motion across the skin, one usually refers to speed and distance moved. Our studies with simulated motion suggest that velocity and path length are not the most fundamental parameters determining cortical responses. What we really mean by velocity is the *temporal frequency* at which adjacent mechanoreceptors are activated. The temporal frequency of stimulation of mechanoreceptors and cortical neurons is the major determinant of their firing patterns. This appears to be characteristic of both simulated moving patterns on the OPTACON and motion of textured surfaces such as gratings and Braille dot arrays (Darian-Smith and Oke, 1980; Darian-Smith et al., 1980; Johnson and Lamb, 1981). Our psychophysical data suggest that discrimination of direction seems to be crucially dependent on *how many* channels are activated sequentially, rather than on their proximity on the skin, or the total distance traversed (Gardner and Sklar, 1986).

We have also demonstrated that spatial location on the skin is not clearly specified in the spike discharge patterns of cortical neurons. Sensitivity within the receptive field tends to be uniform, particularly when slow rates of stimulation are used. Furthermore, many cells are most active at the onset of stimulation, regardless of field location, suggesting that spatial localization may not be accurately reflected in the spike train.

Instead, we suggest that spatial localization may be encoded by means of the dendritic site of activation from particular input channels. Figure 3-7 shows a model we have proposed for direction selectivity in SI cortex (Warren et al., 1986b). The cortical neuron receives inputs from presynaptic neurons whose receptive fields overlap. The model assumes that each afferent nerve innervates particular dendritic branches, and these are somatotopically organized. In the figure, the stippling patterns on the receptive fields are matched to those of the preferred dendrites. Thus, spatial localization might be related to the specificity of current flows in particular dendritic branches. Activity should shift across the dendritic tree as different portions of the receptive field are stimulated.

Inhibition is also postulated to be distributed in a spatially organized fashion on the dendritic tree. Inhibitory synapses are placed on the proximal dendritic branches, so that they can selectively regulate current flow through that branch by shunting current through inhibitory channels. Unlike inhibitory postsynaptic potentials (IPSPs), which are traditionally assumed to inhibit the entire cell, these inhibitory contacts can regulate current flow through some dendrites while sparing other input channels. The spread of inhibitory contacts and the degree of lateral

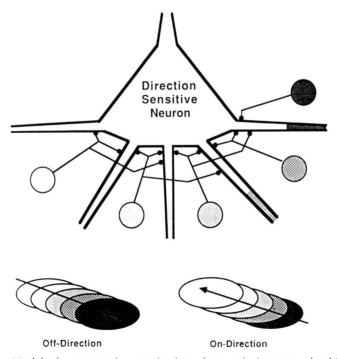

Off-Direction   On-Direction

Figure 3-7 Model of somatotopic organization of synaptic inputs to dendritic tree of cortical neurons. Excitatory inputs are placed on distal dendritic branches, while inhibitory interneurons terminate on the proximal dendrites. Stippling indicates receptive field of input neurons. The degree of direction selectivity depends on spatial distribution of shunting inhibition. [Reproduced with permission from Warren et al., (1986b).]

asymmetry can confer motion or direction selectivity on the cortical cell. Similar models of shunting inhibition have been proposed by Koch and co-workers (1982, 1983) for direction selectivity in the visual system. Shunting inhibition can diminish excitability of portions of the dendritic tree with a time course similar to the reduced excitation demonstrated with OPTACON stimuli. Effects of inhibition tend to be diminished as stimuli are spaced further apart on the skin, and therefore on the dendritic tree, leading to greater responsiveness to widely spaced stimuli.

The topographic organization of somatosensory cortex in primates has been documented on the single-cell level by Gerhard Werner and colleagues (Werner and Whitsel, 1968; Whitsel et al., 1969; McKenna et al., 1982), and by Merzenich and Kaas's laboratories (reviewed in Kaas et al., 1981). We now propose that, in addition to these macrocolumnar maps of the body surface in each cytoarchitectural area of the cortex, individual neurons may have a fine-grain map of their receptive field spread over their dendritic tree. We hope our future studies with more complicated spatial patterns on the OPTACON will reveal more of the spatial topography of cortical neurons, as well as the discriminative mechanisms used to perceive shapes as complicated as letters of the alphabet.

ACKNOWLEDGEMENTS

We thank Dr. Daniel Gardner for many helpful criticisms of this manuscript. We are most grateful to Jane Davis, Paula Genduso, and James Santoro for skilled technical assistance. This research was supported by U.S. Public Health Service Research Grants NS-11862 and NS-17973 from the National Institute of Neurological and Communicative Disorders and Stroke.

REFERENCES

Barlow, H. B., and Levick, W. R. (1965). The mechanism of directionally selective units in rabbit's retina. *J. Physiol. (Lond.)*.178:477–504.

Bliss, J. C., Katcher, M. H., Rogers, C. H., and Shepard, R. P. (1970). Optical-to-tactile image conversion for the blind. *IEEE Trans. MMS* 11:58–65.

Constanzo, R. M., and Gardner, E. P. (1980). A quantitative analysis of responses of direction-sensitive neurons in somatosensory cortex of alert monkeys. *J. Neurophysiol.* 43:1319–1341.

Craig, J. C. (1985). Tactile pattern perception and its perturbations. *J. Acoust. Soc. Amer.* 77:238–256.

Craig, J. C., and Sherrick, C. E. (1982). Dynamic tactile displays. In: W. Schiff and E. Foulke (eds.). *Tactual Perception: A Sourcebook,*pp. 209–233. Cambridge: Cambridge University Press.

Darian-Smith, I., Davidson, I., and Johnson, K. O. (1980). Peripheral neural representation of spatial dimensions of a textured surface moving across the monkey's finger pad. *J. Physiol. (Lond.)* 309:135–146.

Darian-Smith, I., and Kenins, P. (1980). Innervation density of mechanoreceptive fibres supplying glabrous skin of the monkey's index finger. *J. Physiol. (Lond.)* 309:147–155.

Darian-Smith, I., and Oke, L. E. (1980). Peripheral neural representation of the spatial frequency of a grating moving across the monkey's pad. *J. Physiol. (Lond.)* 309:117–133.

Ferrington, D. G., and Rowe, M. (1980). Differential contributions to coding of cutaneous vibratory information by cortical somatosensory areas I and II. *J. Neurophysiol.* 43:310–331.

Gardner, E. P. (1984). Cortical neuronal mechanisms underlying the perception of motion across the skin. In: C. Von Euler, O. Franzen, U. Lindblom, and D. Ottoson (eds.). *Somatosensory Mechanisms,*pp. 93–112. London:Macmillan Press.

Gardner, E. P., and Costanzo, R. M. (1980a). Spatial integration of multiple-point stimuli in primary somatosensory cortical receptive fields of alert monkeys. *J. Neurophysiol.*43:420–443.

Gardner, E. P., and Costanzo, R. M. (1980b). Temporal integration of multiple-point stimuli in primary somatosensory cortical receptive fields of alert monkeys. *J. Neurophysiol.* 43:444–468.

Gardner, E. P., and Costanzo, R. M. (1980c). Neuronal mechanisms underlying direction sensitivity of somatosensory cortical neurons in alert monkeys. *J. Neurophysiol.* 43:1342–1354.

Gardner, E. P., and Sklar, B. F. (1986). Factors influencing discrimination of direction of motion on the human hand. *Soc. Neurosci. Abstr.* 12:798.

Hubel, D. H., and Wiesel, T. N. (1968). Receptive fields and functional architecture of monkey striate cortex. *J. Physiol. (Lond.)* 195:215–243.

Hyvarinen, J., and Poranen, A. (1978). Movement-sensitive and direction and orientation-selective cutaneous receptive fields in the hand area of postcentral gyrus in monkeys. *J. Physiol. (Lond.)* 283:523–537.

Iwamura, Y., and Tanaka, M. (1978). Postcentral neurons in hand region of area 2: their possible role in the form discrimination of tactile objects. *Brain Res.* 150:662–666.

Johnson, K. O., and Lamb, G. D. (1981). Neural mechanisms of spatial tactile discrimination: neural patterns evoked by Braille-like dot patterns in the monkey. *J. Physiol. (Lond.)* 310:117–144.

Kaas, J. H., Nelson, R. J., Sur, M., and Merzenich, M. M. (1981). Organization of somatosensory cortex in primates. In: F. O. Schmitt, F. G. Worden, G. Adelman, and S. G. Dennis (eds.). *The Cerebral Cortex,* pp. 237–261. Cambridge: MIT Press.

Koch, C., Poggio, T., and Torre, V. (1982). Retinal ganglion cells: a functional interpretation of dendritic morphology. *Phil. Trans. Roy. Soc. Lond. Ser. B* 298:227–264.

Koch, C., Poggio, T., and Torre, V. (1983). Nonlinear interactions in a dendritic tree: localization, timing, and role in information processing. *Proc. Natl. Acad. Sci. USA* 80:2799–2802.

McKenna, T. M., Whitsel, B. L., and Dreyer, D. A. (1982). Anterior parietal cortical topographic organization in macaque monkey: a reevaluation. *J. Neurophysiol.* 48:289–317.

Mountcastle, V. B., Talbot, W. H., Sakata, H., and Hyvarinen, J. (1969). Cortical neuronal mechanisms in flutter-vibration studied in unanesthetized monkeys. *J. Neurophysiol.* 32:452–484.

Newsome, W. T., Mikami, A., and Wurtz, R. H. (1986). Motion selectivity in macaque visual cortex. III. Psychophysics and physiology of apparent motion. *J. Neurophysiol.* 55:1340–1351.

Phillips, J. R., and Johnson, K. O. (1981). Tactile spatial resolution. II. Neural representation of bars, edges and gratings in monkey primary afferents. *J. Neurophysiol.* 46:1192–1203.

Talbot, W. H., Darian-Smith, I., Kornhuber, H. H., and Mountcastle, V. B. (1968). The sense of flutter-vibration: comparison of the human capacity with response patterns of mechanoreceptive afferents from the monkey hand. *J. Neurophysiol.* 31:301–355.

Vallbo, A. B., and Johansson, R. S. (1978). The tactile sensory innervation of the glabrous skin of the human hand. In: G. Gordon (ed.). *Active Touch,* pp. 29–54. Oxford:Pergamon Press.

Verrillo, R. T. (1966). A duplex theory of mechanoreception. In: D. R. Kenshalo (ed.). *The Skin Senses,* pp. 139–159. Springfield, IL:Charles C Thomas.

Warren, S., Hamalainen, H., and Gardner, E. P. (1986a). Objective classification of motion- and direction-sensitive neurons in primary somatosensory cortex of awake monkeys. *J. Neurophysiol.* 56:598–622.

Warren, S., Hamalainen, H., and Gardner, E. P. (1986b). Coding of the spatial period of gratings rolled across the receptive fields of somatosensory cortical neurons in awake monkeys. *J. Neurophysiol.* 56:623–639.

Werner, G., and Whitsel, B. L. (1968). Topology of the body representation in the somatosensory area 1 of primates. *J. Neurophysiol.* 31:856–869.

Whitsel, B. L., Dreyer, D. A., Hollins, M., and Young, M. G. (1979). The coding of direction of tactile stimulus movement: correlative psychophysical and electrophysiological data. In: D. R. Kenshalo (ed.). *Sensory Functions of the Skin of Humans,* pp. 79–107. New York: Plenum.

Whitsel, B. L., Petrucelli, L. M., and Werner, G. (1969). Symmetry and connectivity in the map of the body surface in somatosensory area II of primates. *J. Neurophysiol.* 32:170–183.

Whitsel, B. L. Roppolo, J. R., and Werner, G. (1972). Cortical information processing of stimulus motion on primate skin. *J. Neurophysiol.* 35:691–717.

4

The Vibrissa/Barrel Cortex as a Model of Sensory Information Processing

DANIEL J. SIMONS, GEORGE E. CARVELL,
AND PETER W. LAND

Our work on the somatosensory system is based on several premises. First, subsets of cortical neurons are linked to form local networks that transform incoming information. Second, these subsets interact with other neuron ensembles according to definable and experimentally identifiable rules. These rules include a highly nonrandom pattern of intracortical connections, such as those demonstrated in primate visual cortex (see, e.g., Chapter 10). A third assumption is that operations performed by such linked networks constitute a neural representation or percept of the organism's corporeal and extrapersonal space. These ideas have been elegantly elucidated by Mountcastle (1979) in his theory of cortical columnar organization and distributed systems. We emphasize the concept of local neuronal networks. This reflects an experimental approach in which it is increasingly useful for us to regard a functional cortical column as being comprised of a number of distinguishable, highly interrelated compartments.

We are studying a component of the rodent somatosensory system that processes tactile information from the large facial whiskers. As has been shown first in mice by Thomas Woolsey (1967) and later in rats by Carol Welker (1971, 1976), the face region within the first somatic sensory cortex contains identifiable aggregates of layer-IV cells that are related, one to one, to individual vibrissae on the contralateral face (see Figure 4-1). Woolsey and Van der Loos (1970) called these cellular aggregates "barrels." Barrels are structural, metabolic, neurochemical, and functional units. A layer-IV barrel is just one component of a larger module that extends throughout the thickness of the somatosensory cortex. Figure 4-2 shows some essential features of this organization. It is a photomicrograph of an oblique coronal section through this cortex, which has been stained for the mitochondrial enzyme cytochrome oxidase; metabolically active regions stain darkest. In layer IV, the centers of the barrels are more metabolically active than the barrel sides or the septa between barrels. Regions of comparatively high metabolic activity can also be observed in the lower part of layer V and in layer VI, where they form columns that are in register with the overlying layer-IV barrels (Land and Simons, 1985a).

In layer IV and the deeper layers, regions of heightened metabolic activity are

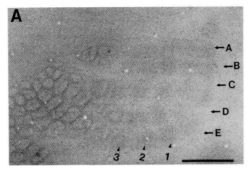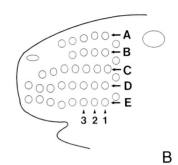

Figure 4-1   Cytoarchitectural organization of the rat vibrissa cortex. (A) A tangential 80-μm section through lamina IV of the right first somatic sensory cortex stained with thionine. (Orientation: lateral, left; anterior, bottom. Scale = 1 mm.) (B) A schematic drawing of the left face. Large sinus hairs, called vibrissae or whiskers, are organized in five rows containing four to seven whiskers. Barrels in the contralateral hemisphere are organized and named correspondingly; they are characterized by a relatively homogeneous, densely packed distribution of granule cells and lack well-defined sides. The more clearly definable barrels at the left of panel A correspond to the small sinus hairs on the more anterior aspects of the face. Barrels related to the large vibrissae are better delineated in sections stained for cytochrome oxidase (see Figure 4-2, and Land and Simons, 1985a).

coextensive with zones of thalamic input from the ventrobasal complex, the principal relay for discriminative touch. Moreover, we have found that the cytochrome oxidase-rich regions in layer VI contain corticothalamic neurons that are clustered in patches whose spatial organization reflects that of the barrels (unpublished observations). In mice, the apical dendrites of the thalamic projecting cells of layer VI extend into the barrels, where they are contacted by thalamocortical axons (White and Hersch, 1982). Also, axon collaterals of corticothalamic cells synapse on barrel neurons (White and Keller, 1987). These findings suggest a central columnar core that is intimately associated with the ventrobasal (VB) thalamus.

Some major afferent and efferent fiber systems, other than those associated with the thalamic relay nucleus, appear to be organized in a *complementary* fashion. Callosal projections terminate largely within the interbarrel septa in layer IV and in the supra and infragranular laminae superficial and deep to them (Olavarria et al., 1984). Afferents from the medial division of the posterior thalamic nucleus (POm) also terminate preferentially within the septa (Koralek et al., 1985; Lu and Lin, 1985). Finally, evidence from studies in mice indicates that corticotectal and corticobulbar neurons in layer V are located preferentially in register with the barrel sides and septa (Crandall et al., 1983). Apical dendrites of these neurons ascend through the barrel sides or septa, avoiding the barrel centers (Escobar et al., 1986). Such cells may be part of an integrative zone surrounding the column's core (see Land and Simons, 1985a, and the following).

FUNCTION OF THE RODENT WHISKER CORTEX

Physiological studies during the last 10 years indicate that an important function of the rodent primary somatosensory cortex is to integrate information arising

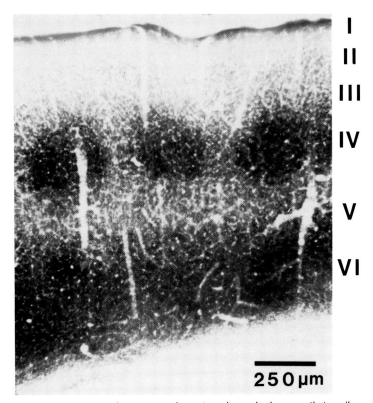

I

II

III

IV

V

VI

250 µm

Figure 4-2 Cytochrome oxidase–stained section through the rat vibrissa/barrel cortex. Three metabolically active (*dark staining*) foci are visible in layer IV and lower layer III; these correspond to the cytochrome oxidase–rich centers of barrels representing three adjacent vibrissae in a row of whiskers on the contralateral face. Metabolically active regions are also visible in lower layer V and layer VI, forming a columnlike pattern with the overlying layer IV barrels (see Land and Simons, 1985a).

from different, individual whiskers (Simons, 1978; Simons and Woolsey, 1979; Ito, 1985; Chapin, 1986; Armstrong-James, 1975; see also Lamour et al., 1983). Receptive fields of layer-IV cells are quite small and often restricted to the single whisker that corresponds anatomically to the barrel in which the cell is located. We call this anatomically appropriate vibrissa the principal whisker (PW). Within a vertical electrode penetration from pia to white matter, all drivable cells are activated by stimulating the PW. Such findings demonstrate that cells are organized functionally into vertical columns, of which the layer-IV barrels are morphological correlates. Important supporting evidence is provided by 2-deoxyglucose studies in alert behaving animals (Durham and Woolsey, 1977; Kossut and Hand, 1984).

Interestingly, repetitive stimulation of all the vibrissae produces heaviest glucose use in laminae IIIb, IV, Vc, and VI, a pattern similar to that seen in our cytochrome oxidase material (Gonzalez and Sharp, 1985). A second important feature of the vibrissa cortex is that receptive fields of layer V, and to a lesser extent of layer III, typically encompass many whiskers in addition to the PW.

The modular nature of the columnar organization thus provides a framework within which information is manipulated and transformed, for example, as the synthesis of multiwhisker receptive fields.

The mystacial vibrissae of rats and some other rodents consist of an array of 25 to 30 discrete tactile organs that is swept back and forth through the sensory environment (Welker, 1964; Wineski, 1983), each hair in turn contacting a stationary or moving object within the animal's immediate environment. In an attempt to mimic this situation partially in the laboratory, we have developed an array of independently controllable piezoelectric stimulators that can deflect groups of whiskers in specific spatial and temporal combinations (Simons, 1983). The stimulator array is controlled by a laboratory computer that also acquires, stores, and analyzes the neural data. Glass micropipettes are used to record the extracellular or intracellular responses of single neurons in pharmacologically immobilized rats. The animals are maintained either in the absence of general anesthesia or, more recently, with a slow infusion of fentanyl, a synthetic opiate receptor agonist. This eliminates the well-known effects of the more commonly used general anesthetics on the dynamic properties of somatosensory cortical neurons.

The integrative function of the rodent somatosensory cortex is nicely revealed when two or more whiskers are moved in combination (Simons, 1985). After a whisker is displaced, cortical unit discharges to subsequent deflections of the same or adjacent whiskers are reduced in a time-dependent fashion. Response suppression is most pronounced at short interdeflection intervals (10 to 20 ms separations of the paired whisker movements), and decreases during the 50 to 100 ms following the initial whisker deflection. Intracellular recordings demonstrate the presence of IPSPs that correspond to the period of this response suppression (Carvell and Simons, 1988; see also Hellweg et al., 1977). Since afferent input to the cortex is thought to be excitatory, we attribute a cortical origin to these IPSPs. These findings in the rodent somatosensory cortex are generally consistent with those from other species, demonstrating that somatosensory stimuli evoke an initial excitation followed by a longer-lasting inhibition within central neurons receiving inputs from ascending dorsal column/medial lemniscal pathways (e.g., Mountcastle et al., 1957; Mountcastle and Powell, 1959; Gardner and Costanzo, 1980).

In our experiments, response suppression has been observed for every cortical neuron studied. For a given unit, however, the presence and degree of suppression depends on certain spatial features of the multiwhisker stimulus. One such factor is which whisker is deflected first. Thus, the PW strongly inhibits responses to subsequent deflections of adjacent (i.e., noncolumnar) whiskers; typically, the PW also elicits the strongest excitatory responses (see also Simons, 1978; Chapin, 1986). By contrast, deflection of a noncolumnar whisker may or may not affect responses to subsequent deflections of the PW. Particularly in the deeper layers of the cortex, excitatory and inhibitory subregions of a cell's receptive field can be distributed asymmetrically around the columnar whisker. These asymmetries often parallel the differential responsiveness of the neurons to bending of the individual whiskers in particular directions. In these cases receptive fields display a distinct spatial orientation, and the units respond selectively to stimuli we would like to think mimic object movement through the vibrissal field. In this regard, data on how the individual whiskers are displaced when actively "whisked" against an object are sorely lacking. Descriptive data of fundamental importance for designing neurophysiological experiments could be obtained using high-speed, high-resolution motion pictures to observe whisking behavior in controlled settings. The extension of such an approach to the study of animals trained to make vibrissal discriminations would provide new and valuable data on the informational capacity of the whisker system. Psychophysical techniques for studying vibrissal behavior are now available (Huston and Masterton, 1986).

INFORMATION PROCESSING IN THE LAYER–IV BARRELS

Data from single-cell recordings suggest the existence of specific excitatory and inhibitory interactions among cells located within their parent cortical column and among neighboring columns. These interactions must be elucidated in further studies; even at our present level of understanding, however, it is clear that the layer-IV barrels play a key role in the function of this cortex. First, as noted earlier, the barrels appear to serve as single-whisker modules for the synthesis of multiwhisker receptive fields in other cortical layers. Second, the neuronal circuitry within each barrel may help to transform afferent information from the thalamus into a temporal code that underlies the operations performed by the rest of the cortical column.

Anatomically, the barrels are organized with respect to a highly segregated, whisker-related pattern of specific thalamocortical fibers that terminate principally in the barrel centers or hollows (Lorente de Nò, 1922; Killackey and Leshin, 1975; Woolsey and Dierker, 1978). The dendrites of barrel neurons are oriented centrally as if to maximize contact with this incoming fiber system (Woolsey et al., 1975; Steffan, 1976; Simons and Woolsey, 1984). Barrel neurons can be classified into two *broadly defined* groups according to a number of qualitative and quantitative morphological criteria (Woolsey et al., 1975; Harris and Woolsey, 1983; Simons and Woolsey, 1984). Class I cells have small somata and spinous dendrites, whereas the cell bodies of class II neurons are larger and their dendrites are smooth or beaded. Axons of class II cells are longer and have more branches than class I cells, and axons of both cell types ramify extensively, but not exclusively, in layer IV and within their parent barrel. Class I and II barrel neurons are similar to the spiny stellate and aspinous or sparsely spined nonpyramidal neurons that have been described in a variety of species and cortices. In the barrels, as elsewhere, the former are thought to be excitatory interneurons and the latter GABAergic and inhibitory (e.g., LeVay, 1973; Somogyi, 1978; Houser et al., 1983). GABAergic cells probably account for at least 15 percent of the barrel neuron population (Lin et al., 1985; Fairén et al., 1986).

Both spinous and smooth barrel neurons are contacted monosynaptically by afferents from the thalamic ventrobasal complex (White, 1978). The former receive thalamocortical synapses on their dendritic spines; the latter are contacted more proximally, on their somata and dendritic shafts. Available evidence is consistent with the idea that spiny and smooth barrel neurons are substantially interconnected, with cells of each type contacting other cells of the same and of the other class (White and Rock, 1980, 1981; Harris and Woolsey, 1983; Benshalom and White, 1986). Indeed, together with thalamocortical synapses, the axons of these two cell types probably account for the majority of contacts within the barrel neuropil (see Land et al., 1986; Lapenko and Podladchikova, 1983). Thus, a barrel represents a definable network of cells interposed between sources of input and output of the rodent somatosensory cortex. A critical task for the future is the quantitative description of the synaptology of the barrels, particularly in terms of contacts made among individual cells identified on the basis of their morphology and their putative neurotransmitter.

Functionally, the barrel circuitry helps establish a robust temporal sequence of stimulus-evoked excitation and inhibition. This is nicely illustrated by a population response profile, as shown in Figure 4-3B. This peristimulus time histogram was constructed by a bin-by-bin accumulation of the extracellularly recorded responses of 210 individual barrel neurons to standard, controlled deflections of their PWs. The profile shows that activity in the barrels is sharply reduced immediately after the (excitatory) ON response, and over a period of 75 msec it gradually returns to and then transiently exceeds prestimulus levels. This reduction of spike activity parallels the time course of IPSPs observed with intracellular recordings from middle-depth neurons in this cortex (Figure 4-3A; see also Carvell and Simons, 1988). Moreover, it coincides with the period of response suppression observed when whiskers are deflected sequentially (see preceding). Of further interest is the presence, during the later phase, of the stimulus of a second and possibly third cycle of excitation–inhibition, which also has a periodicity of approximately 75 msec. Similar patterned activity follows the OFF response. This damped oscillatory behavior is not an artifact produced by ringing of the mechanical stimulator used to deflect the vibrissae, nor is it likely solely to reflect intrinsic membrane properties of the cells. Our working hypothesis is rather that it emerges from the network properties of the barrel circuitry. This hypothesis is supported indirectly by our finding that the temporal pattern of excitation–inhibition is markedly altered in barrels deprived of functional input by chronic trimming of the vibrissae in young animals (Simons and Land, 1987). The period of cyclical inhibition–excitation evoked by stimulating the regrown whisker is reduced by roughly a factor of 2.

Inhibition with the barrels is almost certainly mediated by smooth or sparsely spined cells that use GABA as a neurotransmitter (Lin et al., 1985; Lyon and Connors, 1985; Chmielowska et al., 1986; Fairén et al., 1986; Keller and White, 1986). Smooth cells have axons that ramify profusely within the barrel centers and make symmetric synapses; these presumably inhibitory profiles are observed on both spinous and aspinous cells. In specimens from rats stained immunocytochemically for glutamic acid decarboxylase, immunoreactive terminals are especially prominent in the barrel centers and sides but sparse in the septa between barrels (Lyon and Connors, 1985; Land, unpublished observations). GAD-immunoreactive terminals are observed on both GAD-positive and GAD-negative somata. These findings indicate that there is a rich inhibitory component to the barrel circuitry, and at least some inhibitory neurons may themselves be subject to inhibition. As noted earlier, smooth barrel neurons receive monosynaptic inputs from thalamocortical axons on their cell bodies and proximal dendrites. These

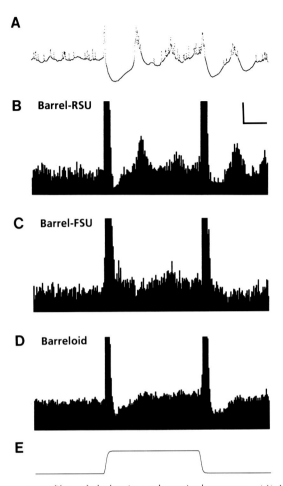

A

B Barrel-RSU

C Barrel-FSU

D Barreloid

E

Figure 4-3 Response profiles of thalamic and cortical neurons. (*A*) Intracellularly re-corded membrane potential changes of a single regular-spike layer IV cell averaged over 40 stimulus presentations. The whisker was deflected randomly in eight angular directions (in 45-degree increments relative to the horizontal alignment of whisker rows), and the battery was repeated five times. The stimulus waveform is shown in panel *E*; the stim-ulator was attached 8 to 10 mm from the base of the whisker, which was deflected 1 mm at 135 mm/sec. Peristimulus time histograms in panels *B* through *D* are population profiles constructed by a bin-by-bin accumulation of extracellularly recorded responses elicited by stimulating each cell's maximally excitatory vibrissa (i.e., principal whisker responses). For each cell the battery of eight stimuli was repeated 10 times. Data are expressed as the observed probability that an action potential occurred within each 2-ms bin. (*B*) Population profiles based on responses of 210 regular-spike units (RSUs) recorded in the barrels; (*C*) responses of five selected fast-spike units (FSUs) in the barrels; and (*D*) responses of 135 thalamic (barreloid) cells. The vertical axis is expanded to demonstrate the modulation of spike activity in each neuron population. The close correspondence between the RSU-barrel profile and the intracellular record in panel *A* indicates that the population profiles reflect underlying membrane potential changes of the individual neu-rons. Vertical scale in panel *B* applies as follows: panel *A*—7.8 mV; panel *B*—probability of 0.002; panel *C*—probability of 0.06; panel *D*—probability of 0.01. Horizontal scale in *B* = 50 ms for all panels.

cells are therefore likely to be strongly driven by the afferent input to the barrels and are well-situated for exerting potent feedforward inhibitory effects on the barrel circuitry. Interconnections among smooth and spiny neurons could provide a mechanism for recurrent (feedback) inhibition and excitation resulting in the cyclic pattern of activity observed in the barrel neuron population.

There is increasing evidence that smooth barrel neurons can be distinguished electrophysiologically by the rapid time course of their action potentials. Such potentials, which we have termed "fast-spikes," have an extracellularly recorded duration of approximately 750 μsec, compared with regular-spike potentials that last about twice as long (Simons, 1978; Simons and Woolsey, 1979; see also Mountcastle et al., 1969). Intracellular recordings in vitro indicate that the short duration of the fast-spikes is due to rapid membrane repolarization, perhaps by a rapidly activating and inactivating potassium conductance (McCormick et al., 1985). Most significantly, these studies provide direct evidence for the hypothesis that fast-spike units are discharged by smooth barrel cells (see Simons and Woolsey, 1984). This identity is consistent with some of the known physiological and anatomical properties of these neurons. Thus, fast-spike units respond especially vigorously to whisker displacements and, unlike most regular-spike units, can be entrained by relatively high-frequently vibrations of the vibrissae (see also Mountcastle et al., 1969; Ferrington and Rowe, 1980). Fast-spike units also display unusually high levels of spontaneous activity. In accord with this, cells morphologically similar to smooth barrel neurons are highly reactive for cytochrome oxidase, a mitochondrial metabolic enzyme (Land and Simons, 1985a). These physiological observations indicate that fast-spike/smooth barrel neurons are especially well driven by thalamic axons; this is consistent with the findings that thalamocortical synapses are made on the soma and proximal dendrites of these cells.

A critical and unresolved issue is how smooth barrel neurons produce relatively long periods of stimulus-evoked inhibition. Unitary IPSPs appear to last only a short time, perhaps on the order of 4 to 5 msec (Connors et al., 1982; see also Kriegstein and Connors, 1986). An obvious possibility is that the inhibitory neurons discharge action potentials throughout the postexcitatory inhibition period. Excluding their initial on and off responses, the activity profile of these cells would thus be somewhat reciprocal to that shown in Figure 4-3B, which was constructed entirely of regular-spike unit responses. Figure 4-3C shows a population profile constructed from five fast-spike units that showed such responses. Fast-spike units are difficult to find and to record from for extended periods, and therefore our sample is limited. Nevertheless, although the five fast-spike units illustrated in Figure 4-2C discharge during the inhibitory period, seven others (not shown) did not. In this regard, it should be noted that sustained or tonic responses in inhibitory neurons may not be required to elicit a long-lasting hyperpolarization in the postsynaptic cell. Temporal summation of short unitary IPSPs could hyperpolarize the postsynaptic cell sufficiently to induce a longer-lasting potassium current (see, e.g., Kreigstein and Conners, 1986). Such temporal summation would presumably be enhanced by the ability of the presynaptic FSUs to repolarize quickly and respond in high-frequency bursts (Simons, 1978; McCormick et al., 1985). The ability to identify the discharges of smooth barrel neurons with extracellular recordings greatly facilitates the elucidation of a barrel's functional wiring diagram. Moreover, it may eventually be possible to label smooth and spiny barrel neurons selectively, so that simultaneous intracellular recordings can be used in vitro to study the synaptic physiology of the barrel circuitry directly (e.g., Katz et al., 1984).

THALAMOCORTICAL RELATIONS: BARRELOID-BARREL RESPONSE TRANSFORMATION

The face region of the rodent thalamic ventrobasal complex, like that of the primary somatosensory cortex, contains whisker-related aggregates of neurons called *barreloids* (Van der Loos, 1976; Land and Simons, 1985b). Cells within individual barreloids project preferentially, though not exclusively, to the corresponding cortical barrel. Small injections of horseradish peroxidase into the center of a cortical barrel yield retrograde labeling of cell bodies within the appropriate thalamic barreloid (approximately 95 percent of the labeled thalamic cells), whereas injections located near the barrel sides or within the septa between barrels label cells in both the isomorphic barreloid and the adjacent barreloids (86 percent isomorphic for barrel side injections; 50 percent for septal injections; Land et al., 1986; see also Podlachikova and Lapenko, 1982; Lu and Lin, 1985). These results accord with our physiological studies in rats. Approximately 80 to 90 percent of units recorded within the cytochrome oxidase–rich barrel centers have single-whisker receptive fields, determined by using manually applied stimuli, whereas most cells in the interbarrel septa and, to a somewhat lesser extent, in the barrel side have distinctly multiwhisker receptive fields (see also Uhr et al., 1982). Thus, a certain proportion of multiwhisker receptive fields in layer IV could result from direct convergence of inputs from two (or more) barreloids (see also following).

Our cell counts indicate that the C2 barreloid in rats contains roughly 220 neurons. There are approximately 2000 neurons within the cortical C2 barrel of *mice* (Lee and Woolsey, 1975; see also Curcio and Coleman, 1982). Since rat barrels are considerably larger and more densely populated than corresponding barrels in mice (Welker and Woolsey, 1974), a conservative estimate is that the C2 barrel in rats contains at least 10 times as many neurons as its corresponding barreloid. Moreover, all barreloid cells in rats are thought to project to the barrel cortex (see Saporta and Kruger, 1977; Wells et al., 1982), and all barrel cells receive thalamocortical synapses (White, 1979). A reasonable assumption, therefore, is that *on average* a single barreloid neuron directly contacts at least 10 barrel cells. A consequence of this assumption would be that a given barrel cell receives convergent input from many barreloid neurons.

A second important anatomical feature of the barreloids is that they are virtually devoid of intrinsic inhibitory interneurons, a situation unique among mammalian thalamic relay nuclei (Ralston, 1984; Barbaresi et al., 1986). Inhibitory synapses within the barreloids arise almost exclusively from the thalamic reticular nucleus (see Peschanski et al., 1983). Cells there are contacted by collaterals of thalamocortical axons, by corticothalamic axons, or by both, so that inhibition within a barreloid is feedback only (see Harris, 1987). By comparison, inhibitory neurons are contained within each cortical barrel and provide them with mechanisms for both feedforward and feedback inhibition.

We have recently studied the functional response properties of barreloid neurons using whisker stimuli identical to those employed in our cortical studies. Responses of many thalamic cells to deflection of their "principal" excitatory whisker, unlike those of cortical neurons, are unaffected by prior deflection of an adjacent whisker. When observed, cross-whisker inhibition is weaker in the barreloids, on average, than in the barrels. The cyclic pattern of excitation—inhibition following the on response is also considerably less pronounced (see Figure 4-3D). Spontaneous discharge rates of thalamic neurons are generally six times greater than those of cells in the barrels, a finding consistent with the intense cytochrome oxidase staining of cell bodies in the barreloids (Land and Simons, 1985b). Moreover, as a population, more barreloid cells respond in a slowly adapting fashion than barrel neurons (29 versus 15 percent). Barreloid neurons also discharge more vigorously, during their on reponses, to their "best" or maximally excitatory deflection angle. These differences between barreloid and barrel neurons are generally consistent with anatomical findings that the former lack intrinsic inhibitory cells but the latter do not.

Our hypothesis is that the network properties of the barrels maintain a tonic level of inhibition that suppresses cortical activity and firmly establishes a cyclic pattern of stimulus-evoked excitation and inhibition. The periodicity of this pattern and the time course of its decay are determined by anatomical interconnections among the barrel cells and also by the dynamic membrane and synaptic properties of the neurons themselves. Such constraining features of the network are presumably affected by neonatal sensory deprivation (Simons and Land, 1987). In addition to abnormal patterns of stimulus-evoked excitation and inhibition, neonatally deprived barrel neurons display abnormally vigorous on and off responses and abnormally high levels of spontaneous activity. It is interesting to speculate that the barrel circuitry normally functions to partially offset the barrage of excitation resulting from the convergence of numerous thalamocortical axons onto individual barrel cells. Indeed, such inhibitory network properties may be obligatory in afferent systems characterized by a high degree of convergent inputs; perhaps the lack of inhibitory interneurons within the rodent VB nucleus reflects the absence of significant convergence from second-order neurons in the brainstem principal sensory nucleus of V.

The convergence–divergence factor suggested by the barreloid-to-barrel topography may play a significant role in the response transformation between thalamus and cortex. Compared to cells in the barreloids, barrel neurons generally respond less selectively to initial movements of the whisker hairs in different angular directions, that is, their on responses are less "tuned." Because barrel neurons also respond less selectively to hair movements in opposing angular directions they are more likely to discharge to both stimulus onset and offset. Many barreloid neurons respond in an "on or off" fashion. Such cells display on responses to one angle of movement and off but not on responses to movements in the opposite direction; the off response in these cases is elicited by return of the whisker to its neutral position. The functional differences between these cell populations are consistent with the idea that a barrel neuron receives convergent inputs from numerous barreloid cells, each of which responds to a given whisker stimulus somewhat differently.

A surprising result of our barreloid studies is that almost 70 percent of thalamic cells are strongly excited by more than one whisker; this finding contrasts with previous observations based on recordings in barbiturate anesthetized animals (Waite, 1973; Verley and Onnen, 1981; Vahle-Hinz and Gottschaldt, 1983). Multiwhisker inputs onto barrel neurons are present but weak; in most cases they can be discerned in extracellular recordings only when responses are averaged over many trials. With intracellular recordings from presumed barrel neurons, they are visible as subthreshold postsynaptic excitatory potentials (EPSPs) (Carvell and Simons, 1988). This apparent paradox—many strongly multiwhisker receptive fields in the presynaptic population and few in the postsynaptic population—may reflect the divergent nature of the thalamocortical input, whereby a single thalamic cell contacts numerous barrel neurons. The influence of an individual barreloid neuron would therefore be diminished in some proportion to the number of other afferent cells that are also presynaptic to the barrel neuron. In the case of spiny stellate cells, the effects of this divergence are presumably accentuated by the fact that thalamocortical synapses occur distally on the dendrites (e.g., Rall, 1967; Barrett and Crill, 1974). Only the most numerous and synchronously active inputs would cause the cell to reach threshold for action potential discharge (Sheperd et al., 1985). Such inputs are presumably manifest in the responses of regular-spike neurons to deflection of their barrel's PW. Inputs from a given "inappropriate" whisker, which are proportionally less numerous, would be less effective in firing the cell. According to this scheme, an individual multiwhisker barreloid cell should have greater influence on smooth barrel cells because thalamocortical synapses are made on their somata and proximal dendritic shafts. In fact, fast-spike units in layer IV are more likely to have strongly multiwhisker receptive fields than regular-spike units, as determined by manually applied stimuli (see Simons, 1978). These findings may account, in part, for the surround inhibition observed in the barrels.

TOWARD A MODEL OF SOMATOSENSORY CORTEX

Presumably one function of the sensory areas of the cerebral cortex is to interpret, at least in a preliminary way, a flow of sensory events distributed across the receptor surface during exploratory and discriminative behaviors. Findings from numerous anatomical and physiological studies in a variety of species and cortices suggest two *general* underlying principles of intrinsic cortical organization: (1) a columnar organization characterized by vertical interconnections among cells distributed in vertical register with each other that preserves certain features of a peripheral stimulus (e.g., which whisker is stimulated) and (2) a horizontal organization largely correlated with the laminar pattern that provides for temporal and spatial integration of inputs from closely adjacent parts of the receptor surface (e.g., sequential deflection of whiskers). *Specific* organizational details clearly vary among systems and species (see, e.g., Humphrey and Hendrickson, 1983;

Sretavan and Dykes, 1983; Sur et al., 1984). Nevertheless, determining how information is integrated within and among identified populations of cortical cells should provide important and broadly applicable clues about how cortex "works." Indeed, functional interrelationships among ensembles of cortical cells form the fundamental framework on which experimentally testable models of cerebral cortical function must be based (Mountcastle, 1979).

Available physiological data indicate that one important function of the rodent primary somatic sensory cortex is to integrate information arising from temporally and spatially patterned deflections of the mystacial vibrissae. An emerging model of this cortex is that it is comprised of 25 to 30 individual columns, each of which is related principally to a specific facial vibrissa and contains a central core surrounded by integrative zones where individual columns interface. Our working hypothesis is that not only do neurons display laminar-dependent response properties, but functional differences among neurons within a lamina can also be correlated with their location in the central core of the column or in the zone that surrounds it.

A related assumption is that these various compartments are anatomically linked in a highly nonrandom fashion; that is, cells located in different compartments of a cortical column distribute their axons or axon collaterals to specific regions within their parent or neighboring columns, or both. As a starting point, at least four specific hypotheses can be identified: (1) non-layer-IV cells in the central core, as defined in part by cytochrome oxidase staining, display functional properties that most closely resemble those of cells in the barrels, that is, relatively small or sharply focused, nonspatially oriented receptive fields; (2) neurons in zones where columns interface, specifically those in the septa between barrels and the pyramidal cells superficial and deep to them, display more complicated, spatially integrated properties, such as large excitatory receptive fields with asymmetrically distributed inhibitory side regions; (3) neurons in layers IV, III, and V receive axon projections from progressively more widespread regions within the barrel cortex; and (4) pyramidal neurons with asymmetrically (i.e., spatially) organized receptive fields distribute their axons or axon collaterals to neighboring columns asymmetrically (e.g., Gilbert and Wiesel, 1983).

The barrel in layer IV represents a well-defined network of interconnected cells that appears to perform an initial transformation of afferent input to the cortex. Our current hypothesis is that a great deal of this response transformation can be accounted for in terms of (1) the convergent and divergent parallelism between a thalamic barreloid and its corresponding cortical barrel and (2) the network properties of the barrel circuitry. More specifically, factors of fundamental importance are the approximately 10:1 ratio of barrel to barreloid cells, the differential pattern of thalamocortical synapses on the two principal cell types of the barrels, and the time-dependent inhibition provided by the fast-spike/smooth stellate neurons. Our motivation for attempting to formulate the operations of a cortical barrel in terms of such a concise topography is that these factors can be readily incorporated into connectionist network models constrained by neurobiological parameters that are already known or knowable in the near future (see Chapter 17). Though still largely untested, such approaches may be a source of new and powerful tools for addressing issues of perception and object representation in the cerebral cortex. The synthesis of computer science and experimental neurobiology may provide sensory physiologists with a language that formally describes their neurophysiological data and a theoretical basis for generating specific hypotheses about cortical function that can be tested in the neurobiology laboratory.

ACKNOWLEDGMENTS

We thank Glenn Fleet for expert technical assistance and Theresa Harvey for typing the manuscript. This work is supported by National Institute of Health Grant NS 19950.

REFERENCES

Armstrong-James, M. (1975). The functional status and columnar organization of single cells responding to cutaneous stimulation in neonatal rat somatosensory cortex SI. *J. Physiol. (Lond.)* 246:501–538.

Barbaresi, P., Spreafico, R., Frassoni, C., and Rustioni, A. (1986). GABAergic neurons are present in the dorsal column nuclei but not in the ventroposterior complex of rats. *Brain Res.* 382:305–326.

Barrett, J. N., and Crill, W. E. (1974). Influence of dendritic location and membrane properties of the effectiveness of synapses on cat motoneurones. *J. Physiol. (Lond.)* 293:325–345.

Benshalom, G., and White, E. L. (1986). Quantification of thalamocortical synapses with spiny stellate neurons in layer IV of mouse somatosensory cortex. *J. Comp. Neurol.* 253:303–314.

Carvell, G. E. and Simons, D. J. (1988). Membrane potential changes in rat SmI cortical neurons evoked by controlled stimulation of the mystacial vibrissae. *Brain Res.* 448:186–191.

Chapin, J. K. (1986). Laminar differences in sizes, shapes, and response profiles of cutaneous receptive fields in the rat SI cortex. *Exp. Brain Res.* 62:549–559.

Chmielowska, J., Stewart, M. G., Bourne, R. C., and Hamori, J. (1986). γ-Aminobutyric acid immunoreactivity in mouse barrel field: a light microscopical study. *Brain Res.* 368:371–374.

Connors, B. W., Gutnick, M. J., and Prince, D. A. (1982). Electrophysiological properties of neocortical neurons in vitro. *J. Neurophysiol.* 48:1302–1320.

Crandall, J. E., Kordi, M., and Caviness, V. S. (1983). Grid pattern of layer V efferent neurons of mouse barrel field cortex. *Neurosci. Abstr.* 9:38.

Curcio, C. A., and Coleman, P. D. (1982). Stability of neuron number in cortical barrels of aging mice. *J. Comp. Neurol.* 212:158–172.

Durham, D., and Woolsey, T. A. (1977). Barrels and columnar cortical organization: evidence from 2-deoxyglucose (2-DG) experiments. *Brain Res.* 137:169–174.

Escobar, M. I., Pimienta, H., Caviness, V. S., Jacobson, M., Crandall, J. E., and Kosik, K. S. (1986). Architecture of apical dendrites in the murine neocortex: dual apical dendritic systems. *Neurosci.* 17:975–989.

Fairén, A., Cobas, A., and Fonesca, M. (1986). Times of generation of glutamic acid decarboxylase immunoreactive neurons in mouse somatosensory cortex. *J. Comp. Neurol.* 251:67–83.

Ferrington, D. G., and Rowe, M. (1980). Differential contributions to coding of cutaneous vibratory information by cortical somatosensory areas I and II. *J. Neurophysiol.* 43:310–331.

Gardner, E. P., and Costanzo, R. M. (1980). Temporal integration of multiple-point stimuli in primary somatosensory cortical receptive fields of alert monkeys. *J. Neurophysiol.* 43:444–468.

Gilbert, C. D., and Wiesel, T. N. (1983). Clustered intrinsic connections in cat visual cortex. *J. Neurosci.* 3:1116–1133.

Gonzalez, M. F., and Sharp, F. R. (1985). Vibrissae tactile stimulation: (14C) 2-deoxyglucose uptake in rat brainstem, thalamus, and cortex. *J. Comp. Neurol.* 231:457–472.

Harris, R. M. (1987). Axon collaterals in the thalamic reticular nucleus from thalamocortical neurons of the rat ventrobasal thalamus. *J. Comp. Neurol.* 258:399–406.

Harris, R. M., and Woolsey, T. A. (1983). Computer-assisted analyses of barrel neuron axons and their putative synaptic contacts. *J. Comp. Neurol.* 220:63–79.

Hellweg, F. C., Schultz, W., and Creutzfeldt, O. D. (1977). Extracellular and intracellular recordings from cat's cortical whisker projection area: thalamocortical response transformation. *J. Neurophysiol.* 40:463–479.

Houser, C. R., Hendry, S. H. C., Jones, E. G., and Vaughn, J. E. (1983). Morphological diversity of immunocytochemically identified GABA neurons in the monkey sensory-motor cortex. *J. Neurocytol.* 12:617–638.

Humphrey, A. L., and Hendrickson, A. E. (1983). Background and stimulus-induced patterns of high metabolic activity in the visual cortex (area 17) of the squirrel and macaque monkey. *J. Neurosci.* 3:345–358.

Huston, K. A., and Masterton, R. B. (1986). The sensory contribution of a single vibrissa's cortical barrel. *J. Neurophysiol.* 56:1196–1223.

Ito, M. (1985). Processing of vibrissa sensory information within the rat neocortex. *J. Neurophysiol.* 54:479–490.

Katz, L. C., Burkhalter, A., and Dreyer, W. J. (1984). Fluorescent latex microspheres as a retrograde neuronal marker for in vivo and in vitro studies of visual cortex. *Nature* 310:498–500.

Keller, A., and White, E. L. (1986). Distribution of glutamic acid decarboxylase-immunoreactive structures in the barrel region of mouse somatosensory cortex. *Neurosci. Lett.* 66:245–250.

Killackey, H. P., and Leshin, S. (1975). The organization of specific thalamocortical projections to the posteromedial barrel subfields of the rat somatic sensory cortex. *Brain Res.* 86:469–472.

Koralek, K. A., Jensen, K. F., and Killackey, H. P. (1985). Projections of the medial portion of the posterior thalamic nuclear group (POM) to the somatosensory cortex of the rat. *Neurosci. Abstr.* 11:905.

Kossut, M., and Hand, P. J. (1984). The development of the vibrissal cortical column: a 2-deoxyglucose study in the rat. *Neurosci. Lett.* 46:1–6.

Kriegstein, A. R., and Connors, B. W. (1986). Cellular physiology of the turtle visual cortex: synaptic properties and intrinsic circuitry. *J. Neurosci.* 6:178–191.

Lamour, Y., Guilbaud, G., and Willer, J. C. (1983). Rat somatosensory (SmI) cortex: II. Laminar and columnar organization of noxious and non-noxious inputs. *Exp. Brain Res.* 49:46–54.

Land, P. W., and Simons, D. J. (1985a). Cytochrome oxidase staining in the rat SmI barrel cortex. *J. Comp. Neurol.* 238:225–235.

Land, P. W., and Simons, D. J. (1985b). Metabolic and structural correlates of the vibrissae representation in the thalamus of the adult rat. *Neurosci. Lett.* 60:319–324.

Land, P. W., Simons, D. J., and Buffer, S. A. (1986). Specificity of thalamocortical connections in the rat somatosensory system. *Neurosci. Abstr.* 12:1434.

Lapenko, T. K., and Podladchikova, O. N. (1983). Study of intercortical connections between groups of neurons in the somatosensory cortex of the rat brain by the method of retrograde horseradish peroxidase transport. *Neirofiziologiya* 15:22–26.

Lee, K. J., and Woolsey, T. A. (1975). A proportional relationship between peripheral innervation density and cortical neuron number in the somatosensory system of the mouse. *Brain Res.* 99:349–353.

LeVay, S. (1973). Synaptic patterns in the visual cortex of the cat and monkey. Electron microscopy of Golgi preparations. *J. Comp. Neurol.* 150:53–86.

Lin, C., Lu, S. M., and Schmechel, D. E. (1985). Glutamic acid decarboxylase activity in layer IV of barrel cortex of rat and mouse. *J. Neurosci.* 5:1934–1939.

Lorente de Nò, R. (1922). La corteza cerebral del raton. *Trab. Lab. Invest. Biol. Madrid* 20:41–78.

Lu, S. M., and Lin, C. (1985). Projection patterns from the ventrobasal complex of the thalamus to the rat barrel cortex. *Neurosci. Abstr.* 11:755.

Lyon, S., and Connors, B. W. (1985). Distribution and forms of GABAergic neurons in rat barrel cortex: a comparative immuno- and histochemical study of GAD, GABA and GABA-transaminase. *Neurosci. Abstr.* 11:751.

McCormick, D. A., Connors, B. W., Lighthall, J. W., and Prince, D. A. (1985). Comparative electrophysiology of pyramidal and sparsely spiny stellate neurons of the neocortex. *J. Neurophysiol.* 54:782–806.

Mountcastle, V. B. (1979). An organizing principle for cerebral function: the unit module and the distributed system. In: F. O. Schmitt and F. G. Worden (eds.). *The Neurosciences: Fourth Study Program*, pp. 21–42. Cambridge: The MIT Press.

Mountcastle, V. B., Davies, P. W., and Berman, A. L. (1957). Response properties of neurons of cat's somatic sensory cortex to peripheral stimuli. *J. Neurophysiol.* 20:374–407.

Mountcastle, V. B., and Powell, T. P. S. (1959). Neural mechanisms subserving cutaneous sensibility, with special reference to the role of afferent inhibition in sensory perception and discrimination. *Bull. John Hopkins Hosp.* 105:201–232.

Mountcastle, V. B., Talbot, W. H., Sakata, H., and Hyvärinen, J. (1969). Cortical neuronal mechanisms in flutter-vibration studies in unanesthetized monkeys. Neuronal periodicity and frequency discrimination. *J. Neurophysiol.* 32:452–484.

Olavarria, J., Van Sluyters, R. C., and Killackey, H. P. (1984). Evidence for the complementary organization of callosal and thalamic connections within rat somatosensory cortex. *Brain Res.* 291:364–368.

Peschanski, M., Ralston, H. J., and Roudier, F. (1983). Reticularis thalami afferents to the ventrobasal complex of the rat thalamus: an electron microscope study. *Brain Res.* 270:325–329.

Podladchikova, O. N., and Lapenko, T. K. (1982). A study of horseradish peroxidase-labelled sources of thalamic projections to the area of vibrissae representation of the rat somatosensory cortex. *Neirofiziiologiya* 14:631–635.

Rall, W. (1967). Distinguishing theoretical synaptic potentials computed for different soma-dendritic distributions of synaptic input. *J. Neurophysiol.* 30:1138–1168.

Ralston, H. J. (1984). Synaptic organization of spinothalamic tract projections to the thalamus, with special reference to pain. In: L. Kruger and J.C. Liebeskind (eds.). *Advances in Pain Research and Therapy*, pp. 183–195. New York: Raven Press.

Saporta, S., and Kruger, L. (1977). The organization of thalamocortical relay neurons in the rat ventrobasal complex studied by the retrograde transport of horseradish peroxidase. *J. Comp. Neurol.* 174:187–208.

Sheperd, G. M., Crayton, R. K., Miller, J. P., Segev, I., Rinzel, J., and Rall, W. (1985). Signal enhancement in distal cortical dendrites by means of interactions between active dendritic spines. *Proc. Natl. Acad. Sci. USA* 82:2192–2195.

Simons, D. J. (1978). Response properties of vibrissa units in the rat SI somatosensory neocortex. *J. Neurophysiol.* 41:798–820.

Simons, D. J. (1983). Multi-whisker stimulation and its effects on vibrissa units in rat SmI barrel cortex. *Brain Res.* 276:178–182.

Simons, D. J. (1985). Temporal and spatial integration in the rat SI vibrissa cortex. *J. Neurophysiol.* 54:615–635.

Simons, D. J., and Land, P. W. (1987). Early experience of tactile stimulation influences organization of somatic sensory cortex. *Nature* 326:694–697.

Simons, D. J., and Woolsey, T. A. (1979). Functional organization in mouse barrel cortex. *Brain Res.* 165:327–332.

Simons, D. J., and Woolsey, T. A. (1984). Morphology of Golgi-Cox-impregnated barrel neurons in rat SmI cortex. *J. Comp. Neurol.* 230:119–132.

Somogyi, P. (1978). The study of Golgi stained cells and of experimental degeneration under the electron microscope: a direct method for the identification in the visual cortex of three successive links in a neuron chain. *Neurosci.* 3:167–180.

Sretavan, D., and Dykes, R. W. (1983). The organization of two cutaneous submodalities in the forearm region of area 3b of cat somatosensory cortex. *J. Comp. Neurol.* 213:381–398.

Steffan, H. (1976). Golgi stained barrel-neurons in the somatosensory region of the mouse cerebral cortex. *Neurosci. Lett.* 2:57–59.

Sur, M., Wall, J. T., and Kaas, J. H. (1984). Modular distribution of neurons with slowly adapting and rapidly adapting responses in area 3b of somatosensory cortex in monkeys. *J. Neurophysiol.* 51:724–744.

Uhr, J. L., Chapin, J. K., and Woodward, D. J. (1982). Variation in receptive field size across layer IV in rat barrelfield cortex. *Neurosci. Abstr.* 852:82.

Vahle-Hinz, C., and Gottschaldt, K.-.M. (1983). Principal differences in the organization of the thalamic face representation in rodents and felids. In: G. Macchi, A. Rustioni, and R. Spreafico (eds.). *Somatosensory Integration in the Thalamus,* pp. 125–145. Amsterdam: Elsevier Science Publishers.

Van der Loos, H. (1976). Barreloids in mouse somatosensory thalamus. *Neurosci. Lett.* 2:1–6.

Verley, R., and Onnen, I. (1981). Somatotopic organization of the tactile thalamus in normal adult and developing mice and in adult mice dewhiskered since birth. *Exp. Neurol.* 72:462–474.

Waite, P. M. E. (1973). Somatotopic organization of vibrissal responses in the ventrobasal complex of the rat thalamus. *J. Physiol. (Lond.)* 228:527–540.

Welker, C. (1971). Microelectrode delineation of fine grain somatotopic organization of SmI cerebral neocortex in albino rat. *Brain Res.* 26:259–275.

Welker, C. (1976). Receptive fields of barrels in the somatosensory neocortex of the rat. *J. Comp. Neurol.* 166:173–189.

Welker, C., and Woolsey, T. A. (1974). Structure of layer IV in the somatosensory neocortex of the rat: description and comparison with the mouse. *J. Comp. Neurol.* 158:437–453.

Welker, W. I. (1964). Analysis of sniffing of the albino rat. *Behaviour* 22:223–244.

Wells, J., Mathews, T. J., and Ariano, M. A. (1982). Are there interneurons in the thalamic somatosensory projection nucleus in the rat? *Neurosci. Abstr.* 8:37.

White, E. L. (1978). Identified neurons in mouse SmI cortex which are postsynaptic to thalamocortical axon terminals: a combined Golgi-electron microscopic and degeneration study. *J. Comp. Neurol.* 181:627–661.

White, E. L. (1979). Thalamocortical synaptic relations: a review with emphasis on projections of specific thalamic nuclei to the primary sensory areas of the neocortex. *Brain Res. Rev.* 1:275–311.

White, E. L. and Hersch, S. M. (1982). A quantitative study of thalamocortical and other synapses involving the apical dendrites of corticothalamic projection cells in mouse SmI cortex. *J. Neurocytol.* 11:137–157.

White, E. L., and Keller, A. (1987). Intrinsic circuitry involving the local axon collaterals of corticothalamic projection cells in mouse SmI cortex. *J. Comp. Neurol.* 262:13–26.

White, E. L., and Rock, M. P. (1980). Three-dimensional aspects and synaptic relationships of a Golgi-impregnated spiny stellate cell reconstructed from serial thin sections. *J. Neurocytol.* 9:615–636.

White, E. L., and Rock, M. P. (1981). A comparison of thalamocortical and other synaptic inputs to dendrites of two non-spiny neurons in a single barrel of mouse SmI cortex. *J. Comp. Neurol.* 195:265–277.

Wineski, L. E. (1983). Movements of the cranial vibrissae in the Golden hamster (Mesocricetus auratus). *J. Zool., Lond.* 200:261–280.

Woolsey, T. A. (1967). Somatosensory, auditory and visual cortical areas of the mouse. *Johns Hopkins Med. J.* 121:91–112.

Woolsey, T. A., and Dierker, M. L. (1978). Computer-assisted recording of anatomical data. In: R. T. Robertson (ed.). *Neuroanatomical Research Techniques,* pp. 47–85. New York: Academic Press.

Woolsey, T. A., Dierker, M. L., and Wann, D. F. (1975). Mouse SmI cortex: qualitative and quantitative classification of Golgi-impregnated barrel neurons. *Proc. Natl. Acad. Sci. U.S.A.* 72:2165–2169.

Woolsey, T. A., and Van der Loos, H. (1970). The structural organization of layer IV in the somatosensory region (SI) of mouse cerebral cortex. *Brain Res.* 17:205–242.

5

Dynamic Processes Governing the Somatosensory Cortical Response to Natural Stimulation

BARRY L. WHITSEL, OLEG V. FAVOROV,
MARK TOMMERDAHL, MATHEW E. DIAMOND,
SHARON L. JULIANO, AND DOUGLAS G. KELLY

For many years neuroscientists have been aware of the possibility that large, extensively interconnected neural networks might behave in ways not predictable from the functional properties of the network's individual elements or from how the elements are interconnected. Such behaviors have been termed "dynamic" and their detection and analysis remain among the most challenging tasks for experimental neuroscientists. In fact, many theorists and experimentalists share the view that systems neuroscience will continued to lag as a discipline (compared with the advancing cellular and molecular neurosciences) until objective, standardized approaches for direct demonstration and systematic analysis of neural network dynamics are developed.

Since the dynamic behaviors of a cerebral cortical network can be elucidated only by information on the time-dependent aspects of its function (Mountcastle, 1966), the most direct approach to studying neural network dynamics would be to use large arrays of recording electrodes to sample simultaneously the activity of as many neurons as possible (Mountcastle, 1986). The principal approach available for analyzing multiple spike-trains involves cross-correlation analysis of pairs or triplets of spike-trains. The method implies functional relations between a pair of neurons whenever excessive or abnormally low numbers of near-coincident firings are detected in the recorded spike-trains. Systematic use of the cross-correlation method allows one to distinguish synaptically mediated interactions among the observed neurons (neural interaction effects) from the tendency for neurons receiving common input to exhibit coordinated changes in firing rate (stimulus coordination effects). In principle, the method permits construction of a logical wiring diagram among the observed neurons. Moreover, when used with an appropriate experimental design, it theoretically allows one to examine whether the wiring diagram changes with stimulus, behavior, or any other laboratory variable.

A major obstacle arises when one tries to use the cross-correlation method to analyze more than a few simultaneously recorded spike-trains. The obstacle is combinatorial because of the nature of the computation (the number of neuron pairs or triplets, multiplied by the several time resolutions needed in the computation). In estimating the wiring diagram within the sector of a living neural network that processes stimulus-evoked input, the number of neurons that must be recorded simultaneously is unrealistically large for most networks. Even the largest number of simultaneously recorded single neurons reported to date (20 neurons; Gerstein et al., 1978; 1983) falls far short of an adequate sample of the responding somatosensory cortical neuronal population. Finally, in the unlikely event (at least with current methods) that a very large number of single neurons are recorded simultaneously, the required computation time would be formidable; perhaps worse, the enormous volume of material requiring interpretation would be overwhelming. Related to this issue is the fact that the recordings in most multineuron recording experiments are blind, in the sense that the observed activities are sampled from neural elements that cannot be assigned a priori to distinguishable functional groupings. A set of neurons is a functional grouping if the neurons' impulses are coordinated to the extent that their temporal relationships are arranged, at least probabilistically, in characteristic patterns (Perkel et al., 1975). Thus, one cannot reduce the combinatorial problem by categorizing the samples on the basis of function before carrying out cross-correlation analyses. Although some computational approaches have been suggested as having potential value in this regard (Gerstein et al., 1978), the combinatorial/sampling problem remains the major deterrent to studies of network dynamics based on cross-correlation analysis. As a result, the number of experimental studies of the time-dependent aspects of cortical function using cross-correlation analysis has been limited, and they have not provided the anticipated insights into cortical dynamics. Their use, however, has led to the important discoveries that the coupling strength between neocortical sensory neuron pairs and triplets can be dynamically regulated (it can change with stimulus conditions; Dickson and Gerstein, 1974; Abeles, 1982; Schneider et al., 1983) and that significant horizontal interactions can occur between sensory neurons in columns separated as much as several millimeters (Ts'o et al., 1986).

Although they do not permit the quantitative analysis of network dynamics promised by the cross-correlation methods, three other approaches have been suggested to reveal aspects of dynamic processes in the somatosensory cortex. The first approach, single-unit recording in behaving animals, was introduced to improve our understanding of the perceptual relevance of somatosensory neuron discharge activity. Such efforts have been characterized as attempts to "define the functional neural circuits" in which the neurons participate, and the data they have generated establish this demanding experimental approach as a valuable method for investigating the plasticity of somatosensory cortical information processing. For example, Hyvarinen and his colleagues (1978, 1982) detected enhancements of the response of S-I and S-II neurons to tactile vibratory stimuli when the monkey subjects attended to the stimulus. Werner (1978) reported improved signaling of stimulus direction by single, directionally sensitive S-I neurons when this aspect of cutaneous stimulation was assigned significant associative value. And, more recently, Chapin and Woodward (1982), Nelson (1984a,b), and Iwamura et al. (1983, 1985) detected and analyzed changes in monkey S-I neuron responsivity that were associated with particular voluntary motor behaviors and occurred during "active" tactile exploration. Taken together, these findings appear to prove beyond any reasonable doubt that the behavior of individual somatosensory cortical neurons can be highly plastic.

By themselves, however, such observations do not lead to a coherent impression of the physiological basis and meaning of such dynamic single-neuron behaviors. There are three principal reasons for this conclusion. First, the available observations cannot eliminate the possibility that a single neuron can contribute to different (perhaps many) functional neural circuits, each involved in a different sensorimotor behavior. Second, such data provide little information on the nature or the locus of the neural mechanisms responsible for the observed plasticity. And third, they do not reveal the relationships between the activity of the neuron selected for study and that of the widely distributed cortical neuron population involved in the subject's behavior. Because of these limitations, an in-depth appreciation of dynamic processes in the somatosensory neocortex (see Edelman and Finkel, 1984) will result only from the use of experimental strategies with the capacity (1) to demonstrate the "emergent properties of large neuron populations" directly, as well as (2) to analyze the contributions of each of the multiple neural projection systems that converge at the level of the somatosensory cerebral cortical cell column.

The second experimental approach to gain insight into dynamic somatosensory cortical processes employs variations of a receptive field (RF) mapping procedure (minimal RF mapping). Recently this approach has been used extensively in examining somatosensory cortical topographical organization. Briefly, the minimal RF mapping observations obtained in recent studies have been interpreted to indicate that somatosensory cortical representations in adult subjects undergo topographic remodeling after a variety of experimental manipulations, including peripheral nerve transection, amputation of digits, syndactyly, differential use of restricted skin surfaces, and restricted cortical lesions (for recent review, see Merzenich, 1987). Minimal RF mapping observations also serve as the basis for the claim that somatosensory topographic maps not only are use-dependent and use-modifiable throughout life, but also can be modified by alterations in behavioral state (Merzenich, 1987). This recent view of somatosensory cortical functional plasticity conflicts with the long-held belief that the adult neocortex is hard-wired, with its operations preshaped mainly by developmental processes. Rather, it is maintained that the basic process underlying cerebral development (i.e., synaptic weight modification) continues to operate throughout life and retains the capacity to be influenced by experience. As a result, somatosensory cortical topographical organization is suggested to be continuously modifiable; that is, there "is an accumulated record of input histories within neocortical fields" that, in turn, determines the representational details of the cortical maps (Merzenich, 1987). The sometimes appreciable idiosyncratic variations in map details detected in minimal mapping studies of large series of normal adult monkeys are thus thought to reflect the different sensory histories of the individual subjects.

A third group of studies providing information about somatosensory cortical dynamic mechanisms has been carried out in our own laboratory. Taken together, the results of these studies have led us to a view that differs radically from the concept of somatosensory cortical dynamics based on the minimal RF mapping studies just described. In the following pages we summarize our experimental observations and describe a model of how somatosensory cortical information processing can be regulated in a moment-to-moment fashion by an intrinsic dynamic mechanism that is sensitive to the spatiotemporal characteristics of afferent input. Unlike the view of somatosensory cortical dynamics based on minimal RF mapping observations, the dynamic behaviors we have detected do not appear to be the result of a process that involves modification of synaptic weights.

SUMMARY OF EXPERIMENTAL OBSERVATIONS

Mapping the Somatosensory Cortical Neural Population that Responds to Tactile Stimuli

To map the somatosensory cortical neural population responsive to tactile stimuli (Favorov and Whitsel, 1988a, 1988b; Favorov et al., 1987a; 1987b; Diamond et al., 1987), two series of experiments were carried out. The first used extracellular single-neuron recording methods and hand-held mechanical stimuli to study the RFs of neurons in two different cortical fields: area 1 of unanesthetized *Macaca fascicularis* monkeys and area 3b of anesthetized and unanesthetized cats. Both in monkey area 1 and cat area 3b, the RF data obtained in near-radial microelectrode penetrations demonstrated that the cutaneous RFs of neurons located within the same cell column can differ strikingly, and the RFs of neighboring neurons commonly differ greatly in size and configuration (Figure 5-1). On the other hand, the RF variations between pairs and neurons within a minicolumn (a

RF variations within S-I cell columns

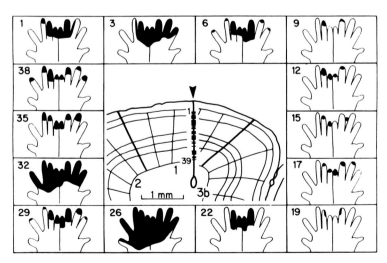

Figure 5-1 Example of a penetration of area 1 in the monkey, in which 39 single units were isolated and their RFs mapped. The location along the electrode track (in center) of each isolated single unit is indicated by a bar, the position of each RF is shown on a figurine. (To conserve space only 14 RFs are shown.)

single radial column of cells) tend to be substantially less than those between pairs of neurons in different minicolumns.

The RF data obtained from arrays of near-radial penetrations led us to suggest that the skin representation in the forelimb region of both area 1 of monkey and area 3b of cat is composed of a mosaic of macrocolumnar neuronal aggregates called *segregates*: the average diameter of a monkey area 1 segregate is 600 μm; for cat area 3b it is 350 μm. Although the RFs of the neurons in a single segregate tend to vary substantially in size and configuration, they all share a single small area on the skin. The boundaries of a segregate can be mapped precisely, because the skin area common to all neurons in one segregate (the segregate RF center) does not overlap the skin area common to the neurons in the adjacent segregate. In other words, some of the neurons located on opposite sides of a segregate boundary (belonging to different segregates) have nonoverlapping RFs. Furthermore, within any given segregate no systematic shift in RF location occurs as the electrode advances through its minicolumns. A systematic RF shift occurs only when the electrode traverses the boundary between neighboring segregates. Our concept of somatosensory cortical organization as a mosaic of segregates is illustrated schematically in Figure 5-2, along with the method used to detect it.

Based on our sample of neurons in the different cortical layers of areas 3b and 1, it seems evident that the RFs of neurons within a single segregate possess a wide variety of sizes and configurations and occupy a wide variety of positions on the skin relative to the segregate RF center. The total skin area providing input to a segregate (the segregate RF which is estimated by the aggregate of the RFs sampled for that segregate) was consistently found to be extensive. Because of the diversity of the RFs of the neurons within a single segregate, the segregate RF is not homogeneous: it is composed of skin regions distinguished by their

Segregate organization in area 1

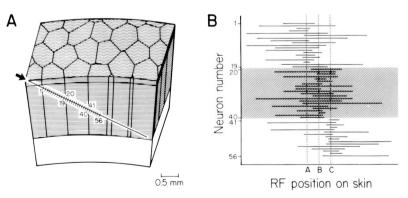

Figure 5-2 (*A*) Positions of 56 single neurons along the electrode track. The cortex is subdivided into a mosaic of 600-μm-wide columns—segregates. (*B*) Position of each neuron's RF on the skin is plotted as a line. The shading distinguishes the groups of neurons belonging to different segregates. RFs shift on the skin in a seemingly random fashion, but in each segregate the oscillations are centered on a single skin point (A, B, and C for the first, second, and third segregates, respectively). Variations in RF size occur in addition to positional oscillations.

strength of representation in the segregate (i.e., the fraction of neurons that include a region in their RFs). Furthermore, the regions of the segregate RF that can be differentiated in this way always exhibit a highly ordered arrangement: we regard this as the spatial context in which the segregate maps its input. As a rule, the skin areas that provide input to neighboring segregates overlap extensively; even segregates separated by large distances (5 to 6 mm) in area 1 or area 3b can receive a substantial common input. Finally, although the arrangement of segregates in both areas 1 and 3b is generally somatotopic, the local relationships among different segregates can deviate significantly from a strictly somatotopic pattern.

A second type of experiment was carried out to evaluate our expectation (based on our findings of extensive overlap in the peripheral input to neighboring segregates and the ability of remote segregates to receive common input) that the activity evoked by most somatic stimuli would be widely distributed in the S-I cortex. To this end, microelectrode penetrations of areas 3b and 1 in unanesthetized monkeys were performed so that they traversed long sequences of cortical cell columns (Figure 5-3). In this type of experiment, exactly the same set of stimulus conditions was employed each time the activity of a single neuron was sampled: the stimulus used was termed the *standard mapping stimulus*. It should be noted that this approach departs radically from the usual approach to single-unit mapping—to vary the stimulus conditions at each recording site until the "best" conditions for evoking discharge activity are discovered. In this experiment the same set of stimulus conditions is used at all recording sites. If a sufficient number of neurons is sampled, this strategy should provide an estimate of the spatial distribution of discharge activity set up in the subject's S-I cortex by the standard mapping stimulus. This estimate can be compared directly with the global spatial pattern of metabolic activity set up in the same cortex by the same stimulus (revealed by the 14C-2-deoxyglucose [2DG] metabolic mapping method).

The representative mapping results shown in Figure 5-3 indicate that even a spatially restricted gentle tactile stimulus (e.g., a unidirectional brushing stimulus to the palmar skin) sets up vigorous activity in neurons located at multiple, disjoint, and widespread locations in both areas 1 and 3b. Based on this outcome, therefore, our expectation (based on our RF mapping studies of area 3b and 1 segregates) that stimulus-evoked activity patterns in S-I should be spatially extensive seems to be supported. Although a relatively large number of single neurons was sampled in penetrations such as the one shown in Figure 5-3, the numbers fall far short of the sampling density required to reconstruct the responding S-I population at high resolution. Thus, we believe only nonneurophysiological mapping methods (e.g., the 2DG metabolic mapping method) can reveal in high resolution the global spatiointensive activity patterns evoked in large neural networks by natural stimuli.

Mapping the Stimulus-Evoked Somatosensory Cortical Network Response with the 2DG Metabolic Mapping Procedure

In an attempt to estimate the global cortical neuroelectrical activity patterns evoked by natural stimuli (Juliano et al., 1981; 1983; Whitsel and Juliano, 1984; Juliano and Whitsel, 1985), 2DG labeling patterns in area S-I of monkeys (*Macaca fas-*

A

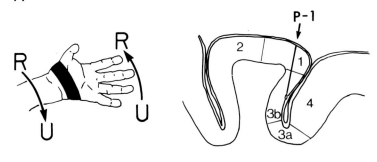

B

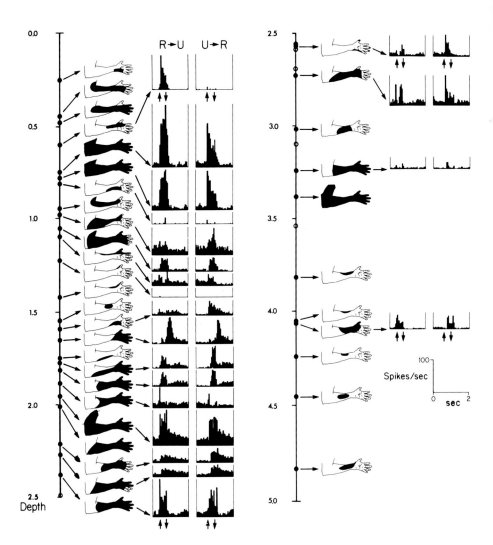

cicularis) were obtained using servocontrolled vertical displacement ("flutter") or constant-velocity brushing stimuli applied repetitively to the skin. In all subjects, most above-background metabolic labeling appeared in the form of intermittent patches, extending radially across laminae II to V. When series of autoradiographs were reconstructed to produce continuous two-dimensional maps, it became evident that the patches aligned to form a complex spatial pattern involving an extensive sector of the S-I cortex. Even in those experiments in which the somatic stimulus was applied to a spatially restricted peripheral field, the spatial pattern of cortical activity consisted of a complex pattern of patches and strips that occupied an extensive sector within S-I. Nevertheless, the 2DG patterns produced in S-I of different animals stimulated with the same stimulus were very similar.

The location of the patches of 2DG labeling within S-I matched well the locus of the representation of the body part stimulated (determined by comparing the labeling distribution with reconstructions of RF mapping data obtained without general anesthesia; see, e.g., Figure 5-4). Such comparisons revealed, however, that the reconstructions based on RF mapping data failed to predict several prominent features of the 2DG labeling patterns; neither the highly periodic strips of activity nor the presence of bordering fields of below-background activity had counterparts in the maps based on neurophysiological mapping data. (The bordering "inhibited" fields are also striplike in configuration, but with widths several times that of the strips of above-background activity; for examples of such fields, see Figures 5-12 and 5-13.) These discrepancies between the 2DG patterns and maps of the distribution of input from the stimulated field—reconstructed from neurophysiological mapping data obtained in conscious animals—could not be attributed to inadequacies in either the 2DG or the neurophysiological mapping methods; thus, it was speculated that they reflect the operation of previously unidentified "dynamic" processes that cause the cortical response to repetitive natural stimulation of the skin, measured by the 2DG method, to deviate from that anticipated from neurophysiological mapping data.

Parallel experiments were carried out to examine the 2DG labeling pattern in S-II in another series of monkeys. In these animals the S-II labeling pattern evoked by a repetitive stimulus was always spatially complex and extensive; nevertheless, the patterns evoked by the same stimulus in different animals were strikingly sim-

Figure 5-3   (*A*) Figure (to *left*): showing skin zone (*blackened region*) stimulated with the standard mapping stimulus—a constant-velocity (10 cm/sec) brushing stimulus that swept across the skin in either of the two opposing directions (*see arrows*); the two directions of motion were presented randomly. On the right is a reconstruction of an electrode track that crossed areas 1 and 3b. (*B*) Dots placed along the vertical lines depict the locations along the microelectrode track at which single neurons were studied in the penetration shown in *A*. The numbers placed along the vertical lines indicate the depth from the initial point of brain contact along the track. For each single neuron studied a figure shows the cutaneous RF. For those neurons that responded to the standard mapping stimulus, a pair of PST histograms show the patterns of spike discharge activity evoked by the two directions of stimulus motion. For each pair of PST histograms, the histogram on the left shows the response to 25 stimuli moving in the radial-to-ulnar direction (R→U), whereas the histogram on the right shows the response to 25 stimuli delivered in the opposite direction (U→R). The scale used for all PST histograms is shown at the bottom right.

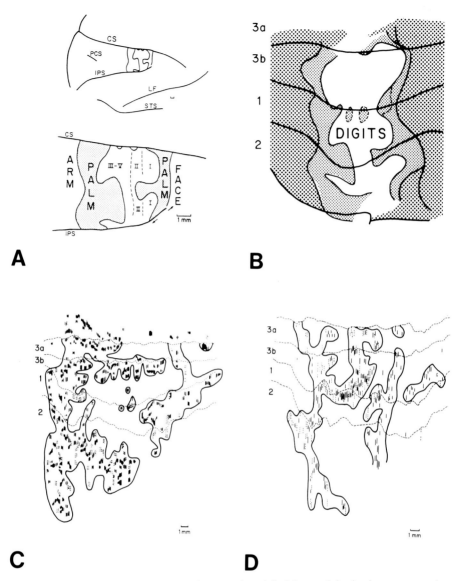

Figure 5-4 (*A* and *B*) Surface maps showing the global form of the body representation in the postcentral gyrus of unanesthetized *Macaca fascicularis* monkeys. The names of body parts placed within regions (e.g., FACE, ARM, etc.) indicate the predominant source of peripheral input to each region. In the unfolded map showing the extent of the anterior parietal body representation (*B*), cytoarchitectural boundaries are indicated by dark, nearly horizontal lines (*C* and *D*). Unfolded surface maps showing the global patterns of distribution of patchlike 2DG labeling in the anterior parietal cortex of *Macaca fascicularis* monkeys. Patches of label are indicated by short line segments; fields of cortex containing patches of label are enclosed by solid lines. Cytoarchitectural boundaries are indicated by interrupted lines. In *C* the stimulus used to evoke the labeling pattern was a brushing stimulus to the contralateral palm (similar to the stimulus described in Figure 5-3); in *D* a flutter vibration stimulus (25 Hz, 1-mm peak-to-peak amplitude) to the volar tip of the contralateral index finger was used.

ilar. Most labeling fell within the cortical territory identified in neurophysiological mapping studies in conscious animals as the S-II representational field for the stimulated skin site; the striplike nature of the labeling pattern within this region, however, was not predicted by the neurophysiological mapping observations. Therefore, just as we suggested for S-I, unknown dynamic processes not detectable in conventional single-unit mapping experiments may have been determining the highly periodic character of the S-II network response to somatic stimulation.

Stimulus Mode and Place Influence the Stimulus-Evoked Somatosensory Cortical Network 2DG Activity Pattern

In analyzing the effects of stimulus mode and place (Juliano et al., 1984; Whitsel and Juliano, 1984; Juliano et al., 1988), as before, somatosensory cortical 2DG activity patterns were produced using precisely controlled skin stimulation (intermittent vertical displacement, constant-velocity brushing, and constant-current electrocutaneous). The parameters of stimulation were varied systematically from one experiment to the next. In sagittal sections the S-I labeling of all subjects appeared as columnlike patches extending from laminae II to V. The density of labeling within a given patch was greatest in lower lamina III and lamina IV. Another prominent feature was that, in area 3b and at times area 1, the patches were superimposed on a continuous band of label in basal laminae III and IV (see examples in Figure 5-11). Surface reconstructions of the labeling patterns in the contralateral S-I cortex indicated that the patches of above-background labeling in adjacent sections aligned to form a mediolaterally oriented pattern of strips. Comparing the patterns of strips in animals stimulated in different ways demonstrated that the labeling pattern in the S-I cortex shifts in an orderly manner, as the peripheral locus and the modality of the somatic stimulation change (Figure 5-4; see also Figures 5-12 and 5-13). It was concluded, therefore, that the striplike S-I cortical network 2DG labeling pattern evoked by somatic stimulation reflects both the place and the mode of somatic stimulation—that is, it is stimulus-specific.

The observations obtained from a series of animals studied using electrocutaneous stimulation permitted rejection of the possibility that intermittent and patchy somatosensory cortical labeling patterns arise because of the differential activation (by the repetitive natural stimulus) of selected mechanoreceptor afferent populations. The patchy pattern of labeled columns, thus, might be attributable to the way different peripheral receptor types project to the cortex: that is, the labeled columns in a 2DG pattern might process input from one mechanoreceptor class within the stimulus field, whereas the intervening unlabeled columns process input from a different, but less well-activated mechanoreceptor class that innervates the same peripheral field. This concept is unlikely to account for the periodic labeling patterns, however, because clear-cut patchy labeling is also evoked when electrocutaneous stimulation is used to evoke 2DG patterns. (The patterns evoked by electrocutaneous stimuli are, in fact, quite similar to those evoked by natural stimulation of the same skin field.) We thus concluded that central dynamic processes, presumably in the cortex itself, are required to account for the intermittent, patchy patterns of 2DG labeling present in the somatosensory cortices of animals subjected to repetitive natural stimuli, since it is known that electrocutaneous stimuli unselectively activate all classes of large-diameter mechanoreceptive afferents.

Dynamic Factors Identified in Combined 2DG and Neurophysiological Mapping Experiments

Our combined experiments which sought to identify the dynamic factors suggested by our earlier 2DG and neurophysiological mapping experiments (see Juliano and Whitsel, 1987; Juliano et al., 1987) involved the collection of two types of data from the same animal; microelectrode recordings were made from small populations of neurons, and 2DG mapping data were also obtained from the same region of the S-I cortex. Two paradigms were used. In the first, standard extracellular recording and peripheral stimulation methods were used to map the RF and submodality properties of the neuron populations at selected S-I loci. Immediately after the microelectrode recordings were completed, the response of the same S-I field to a repetitively delivered and precisely controlled tactile stimulus (either an intermittent vertical displacement stimulus or a constant-velocity unidirectional brushing stimulus) was mapped with the use of the 2DG procedure. The region of skin stimulated during the 2DG procedure and the parameters of the 2DG mapping stimulus were selected to correspond to the RFs and the stimulus requirements of at least some of the neurons sampled in the initial phase of the experiment. Following the experiment, the locations of the recorded neurons were reconstructed, and the RF and response properties of the neurons at each locus were compared with the intensity of 2DG labeling at those same sites.

As expected, at cortical loci where the RF of the neuron population did not include the skin site stimulated in the 2DG experiment, there was no above-background 2DG labeling. Surprisingly, however, the 2DG labeling was above background in only 53 percent of the cortical loci where the neurons had RFs and response properties corresponding to the place and model characteristics of the stimulus used to evoke 2DG labeling. These results were interpreted to indicate that a repetitive stimulus (e.g., those used in the 2DG mapping experiments) "fractionates" the total population of S-I cell columns whose neurons possess RF and submodality properties consistent with those of the applied stimulus: that is, the stimulus activates only a fraction of the total number of cortical columns that receive input from the stimulus field. Although the mechanism of fractionation remains obscure, we favor the idea that it involves a dynamic process that determines which of the cortical columns accessed by thalamocortical input is activated to participate in the representation of the applied stimulus. The operation of such a mechanism in the cortex would have major functional implications: for example, if the proposed fractionation of the available pool of S-I neurons occurred nonrandomly (i.e., if the spatial distribution of cortical columns permitted to participate in an activity pattern is nonrandom), maps prepared by reconstructing samples of single-neuron RF and submodality properties would be very poor predictors of the network response to a somatic stimulus. Not only would this mechanism account for most of the notable discrepancies we have detected between 2DG maps and reconstructions of neurophysiological mapping data, but it would also mean (1) that computations of the volume of representation of a body part based on neurophysiological mapping data overestimate the actual volume of cortex activated by a repetitive natural stimulus, and (2) that volume of somatosensory cortical representation should change as the capacity of the stimulus to recruit dynamic mechanisms changes. (Such alterations might occur, for example, with changes in the attentive state of a subject.) In general, the prominent spatial periodicities in the patterns of metabolic activity evoked by a stimulus led us to suggest that the process underlying neuronal group ("module") selection in the S-I cortex is distributed in a highly regular, nonrandom way over the field receiving stimulus-evoked afferent input.

A second type of combined 2DG and neurophysiological experiment was carried out using a different paradigm. In these experiments the response of the neuron population at each electrode site to exactly the same stimulus used in the 2DG experiment (cutaneous flutter or constant velocity and unidirectional brush strokes) was recorded. In addition, the RF and submodality properties of the neurons at each site were determined using conventional neurophysiological mapping procedures. The findings provided by this second type of experiment were viewed as consistent with those obtained in the first series of experiments. The strength of neural activation recorded at a site often corresponded to the intensity of 2DG labeling, but at some sites the controlled stimulus very effectively evoked neural activity, yet no above-background labeling was apparent. As in the first series of combined experiments, therefore, the sometimes notable discrepancy between 2DG labeling intensity and neurophysiological indices of stimulus effectiveness detected at some cortical loci seemed to be in line with our postulate that dynamic factors (i.e., factors other than single neuron RF and submodality properties) may determine the global properties of the spatial activity pattern evoked in S-I by prolonged repetitive somatic stimulation.

Role of Intrinsic Inhibitory Circuitry in Elaboration of Somatosensory Cortical Response to Peripheral Stimulation

The possibility that GABA-mediated inhibitory processes intrinsic to the cortex might participate in forming somatosensory activity patterns was evaluated directly in studies in which GABA-mediated cortical inhibition was modified by topical applications of the antagonist bicuculline methiodide (BIC; concentrations between 10^{-5} and 10^{-7} M were used) (Juliano and Whitsel, 1987; Tommerdahl et al., in preparation). These studies showed that applying BIC to the somatosensory cortex dramatically altered the size of the stimulus-evoked 2DG patches and led them to be embedded in a field of elevated 2DG uptake. With two-dimensional maps of the labeling pattern it is obvious that the normally separated activity "strips" in the BIC-treated hemispheres tend to fuse, leading to a spatial distribution of labeling much more homogeneous than that obtained in the absence of the drug (Figure 5-5). In an important control experiment, the somatosensory cortex of subjects was treated with BIC, but no somatic stimulation was delivered. Because subjects treated in this way did not exhibit 2DG activity patterns different from those in the same cortical regions in unstimulated drug-free subjects, the homogeneous patterns observed in BIC-treated cortices apparently reflect stimulus-evoked activity, not the effects of direct excitatory actions of BIC.

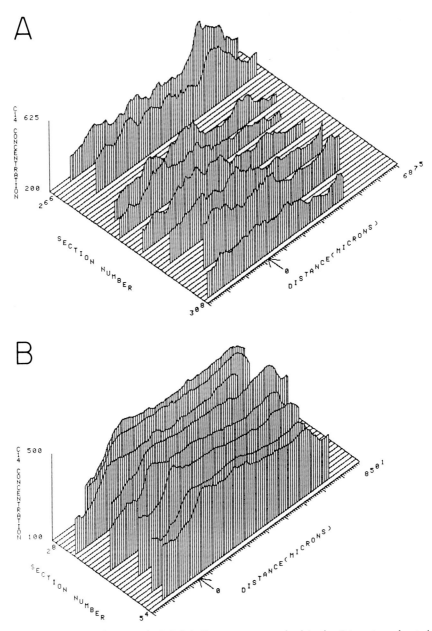

Figure 5-5 Unfolded maps of 2DG labeling patterns evoked in the S-I cortex of cats by flutter vibration stimulation (sinusoidal, 25-Hz, 1-mm peak-to-peak amplitude) of the central pad of the contralateral forepaw. The stimuli were delivered with a probe 5 mm in diameter. (*A*) Map of labeling pattern evoked in S-I of a normal drug-free subject. (*B*) Map of labeling pattern evoked in S-I of subject in which the cortex was pretreated by the topical application (in the CSF) of 10^{-5} *M* bicuculline. Each of the maps consists of multiple histograms; each histogram displays the series of C14 concentration values extracted from a single autoradiographic image of a frontal section 20μm thick. The distance (in micrometers) between any two histograms in a map can be computed by multiplying the difference in section numbers by 20. Sections assigned higher numbers are anterior to those assigned lower numbers. The bin at the extreme left of a histogram indicates the

Identifying Dynamic Factors Through Analysis of Single-Neuron Behavior

Two series of neurophysiological experiments (McKenna et al., 1984; Diamond et al., 1986) have provided evidence that stimulus-evoked somatosensory cortical neuron behavior can change significantly with repetitive stimulation of the skin. The first series of experiments used extracellular and intracellular single-unit recording methods to sample neurons in the upper layers of the cat S-I cortex. Intracellular labeling with horseradish peroxidase (HRP) positively identified the elements sampled. A high proportion of the upper-layer units exhibited a prominent response plasticity: that is, repetitive stimulation of their cutaneous receptive fields led either to progressive enhancement of responsiveness and increased spontaneous activity or to decreased responsiveness and an absence of spontaneous activity (Figure 5-6). Also, for some neurons, a relatively subtle change in stimulus conditions (e.g., in velocity) could reverse the effects of repetitive stimulation; for example, rather than decreasing with repetitive stimulation, the response magnitude might increase. Intracellular recordings obtained from a few upper-layer S-I neurons revealed that the changes in single-neuron responsivity evoked by repetitive tactile stimulation were accompanied by changes in membrane potential or input resistance; this suggests that, at least in these instances, the changes in responsivity were due to mechanisms acting directly on the observed cortical neurons.

A subsequent series of neurophysiological studies was designed to evaluate quantitatively the effects of repetitive somatic stimulation of S-I neurons. Extracellular recording methods were used to sample the activity of single neurons in layers II through VI of the S-I cortex of unanesthetized macaque monkeys and cats. The experiments employed servocontrolled mechanical stimulation of the skin. Changes in the magnitude of the neural response to each stimulus in a series of stimulus presentations (usually 25 to 200) were assessed statistically. Specifically, a Pearson product moment correlation coefficient was computed between response magnitude and the position of the stimulus in the series of repetitive stimuli. Although this statistic assumes a linear relationship between the position of the stimulus in the series and the magnitude of the response it evokes, the correlation coefficient is sensitive to the presence of many nonlinear relationships as well. To test whether the obtained correlation coefficients differed significantly

---

average C14 concentration value computed for a small, column-shaped bin (bin width = 80 μm) located at the medial extreme of the sampled field in the autoradiograph; the bin at the right extreme of each histogram indicates the average C14 concentration value computed for a column of the same size at the extreme lateral edge of the sampled field in the autoradiograph. All bins extended from the pial surface to the white matter. In both A and B the histograms are aligned with respect to that bin (*identified by the arrow*) corresponding to the location of a morphological landmark identifiable in every autoradiograph—that is, the point of maximal curvature in the medial lip of the coronal gyrus. Note the considerable homogeneity in the labeling pattern in the bicuculline pretreated subject (*B*) and the detailed, spatially heterogeneous pattern in the drug-free subject (*A*). The restriction of the field of elevated labeling in the bicuculline pretreated cortex to the same field that contains elevated labeling in the drug-free cortex is also evident.

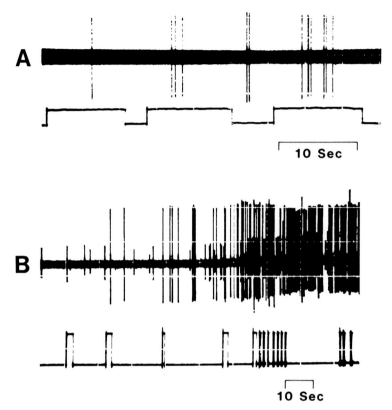

Figure 5-6 (*Upper traces*) Spike trains recorded extracellularly from two different upper layer units in cat S-I cortex responsive to cutaneous stimuli. Unit A was recorded from area 3b; unit B from area 1. (*Lower traces*) Upward deflections indicate periods during which a moving brush stimulus traversed the unit's RF. For each unit all the stimuli moved across the same extent of skin and in the same direction. Stimuli were constant-velocity (0.5 to 2 cm/sec), unidirectional brush strokes. Note that for both units A and B the initial stimulus evoked a weak, short-duration response, and subsequent reapplications of the stimulus led to increases in response magnitude and to longer periods of discharge. The small unit recorded simultaneously with unit B shows a similar tendency.

from zero, the normal deviate corresponding to the coefficient was calculated using Fisher's r to Z transformation. Normal deviates with absolute values greater than that needed to reject the null hypothesis at the $p < 0.01$ level of significance were interpreted as consistent with the idea that amplitude of response and position of stimulus in the stimulus series were significantly correlated—that is, the neuronal response changed significantly in the course of repetitive tactile stimulation.

Data obtained from more than 100 S-I neurons have been analyzed to date. This preliminary analysis allows a number of conclusions, although more detailed analyses (e.g., time series analysis) would provide additional valuable information about the effects of repetitive stimulation on single-neuron response. First, the effects of repetitive stimulation on S-I neuron responsivity is highly reproducible for a given set of stimulus conditions. (A number of neurons have exhibited virtually identical alterations of responsivity during repetitive stimulation in each of many runs conducted over a period of 4 to 6 hours; see Figure 5-7.)

Second, the changes in responsivity exhibited by an individual neuron is not the result of a widespread, uniform change in responsiveness throughout the cortex. This is evident from the fact that, when the same repetitive tactile stimulus is used to study neurons in different cell columns in the same cortical area, some of the neurons sampled exhibit radically different trends in their responsivity. In the experiment illustrated in Figure 5-8, for example, a unidirectional, constant-velocity brushing stimulus was delivered to the monkey's hand as the electrode was driven across a long series of cell columns in areas 1 and 3b. Four of the illustrated neurons exhibited statistically significant negative trends (neurons 1, 3, 4, and 6), two exhibited statistically significant positive trends (neurons 2 and 7),

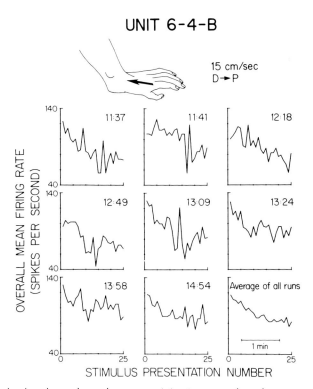

Figure 5-7 Each plot shows how the responsivity (expressed as the mean rate of spike discharge activity evoked during the stimulus period; OMFR) of a single S-I neuron to a brushing stimulus changed during repetitive stimulation. (Each plot shows the data obtained in a "run" consisting of 25 identical stimuli; stimulus conditions were identical in each of the eight runs.) This neuron was studied continuously between 11:36 A.M. (*top left plot*) and 2:54 P.M. (*bottom right plot*); a rest period of at least 5 minutes was allowed to elapse between runs. The plot at the lower right shows the average behavior of the neuron computed over all eight runs. Note that the changes in responsivity are highly reproducible from one run to the next. Each point plots the mean rate of discharge activity evoked during a single stimulus trial. The arrow on the figure at the top indicates the direction of the brushing motion and defines the skin site stimulated.

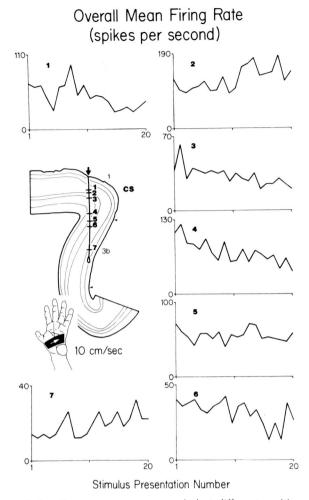

Figure 5-8 Trend plots for seven neurons sampled at different positions along a single microelectrode penetration, which traversed areas 1 and 3b. The numbers and tics placed along the electrode track indicate recording sites where the responsivity of single neurons to repetitive applications of the same stimulus (the standard mapping stimulus) was determined. The format for trend plots is similar to that in Figure 5-7. The ordinate is scaled in mean spikes per second; the figure of a hand shows the direction and location of the brushing stimulus. CS = central sulcus.

and one exhibited no trend (neuron 5). For all the trends illustrated in Figure 5-8, the hypothesis of zero trend could be rejected with a probability of error of ≤0.01.

Third, the sign (+ or −) of the alteration in responsivity produced by repetitive stimulation depends on stimulus conditions (e.g., the velocity or direction of motion or, for some neurons, the region of the RF stimulated; see Figure 5-9).

Fourth, although the time course of the alteration in neuronal responsivity observed with repetitive stimulation is typically rapid (usually reaching a maximum within 10 to 20 stimulus presentations), this time course can also be influenced by changes in stimulus conditions.

Fifth, the stimulus-specific alterations in responsivity can modify a somatosensory cortical neuron's feature extraction capabilities (e.g., we have reversed the directional sensitivity of some S-I neurons with repetitive stimulation.

Sixth, and finally, we have observed that neurons sampled at different times

UNIT 23-2-A

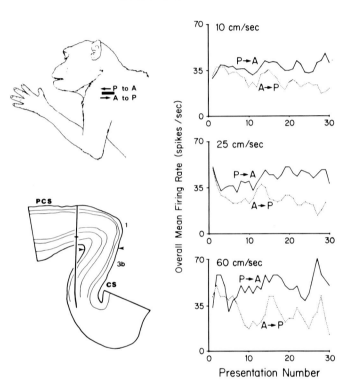

Figure 5-9   Trend plots for an area 1 neuron studied with repetitive stimuli differing in direction and velocity. The format for the plots is similar to that used in Figures 5-7 and 5-8. The filled rectangle and arrows on the figure at the upper left show location and direction of the brushing stimuli used to evaluate the effects of repetitive stimulation. The solid line in each trend plot shows the mean firing rate elicited on each trial by stimuli moving posterior to anterior (P→A) over the RF; the dotted line plots the data obtained using the opposing direction of motion (A→P). In general, it seems evident that, at all the velocities of stimulation used (10, 25, and 60 cm/sec), the response to stimuli applied in the P→A direction is either slightly enhanced or does not change with repetitive stimulation, whereas the response to stimuli moving in the A→P direction declines progressively. As a result, directional sensitivity (computed as the difference in the mean response to stimuli applied in opposing directions) is progressively enhanced with repetitive stimulation. This occurred at all the velocities studied. This is viewed as a demonstration that trend behavior can depend on the parameters of stimulation—that is, for this neuron trend behavior is stimulus-specific.

within the same column-shaped radial S-I aggregate or recorded simultaneously with the same electrode (in the latter case they can be inferred to lie in close proximity) tend to exhibit similar changes in responsivity to repetitive presentations of the same somatic stimulus.

Considered together, the effects of repetitive natural stimulation on single-neuron behavior just described argue strongly for the existence of previously unrecognized dynamic cortical processes that can be recruited into action rapidly, and whose effects summate with repetitive stimulation. Furthermore, when repetitive stimulation is continued for a prolonged period (as in 2DG mapping experiments), the processes underlying the changes in single neuron responsivity are distributed nonrandomly, resulting in the formation of the highly structured patterns of column-shaped neuronal aggregates seen in our 2DG mapping experiments. (The aggregates formed are distinguishable with the 2DG mapping method by the prominent differences in the collective activity of their constituent neurons established by repetitive stimulation; see Figure 5-10.) The mechanisms responsible for either the changes in somatosensory cortical neuronal responsivity or their nonuniform spatial distribution are uncertain, but because of the disruptive

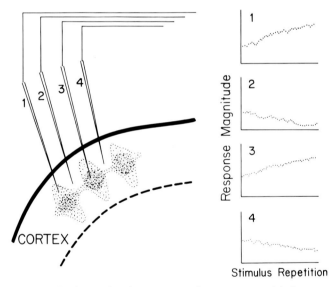

Figure 5-10 Proposed relationship between single-neuron trend behavior and 2DG labeling patterns. Pictured is an array of microelectrodes inserted so as to sample simultaneously four neurons in different S-I cell columns. All of the neurons receive excitatory thalamocortical input evoked by the delivery of a tactile stimulus. Because of the spatiotemporal characteristics of the thalamocortical input triggered by the stimulus, however, a process intrinsic to the cortex is activated that (1) maintains, or even increases, the responsivity of neurons at locations 1 and 3 to successive stimulus presentations and (2) suppresses the responsivity of neurons to stimulus input at locations 2 and 4. Since similar effects are proposed to occur for neurons in all cortical layers (except those in the vicinity of layer IV), a periodic spatial pattern of cortical responsivity develops (indicated by shading: darkly shaded areas are most active; unshaded areas are least active). This periodic pattern is presumed to correspond to that revealed by the 2DG metabolic mapping method.

effect of BIC on 2DG labeling patterns, the mechanisms are assumed to involve, at least in part, the GABAergic cortical intrinsic inhibitory system. Our discovery that the mechanisms are recruited by stimulus repetition (a frequent occurrence in everyday life, e.g., when an object is tactually scanned) raises the intriguing possibility that they may contribute in an important, adaptive way to subjects' somesthetic discriminative capacities. For example, the notable changes in perceptual sensitivity and discriminative capacity that occur with prolonged exposure to natural stimulation could involve contributions from dynamic cortical mechanisms such as those we have discovered.

Spectral Analysis of the Periodicities in the Stimulus-Evoked Somatosensory Cortical 2DG Activity Pattern

The highly periodic nature of stimulus-evoked (2DG) somatosensory cortical activity patterns (Tommerdahl et al., 1985; 1987) suggested to us that spectral image analysis techniques might permit more objective characterization of those patterns than has been possible to date. Thus, we developed an image analysis system that allows digitization, storage, display, and quantitative analysis of 2DG autoradiographic data (see Figure 5-11). In addition, software was developed to generate high-resolution two- and three-dimensional reconstructions of the data series provided by sequences of serial autoradiographs (see Figures 5-12 and 5-13). We have carried out two types of analyses to date. In the first, we sought to quantify the global spatial distribution of labeling within extensive (typically greater than 5 mm^2) cortical sectors. In the second, we used spectral analysis techniques to assess the periodicities in the labeling pattern within relatively small (350 to 1500 μm) sectors of individual high-resolution autoradiographs. These two approaches are subsequently referred to as low- and high-resolution analyses, respectively.

The low-resolution analyses have provided new information about the spatial distribution of stimulus-evoked 2DG labeling in the S-I cortex of cats and monkeys. Of special interest is the finding that periodic, low-frequency spatial variations (<3 cycles/mm) in C14 uptake occur in two locations: along the long axis of the elongated strips of above-background label that characterize 2DG patterns, and in the cortex adjacent to the above-background strips (e.g., see Figures 5-12 and 5-13). We interpret these low-spatial-frequency variations as manifestations of an opponent process mechanism that acts (by means of the intrinsic circuitry of the cortex) to enhance local discrepancies in the pattern of afferent input to adjacent cortical columns, by means of cooperative and competitive lateral interactions. The high-resolution analyses have also been productive, revealing prominent high spatial frequencies (5 to 15 cycles/mm) in the distribution of labeling within individual column-shaped patches. This finding is interpreted as evidence that the population of neuronal elements within the column-shaped territory occupied by a single 2DG patch is also activated in a highly regular and nonuniform manner. Comparison of the periodicities within 2DG labeled area 3B and area 1 patches with those in the same cytoarchitectural regions in Nissl stained sections (compare Figures 5-14 and 5-15) indicates that the within-patch 2DG activation pattern is modular (i.e., made up of a highly structured pattern of active and inactive minicolumns). Still other observations have shown that the within-

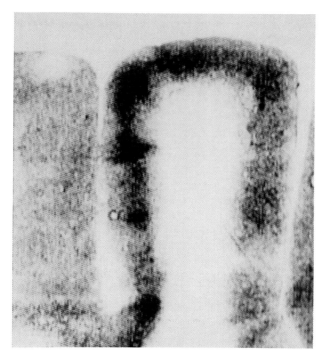

Figure 5-11   Digitized image of a 2DG autoradiograph of a sagittal section through the postcentral gyrus. The subject was stimulated with a flutter vibration stimulus on the volar tip of the index finger of the contralateral hand. The central sulcus (*on the left*) and the intraparietal sulcus (*on the right*) are evident. Also clearly evident are the principal characteristics of stimulus-evoked 2DG labeling patterns; (1) disjoint, column-shaped patches of above-background labeling (*dark profiles*) and (2) a continuous band of elevated labeling confined to the vicinity of the principal thalamocortical recipient layers (base of layer III and layer IV).

patch labeling pattern changes systematically with increasing distance from the patch center. Together, therefore, the available data indicate that high-resolution spectral analysis of 2DG labeling patterns can provide detailed information about the spatiointensive pattern of activity a somatic stimulus evokes within a single cortical column. We expect such analyses will lead to novel and valuable insights into the dynamics of information processing within individual cortical cell columns. For example, they should provide information about the functional wiring diagram among the minicolumns within macrocolumnar neuronal aggregates under controlled peripheral excitatory drive.

DISCUSSION

The model of somatosensory cerebral cortical information processing shown in Figure 5-16 is consistent with all the preliminary observations (both neurophysiological and 2DG mapping data) we have obtained to date. First, it reflects our demonstrations (in RF mapping experiments) that the low-threshold mechanore-

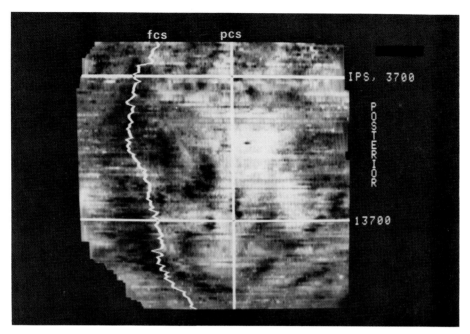

Figure 5-12 Two-dimensional, unfolded reconstruction of 2DG labeling pattern in area S-I of monkey cortex. The pattern was reconstructed from digitized images of more than 200 serial autoradiographs. The anterior is at the left; the lateral at the top. The distance between the two horizontal lines is 10 mm. C14 labeling is greatest in dark regions and least in light regions. Activity was evoked by repetitive brushing stimulation of the skin of the dorsal distal hand. Note the large region lacking label (appears white), posterior to the region of intense labeling between. PCS (lip of posterior bank of central sulcus) and FCS (fundus of central sulcus). IPS = level of the lateral end of the intraparietal sulcus.

ceptors from a restricted region of the forelimb skin project to an extensive sector of the S-I cortex. The fact that the skin field, shown in Figure 5-16, projects to an extensive sector of the S-I cortex is indicated by the widely diverging and overlapping brackets drawn between the different parts of the skin field and the S-I cortical field. Second, the model in Figure 5-16 reflects our consistent finding (in 2DG mapping experiments) that stimulation of any skin region with either natural or electrocutaneous stimuli *never* activates the total S-I cortical region that receives input from the stimulated field, but instead fractionates the S-I territory that receives input from the stimulated skin region into an inhomogeneous, strip-like pattern of activation.

The evidence provided by both our neurophysiological and 2DG experiments leads us to propose an explicit mechanism for the nonrandom fractional activation of the cortical field receiving repetitive stimulus-evoked input. Specifically, we believe the specific thalamic input that reaches the S-I middle layers is regulated by intrinsic cortical dynamic or adaptive mechanisms involving GABAergic inhibitory connections, and these regulatory influences occur chiefly in the vicinity of the principal thalamic recipient layer (layer IV and the basal part of layer III).

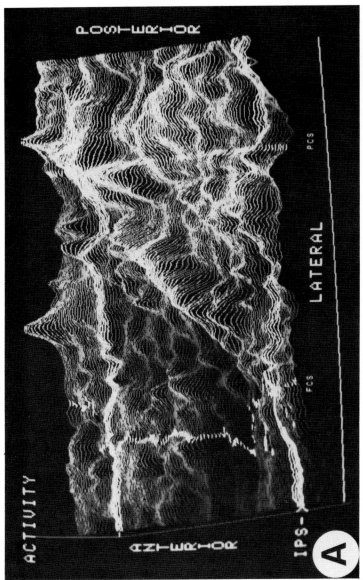

Figure 5-13 Three-dimensional displays of 2DG activity patterns set up in areas 1 and 3b of monkey somatosensory cortex by repetitive tactile stimuli. Each line displays the sequence of C14–2DG concentration values obtained for bins (each approximately 50 μm in diameter and extending across all cortical layers) extracted from the autoradiographic image of a single 20-μm thick section cut in the saggital plane. Displays were constructed with the data from serial autoradiographs. Note that a coherent modular pattern of regions containing high C14 concentrations (*light areas*) is identifiable in each brain.

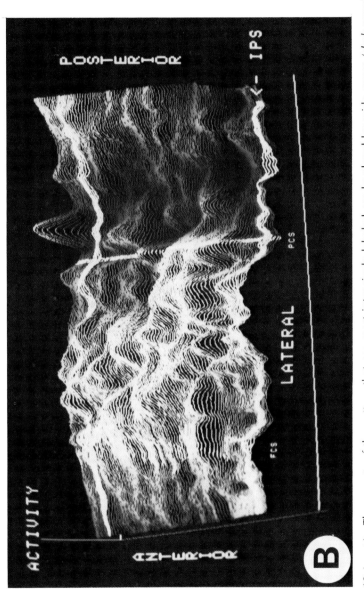

Figure 5-13 (*cont.*) Those parts of the pattern that are most intensively labeled are bordered by wide zones (*dark areas*) containing low concentrations (relative to background) of labeling. The pattern in A was evoked by a flutter vibration stimulus to the volar tip of the contralateral index finger; the pattern in B was evoked by an identical stimulus, but applied in this case to the center of the contralateral thenar eminence. In both subjects the stimulus consisted of repetitive sinusoidal vertical displacements of the skin (25-Hz, 1-mm peak-to-peak amplitude) applied with a probe tip 5 mm in diameter. Abbreviations: IPS, level of lateral terminus of intraparietal sulcus; PCS, location of the posterior lip (point of maximal curvature) of the central sulcus; FCS, location of fundus of the central sulcus.

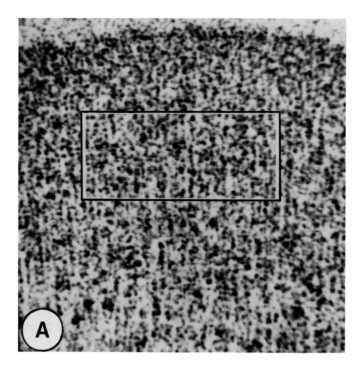

B

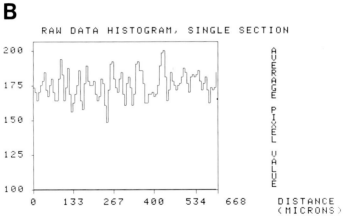

RAW DATA HISTOGRAM, SINGLE SECTION

AVERAGE PIXEL VALUE

DISTANCE (MICRONS)

C

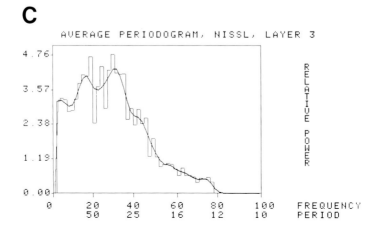

AVERAGE PERIODOGRAM, NISSL, LAYER 3

RELATIVE POWER

FREQUENCY
PERIOD

Furthermore, we propose that when the inhibitory middle-layer neurons in columns receiving strong thalamic input become active, they prevent ventrobasal thalamic excitatory input from reaching cortical neurons located above and below the middle layers in the neighboring columns receiving a weaker excitatory drive (i.e., whenever the intercolumnar intrinsic inhibitory circuitry becomes active it operates as a gate preventing excitatory thalamocortical input from reaching neurons in the superficial and deep layers of neighboring columns). As a result of this gating, the networkwide distribution of activity (indicated by 2DG labeling) in the neuron populations located above and below the middle layers fails to correspond to that in the middle cortical layers. (Compare the middle and nonmiddle layer activity in the stimulus-evoked pattern shown in the center of Figure 5-16; this noncorrespondence between middle and nonmiddle layer activity is evident in the labeling patterns of the autoradiography of Figure 5-11.)

The time-dependency of the dynamic processes that sculpture the spatial pattern of activity in the upper and deep layers of S-I is depicted schematically by the two unfolded cortical sections shown at the bottom of Figure 5-16. According to our model, following a prolonged period without stimulation, the first stimulus delivered to a skin field would evoke activity distributed relatively homogeneously throughout all cortical layers in the entire cortical territory receiving input from that skin field (see hypothetical stimulus number 1 activity pattern at bottom of Figure 5-16). With repetitive presentations of the same somatic stimulus, however, the dynamic cortical processes (presumably involving the intracortical intrinsic inhibitory network and perhaps the nonspecific projection systems that terminate preferentially in the upper and deep layers) would be activated, and a spatiointensive pattern of cortical activity unique to the physical parameters of the stimulus and the behavioral context in which it is delivered would evolve in the supra- and infragranular layers (see hypothetical activity pattern that exists at the time of stimulus number 25 at bottom of Figure 5-16). The hypothesized time-dependency of the steps in generating a somatosensory cortical activity pattern to repetitive stimulation has been reproduced in simulation studies using a mathematical model based on the conceptual model shown in Figure 5-16 (Whitsel and Kelly, 1986).

Figure 5-14 High-resolution spectral analysis of the tangential structure present in layer III of Nissl-stained sections through cytoarchitectural area 3b in monkey. (A) Digitized image of sector of area 3b. The dark regions indicate zones of dense staining, and the light regions correspond to zones containing less stain. Note that the laminae are clearly evident, as are minicolumns and individual cell somas. The area within the boxed region was subjected to spectral analysis—accomplished by dividing the boxed region into radial bins (bin width = 6.67 μm) that extended to the top and bottom limits of the boxed area. (B) Histogram showing the average pixel value for each of the 100 bins contained within the boxed area at the top. Note the relatively high-frequency periodicity in the data series. (C) Average periodogram computed from the data provided by eight serial Nissl-stained sections: in each a sector of layer III in area 3b was sampled in the same manner as described previously. Note that most of the power in the periodogram is present at spatial frequencies between 15 and 35 cycles/mm (reflecting the existence of a repeating structure with a center-to-center spacing of between 28 and 66 μm). The repeating structure is presumed to be the cortical minicolumn.

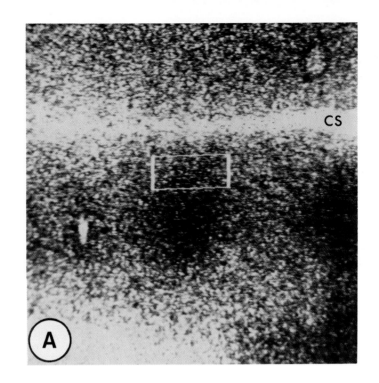

A

CS

B

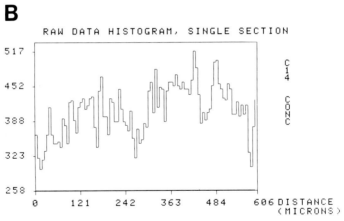

RAW DATA HISTOGRAM, SINGLE SECTION

517

452

388

323

258

C
1
4

C
O
N
C

0   121   242   363   484   606 DISTANCE
(MICRONS)

C

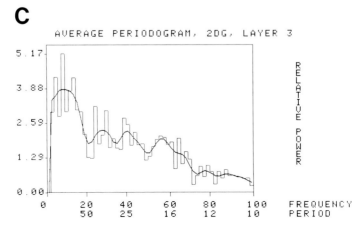

AVERAGE PERIODOGRAM, 2DG, LAYER 3

5.17

3.88

2.59

1.29

0.00

R
E
L
A
T
I
V
E

P
O
W
E
R

0   20   40   60   80   100 FREQUENCY
          50   25   16   12   10   PERIOD

Two properties of the view of somatosensory stimulus representation depicted by the model in Figure 5-16 deserve emphasis. First, it is completely consistent with both the concept of cortical columnar organization and its suggestion that columns are the units of information processing in the neocortex (see Mountcastle, 1978). Second, it offers distinct advantages over a cortex with nonplastic circuitry and one with relatively little redundancy in representation. Given the extent of the S-I cortex that receives input from a skin field (i.e., the extreme amount of "degeneracy" or "redundancy" in representation; Favorov and Whitsel, 1987a,b), and given the existence of dynamic or adaptive processes shaping the pattern of cortical columns or modules activated by a stimulus, the number of different modular patterns that can be set up by cutaneous stimuli engaging a given skin field is likely to be very large. Based on these properties, therefore, the capacity of the cortex to reflect different stimuli by means of different activity "signatures" should also be great (Whitsel and Kelly, 1986). Moreover, it seems appropriate to regard each distinguishable somatosensory cortical signature as a special form of the computational maps, now widely believed to serve a key role in the processing of sensory information in both the auditory and visual systems (Knudsen et al., 1987; Bullock, 1986).

Finally, it is important to recognize that the view of dynamic cortical processes we advocate departs sharply from that proposed in the general theory of neocortical information processing advanced by Merzenich (1987). First, the latter view relies on the assumption that the cortex is functionally two-dimensional (i.e., variations in the RF and response properties of neurons at different levels of a column are so minor they can be ignored); the view we advance in this paper emphasizes laminar variations in RF and response properties, regarding them as meaningful for somatosensory cortical information representation. Second, our view of the mechanisms underlying the moment-to-moment functional plasticity of adult somatosensory cortex, unlike the theory based on minimal RF mapping observa-

---

Figure 5-15 High-resolution spectral analysis of the tangential structure in portions of 2DG autoradiographs corresponding to layer III of area 3b. (A) Digitized image of autoradiograph of cortical areas in the vicinity of central sulcus (CS) of an animal stimulated on contralateral radial hand with a repetitive brushing stimulus. The *boxed* region defines the region subjected to spectral analysis—accomplished by dividing the region within the box into radial bins (bin width = 6.1 μm) that extended to the top and bottom limits of the boxed area. A higher magnification than that shown was used. (B) Histogram showing the average C14 concentration value for each of the 100 bins within the boxed area shown in the autoradiograph at top. Note the presence of both high and low frequencies in the data series. (C) Average periodogram computed from the data provided by three serial autoradiographs. In each periodogram a sector was sampled that corresponded to the part of layer III of area 3b containing the same labeled patch shown at the top. Note that the dominant power in the periodogram in contributed by spatial frequencies lower than 15 cycles/mm. Since, in most cases, these corresponded to periods that were integral multiples of the periodicity attributable to minicolumns, it is suggested that the observed pattern of labeling patches at the level of layer III is "modular"—that is, composed of a nonrandom spatial pattern of active and inactive minicolumns. Such a modular pattern is discernable by eye within the boxed region of the digitized autoradiograph shown in A.

Representation of Skin Field in Cortex

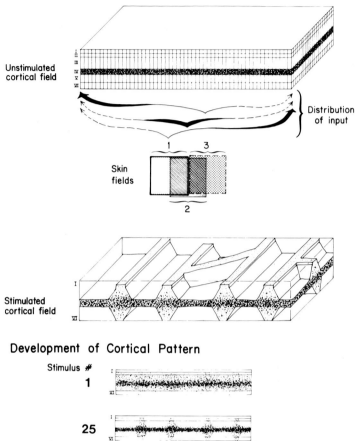

Figure 5-16 (*Top*) A skin region and the S-I cortical region that receives its input. Note that input from all regions of the skin field accesses the entire cortical field illustrated. (This extensive divergence of peripheral input onto S-I is indicated by the diverging arrows between the skin and cortex.) The schematic view of the cortex at the top of the figure shows the distribution of metabolic activity (the shading indicates the degree of metabolic activity) in the resting (unstimulated) cortex: under this condition metabolic activity is low in all layers with the exception of layer IV. (*Middle*) The inhomogeneous spatial pattern of metabolic labeling set up in S-I by stimulation of the skin field that provides its input (Note the striplike pattern of above-background labeling). (*Bottom*) Postulated time course of development of the striplike pattern of activity. (Two endviews of the S-I cortex representing the skin field stimulated are illustrated: the activity evoked by the first stimulus is labeled 1; the activity evoked by the 25th stimulus is labeled 25.

tions, proposes the participation of a mechanism (thalamocortical gating) unlike that generally believed to operate in neocortical development (synaptic weight modification). The possibility that the two mechanisms are not mutually exclusive, but may cooperate in (1) the moment-to-moment regulation of the response of the normal adult somatosensory cortex to peripheral stimulation and (2) the topographic remodeling of the somatic cortex in adults after prolonged and drastic manipulations of afferent activity patterns, cannot be dismissed on the basis of the available evidence.

In the beginning of this chapter, we stated that for many years theorists and experimentalists have predicted and searched for dynamic neural network behaviors. Although the volume of sensory neurophysiological experimentation carried out since the introduction of modern single-unit recording approaches is now enormous, our understanding of the dynamic mechanisms involved in somatosensory cortical functional plasticity in adult subjects remains abysmally deficient. We are optimistic, however, that the recent surge of interest in somatosensory dynamic mechanisms (sparked principally by the work of Merzenich and his colleagues on the cerebral cortex and by Wall and co-workers on the spinal cord; see Merzenich, 1987, and Wall, 1985, for reviews) will ultimately lead to discoveries that greatly modify and extend current ideas about information representation in mature sensory neuronal networks and its sensitivity to experience, pharmacological manipulations, and disease processes. Although continued fundamental neuroscientific studies are certainly needed, truly significant advances in our appreciation of the emergent properties of such large assemblies of neurons as exist in the sensory areas of the neocortex will come only with models and technologies that allow experimentalists to observe and study the coordinated behaviors of large numbers of elements whose properties, connectivities, and input repertoires accurately resemble those of the neurons in the sensory cortices of functioning subjects (see Mountcastle, 1986, for recent review).

Hopefully, the continuing rapid pace of developments in computer science and technology has brought us closer to the time when realistic models of the sensory neocortex can be implemented and studied. Until that time, experimentalists interested in cortical dynamics must continue their efforts to (1) provide increasingly more detailed and accurate accounts of the conditions and agents that influence cortical neuron behavior and (2) obtain higher-resolution descriptions of the organization of the anatomical paths projecting to, within, and from sensory cortical networks. In fact, the need for such detailed information is real and immediate, for the quality of the insights into neural network dynamics provided by the modeling and simulation studies we anticipate depends, to a large extent, on the accuracy and completeness of the available descriptions of cortical connectivity and cortical neuron behavior. The prospect that studies of realistic neural network models will soon be feasible should make the next few years exciting and productive for both theorists and experimentalists interested in sensory neocortex. Hopefully, they will ultimately prove to be formative for systems neuroscience.

ACKNOWLEDGMENTS

The authors gratefully acknowledge the skilled technical contributions of Elizabeth Cox and Chang-Joong Lee to the acquisition and spectral analysis of 2DG autoradiographic images. Calvin Wong and Ms. Cox assisted in preparation of the manuscript and generation of illustrations. Dr. R. Baker made significant contributions to the development of software enabling the acquisition, storage, and display of quantitative 2DG autoradiographic data. Funding for the research was provided by NIDR Program Project Grant DE 07509. The research interests, approaches and perspectives of Barry L. Whitsel continue to be influenced by the ideas and philosophy of his former mentor and colleague, Professor Gerhard Werner. He is indebted to Professor Werner for his kindness, and for communicating his enthusiasm and commitment to excellence in neuroscience research.

REFERENCES

Abeles, M. (1982). *Local Cortical Circuits: An Electrophysiological Study*. Studies of Brain Function, Vol. 6. New York: Springer-Verlag.

Bullock, T. H. (1986). Some principals in the brain analysis of important signals: mapping and stimulus recognition. *Brain Behav. Evol.* 28:145–156.

Chapin, J. K., and Woodward, D. J. (1982). Somatic sensory transmission to the cortex during movement: gating of single cell responses to touch. *Exp. Neurol.* 78:654–669.

Diamond, M., Favorov, O., Tommerdahl, M., Kelly, D., and Whitsel, B. (1986). The responsivity of S-I cortical neurons changes systematically with repetitive tactile stimulation. *Neurosci. Abs.* 12:1431.

Diamond, M., Favorov, O., and Whitsel B. (1987). The body surface is represented in S-I by a mosaic of segregates. *Neurosci. Abs.* 13:471.

Dickson, J., and Gerstein, G. (1974). Interactions between neurons in the auditory cortex of the cat. *J. Neurophys.* 37:1239–1261.

Edelman, G., and Finkel, L. H. (1984). Neuronal group selection in the cerebral cortex. In: G. Edelman, W. Cowan, and E. Gall (eds.). *Dynamic Aspects of neocortical Function*, pp. 653–695. New York: John Wiley & Sons.

Favorov, O., Diamond, M., and Whitsel, B. L. (1987a). Evidence for a mosaic representation of the body surface in Area 3b of the somatic cortex of cat. *Proc. Natl. Acad. Sci.* 84:6606–6610.

Favorov, O., Diamond, M., and Whitsel, B. (1987b). Evidence for discontinuous body surface representation in S-I. *Neurosci. Abs.* 13:470.

Favorov, O., and Whitsel, B. L. (1988a). Spatial organization of the peripheral input to Area 1 cell columns: I. The detection of "segregates." *Brain Res. Rev.*, 13:25–42.

Favorov, O., and Whitsel, B.L. (1988b). Spatial organization of the peripheral input to Area 1 cell columns: II. The forelimb representation achieved by a mosaic of segregates. *Brain Res. Rev.*, 13:43–56.

Gerstein, G. L., Bloom, M. J., Espinosa, I. E., Evanczuk, S., and Turner, M. R. (1983). Design of a laboratory for multineuron studies. In: J. C. Eccles (ed.). Systems, Man, and Cybernetics: IEEE Transactions, vol. 13, pp. 668–675. New York: Springer-Verlag.

Gerstein, G., Perkel, D., and Subramanian, K. (1978). Identification of functionally related neural assemblies. *Brain Res.*, 140:43–62.

Hyvarinen, J. (1982). *The Parietal Cortex of Monkey and Man*. Studies of Brain Function, vol. 8. New York: Springer-Verlag.

Hyvarinen, J., Poranen, A., and Jokinen, Y. (1978). Influence of attentive behavior on neuronal responses to vibration in primary somatosensory cortex of the monkey. *J. Neurophys.* 43:870–882.

Iwamura, Y., Tanaka, M., Saramoto, M., and Hirosaka, O. (1983). Converging patterns of finger representation and complex response properties of neurons in area 1 of the first somatosensory cortex of the conscious monkey. *Exp. Brain Res.* 51:327–337.

Iwamura, Y., Tanaka, M., Sakamoto, M., and Hikosaka, O. (1985). Vertical neuronal arrays in the postcentral gyrus signaling active touch: a receptive field study in the conscious monkey. *Exp. Brain Res.* 58:412–420.

Juliano, S. L., Cheema, S. S., and Whitsel, B. L. (1984). Factors determining patchy metabolic labeling in the somatosensory cortex of cats. *Neurosci. Abs.* 10:946.

Juliano, S. L., Hand, P. J., and Whitsel, B. L. (1981). Patterns of increased metabolic activity in somatosensory cortex of monkeys *Macaca Fascicularis,* subjected to controlled cutaneous stimulation: a 2-deoxyglucose study. *J. Neurophys.* 46:1260–1284.

Juliano, S. L., Hand, P. J., and Whitsel, B. L. (1983). Patterns of metabolic activity in cytoarchitectural area SII and surrounding cortical fields of the monkey. *J. Neurophys.* 50:961–980.

Juliano, S. L., and Whitsel, B. L. (1985). Metabolic labeling associated with index finger stimulation in monkey S-I: between animal variability. *Brain Res.* 342:242–251.

Juliano, S. L., and Whitsel, B. (1987). A combined 2-deoxyglucose and neurophysiological study of primate somatosensory cortex. *J. Comp. Neurol.* 263:514–525.

Juliano, S. L., Whitsel, B., Tommerdahl, M., and Cheema, S. (1988) Determinants of patchy metabolic labeling in the somatosensory cortex of cats: a possible role for intrinsic inhibitory circuitry. *J. Neurosci.* in press.

Knudsen, E. I., du Lac, S., and Esterly, S. D. (1987). Computational maps in the brain. *Annu. Rev. Neurosci.* 10:41–65.

McKenna, T. M., Light, A. R., and Whitsel, B. L. (1984). Neurons with unusual response and receptive-field properties in upper laminae of cat S-I cortex. *J. Neurophys.* 51:1055–1076.

Merzenich, M. M. (1987). Dynamic neocortical processes and the origins of higher brain functions. In: *The Neural and Molecular Basis of Learning: Dahlem Konferenzen,* pp. 337–358. New York: John Wiley & Sons.

Mountcastle, V. B. (1966). The neural replication of sensory events in the somatic afferent system. In: J. C. Eccles (ed.). *Systems, Man, and Cybernetics: IEEE Transactions,* vol. 13, pp. 676–682. New York: Springer-Verlag.

Mountcastle, V. B. (1978). An organizing principle for cerebral function: the unit module and the distributed system. In: G. M. Edelman and V. B. Mountcastle (eds.). *The Mindful Brain,* pp. 7–51. Cambridge: The MIT Press.

Mountcastle, V. B. (1986). The neural mechanisms of cognitive functions can now be studied directly. *Trends Neurosci.* 10:505–508.

Nelson, R. J. (1984a). Responsiveness of monkey primary somatosensory neurons to peripheral stimulation depends on "motor-set." *Brain Res.* 302:143–148.

Nelson, R. J. (1984b). Sensorimotor cortex responses to vibrotactile stimuli during the initiation and execution of hand movement. In: A. Goodwin and I. Darian-Smith (eds.). *Hand Function and the Neocortex.* Berlin: Springer-Verlag.

Perkel, D. H., Gerstein, G. L., Smith, M. S., and Tatton, W. G. (1975). Nerve impulse patterns: a quantitative display technique for three neurons. *Brain Res.* 100:271–296.

Schneider, J., Eckhorn, R., and Reitboeck, H. (1983). Evaluation of neuronal coupling dynamics. *Bio. Cybern.* 46:129–134.

Tommerdahl, M., Baker, R., Whitsel, B., and Juliano, S. (1985). A method for reconstructing patterns of somatosensory cerebral cortical activity. *Biomed. Sci. Instrument.* 21:93–98.

Tommerdahl, M., Whitsel, B. L., Cox, E. G., Diamond, M., and Kelly, D. G. (1987).

Analysis of the periodicities in somatosensory cortical activity patterns. *Neurosci. Abs.* 13:470.

Ts'o, D. Y., Gilbert, C. D., and Wiesel, T. N. (1986). Relationships between horizontal interactions and functional architecture in cat striate cortex as reealed by cross-correlational analysis. *J. Neurosci.* 6:1160–1170.

Wall, P. D. (1985). Future trends in pain research. *Phil. Trans. R. Soc. Lond.* 308:393–401.

Werner, G. (1978). Static and dynamic components of object representation in the central nervous system. In: M. A. B. Brazier and H. Petsche (eds.). *Architectonics of the Cerebral Cortex,* pp. 335–355. New York: Dowen Press.

Whitsel, B. L., and Juliano, S. L. (1984). Imaging the responding neuronal population with 14-C-2-DG; the somatosensory cerebral cortical signature of a tactile stimulus. In: C. Von Euler, O. Franzen, U. Lindblom, and D. Ottoson (eds.). *Somatosensory Mechanisms: Wenner-Gren Symposium Series,* pp. 61–80. London: Macmillan.

Whitsel, B. L., and Kelly, D. G. (1986). Knowledge acquisition ("learning") by the somatosensory cortex. In: J. Davis, E. G. Wegman, and R. Newburgh (eds.). *Brain Structure and Function.* Boulder, CO: Westview Press.

6

Multielectrode Exploration of Somatosensory Cortex Function in the Awake Monkey

ALBERT T. KULICS AND LAWRENCE J. CAULLER

Cooperative activity in dynamically organized and distributed cortical neuronal pools may be the basis of higher functions such as perception or willed movement (Mountcastle, 1984). This idea is supported by research in humans that reflects relatively global aspects of neuronal function such as evoked potential (Galambos and Hillyard, 1981) and positron emission tomography studies (Roland, 1982, 1983; Roland et al., 1980). Ideas about the cortical module (Szentagothai, 1978) and neuronal ensembles (Eccles, 1984), and how such activity might become organized (Edelman and Finkel, 1984), have stimulated thinking on the representation and expression of psychological functions by cerebral cortex. Despite the growing evidence that cell population behavior determines function, current research in cortical mechanisms is dominated by neurophysiologic techniques directed toward the study of individual cellular units. At this level of analysis, the cooperative nature of cortical activity and the rules governing such behavior are not apparent.

This chapter describes our attempt to break away from a reductionist experimental orientation and gain a new perspective on the cortical determinants of somatic sensation. We have used a multielectrode that permits trial-by-trial recordings from numerous (up to 15) cortical depths simultaneously in awake or performing monkeys. The multielectrode recordings provide surface-to-depth response profiles of both field potential (FP) and multiple-unit activity (MUA) evoked by touch stimulation. In addition, on the basis of the FP profiles, the generator currents within cortex are described in the form of current source-density (CSD) depth profiles. Covariation among the FP, MUA, and CSD measures is detected by statistical analysis of trial-by-trial responses.

The three measures afford different but complementary views of cortical activity (Kulics and Cauller, 1986). Combined, they coherently reflect the operation of a single underlying neural process in postcentral gyrus elicited by touch stimulation of the hand.

THE CORTICAL NEURAL PROCESS UNDERLYING TOUCH SENSATION

In the awake monkey, an abrupt skin stimulus evokes a stereotypical response in postcentral gyrus that is much different from the response in sleeping (Cauller and

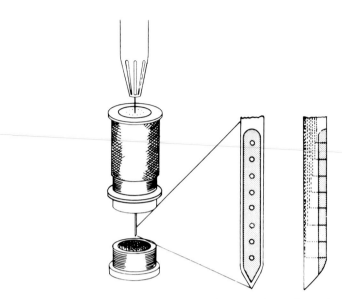

Figure 6-1 The multielectrode inserted in the microdrive above the recording chamber. To the right are frontal and lateral expanded views of the multielectrode tip, showing eight recording tips spaced 400 μm apart embedded in epoxy resin (*shaded area*). [From Kulics and Cauller (1986).]

P12R

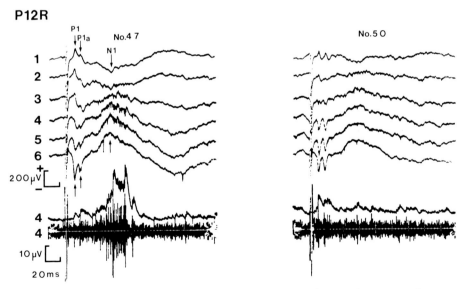

Figure 6-2 Simultaneous surface (1) to depth (6) FP recordings (using 25-μmdiameter tips spaced 200 μm apart, in this case) from two trials (left to right) in an awake monkey. FP components are indicated by arrows. They are P1 (12 msec), P1a (20 msec) N1 (50 msec), and P2 (120 msec), which is not labeled. Below the FPs, the MUA from one level is displayed with its integral just above. Note that the decrease in FP amplitude of around 50 msec (N1) between trials is accompanied by decreased MUA during the same time interval. [From Kulics and Cauller (1986).]

Kulics, in preparation) or anesthetized subjects (Woolsey et al., 1942; Amassian et al., 1964). In the unconscious monkey, cortical unit and FP activity are restricted largely to an early (i.e., approximately 12 msec) poststimulus interval. With the multielectrode (Figure 6-1) in the awake monkey, four FP components (Figures 6-2 and 6-3) lasting approximately 200 msec have been described. Coincident with the first three of these components, evoked MUA discharges whose vigor covaries with each respective component's amplitude are seen intracortically (Figures 6-2 and 6-3). Based on analysis of CSD profiles, the first two components, P1 and P1a, appear to reflect deep-lying current sink activity (see Kulics and Cauller, 1986, for examples), while N1 reflects superficially located current sink activity (e.g., Fig. 3, CSD, level 2). The MUA associated with P1 or P1a tends to be concentrated at the level of each FP component's respective intracortical sink, whereas that of N1 occurs below the superficial sink in the supra- or infragranular cortical layers (Figure 6-4). No MUA has been observed in conjunction with P2, and its generator is as yet unknown. A more detailed description

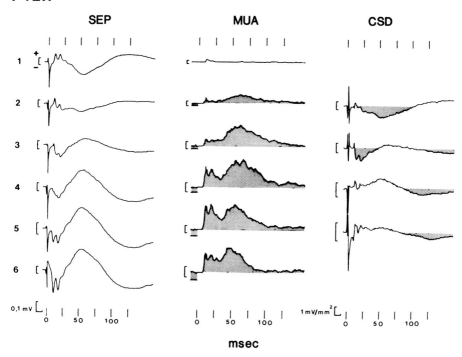

Figure 6-3 Depth profiles of average (N = 50) FP (labeled as SEP), MUA, and CSD from simultaneous recordings during a single penetration through the crown of postcentral gyrus. Figure 6-2 shows single trials on which the averages are based while Figure 6-4 shows histological details. Relative recording depths are indicated by numbers as in Figure 6-2. Brackets around each waveform show ±1 SD of the prestimulus baseline amplitude at each level. Shaded bars below each MUA integral show relative amplitude of spontaneous background activity. Current sinks are shaded in the CSD profiles. Levels 1 and 6 of the CSD are lost during its calculation. [From Kulics and Cauller (1986).]

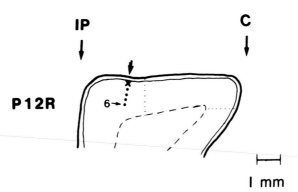

Figure 6-4   Drawing of a histological section containing the electrode penetration represented in Figures 6-2 and 6-3. Central (C) and intraparietal (IP) sulci are indicated. Faint dotted lines delimit cytoarchitectonic regions 3b, 1, and 2. An arrow shows the penetration site. Dots locate recording levels, and the X indicates the one nearest the FP null-potential point and N1 sink, approximately 300 μm below the cortical surface. In this example, MUA during N1 appeared to be maximal near level 4 (see Figure 6-3, MUA), located in the supragranular cortical layers [From Kulics and Cauller (1986).]

of the FP, MUA, and CSD findings has been published (Kulics and Cauller, 1986). Recently, our recording capabilities have been expanded to sample 12 depths simultaneously, with a resultant increase in our ability to resolve sink-source relations and MUA locations (Figure 6-5).

Neural activity observed with our electrodes during P1, P1a, and N1 suggests the operation of a cortical excitatory process mediated by pyramidal cells. The elongated, vertically oriented pyramidal cells are thought to be responsible for the current dipoles underlying the observed FP (Mitzdorf, 1985). The correlation between MUA discharge vigor and FP component amplitudes suggests that the same cell populations are responsible for both phenomena. Since pyramidal cells are the predominant neural element in cerebral cortex (Feldman, 1984) and are thought to exert the main excitatory effect (Szentagothai, 1975), the activity we routinely observe probably largely reflects pyramidal cell excitation. The model in Figure 6-6 summarizes our observations together with the preceding conclusions.

In conjunction with pyramidal cells, other excitatory or inhibitory interneurons such as stellate or basket cells may be activated. However, in contrast with the pyramidal cell, these Golgi type II elements are unlikely to generate current sinks or sources that are observable extracellularly. The radial dendritic geometry surrounding the cell somas, and the scattered, partially superimposed distribution of these cells lead to cancellation of the current sources and sinks they generate (Mitzdorf, 1985). Thus, even though Golgi type II units probably do not contribute to the observed FP, their activity may be reflected in the MUA, along with pyramidal cell discharges. Detailed single-unit analysis appears necessary to resolve the cellular composition of evoked MUA fully.

Mitzdorf (1985) postulates that the current sinks observed under these conditions reflect predominantly excitatory postsynaptic potentials (EPSPs). We believe this is the best explanation for the N1 component. The current sink responsible for N1 is located near the layer I–II border of cortex, a region that contains

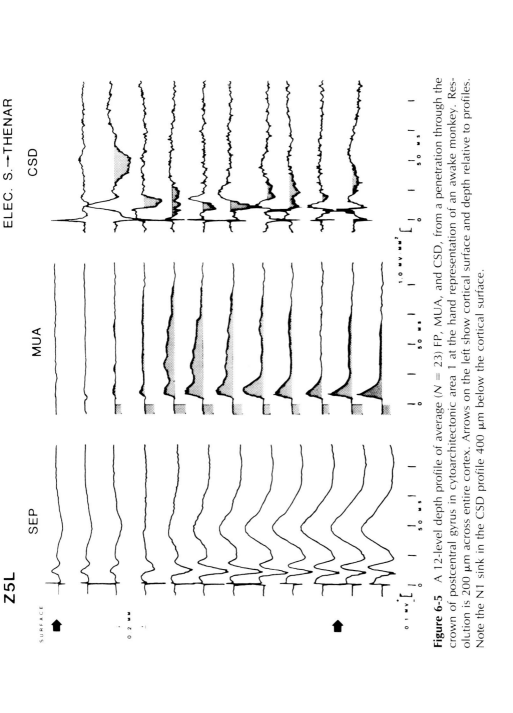

Figure 6-5 A 12-level depth profile of average ($N = 23$) FP, MUA, and CSD, from a penetration through the crown of postcentral gyrus in cytoarchitectonic area 1 at the hand representation of an awake monkey. Resolution is 200 µm across entire cortex. Arrows on the left show cortical surface and depth relative to profiles. Note the N1 sink in the CSD profile 400 µm below the cortical surface.

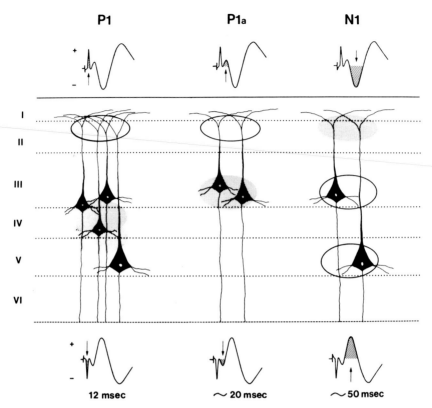

Figure 6-6 The model of postcentral gyrus cortical function suggested by the present results. Cortical layers are indicated by Roman numerals. FPs reflect cortical surface (*above*) and depth (*below*) recordings at the times indicated. Touch stimulation elicits P1 at 12 msec, which results from a deep current sink (*shaded ellipses*) and superficial current source (*clear ellipse*). Simultaneously, pyramidal cell discharges (*in black*) occur in a number of cortical layers. During P1a, at around 20 msec, a more superficial sink appears to be accompanied by cell discharges at the same level. At the time of N1, a large superficial sink develops at the layer I–II border that correlates with cell discharges and current sources in deeper layers. Activity during the N1 interval then propagates to adjacent cortical regions by an unknown mechanism. [From Kulics and Cauller (1986).]

numerous excitatory synaptic contacts on the apical dendrites of the deeper-lying pyramidal cells (Szentagothai, 1978). The development of the superficially lying extracellular current sink probably reflects excitation at synapses on these apical dendrites. Just as important, from the standpoint of the model, complementary deep sources develop simultaneously with the sink above. These sources reveal the general location of the cell bodies in the population whose dendrites are undergoing excitation.

Once initiated, this excitatory process underlying N1 propagates actively in a wavelike fashion through broad regions of postcentral gyrus. (Kulics, 1982; Kulics and Cauller, 1986). We are currently trying to establish whether this activity propagates from postcentral gyrus to nearby functionally related cortical areas.

Other recent observations suggest that summated somatic action potentials, and not the EPSPs themselves, may be responsible for generating the deep-lying sinks seen during P1 and P1a. The somatic action potential from a pyramidal cell, when summated with other nearby ones, can produce a sizable extracellular negative current field, as is well known from work on the hippocampus (Andersen et al., 1971). During the course of N1, recording tips lying in layers demonstrating strong spontaneous and driven cellular activity often show small, sharp, negative deflections superimposed on the normally smooth deep-positive potential waveform. As illustrated in Figure 6-7, the occurrence of these "serrations" coincide temporally with intense MUA bursts. The most likely explanation is that a more or less synchronous discharge in a small population of closely related pyramidal cells will lead to summation of their somatic action potentials and development of a small-current sink in that area. This sink is superimposed on the dominant

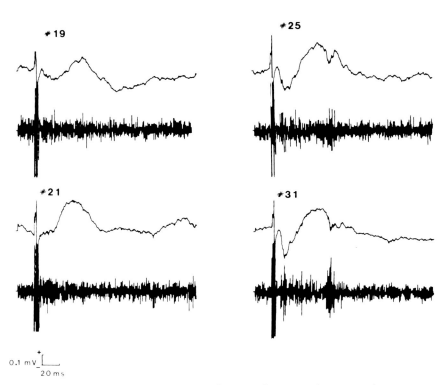

Figure 6-7 Four trials from a sequence showing the FP and associated MUA at one recording tip deep within cerebral cortex. Note the sharp downward deflection (serration) on the descending portion of the positive wave component in the two examples to the right-side and the burst of MUA at that time, shown below each FP. For comparison, two trials are shown on the left in which no serration or MUA burst exists. MUA bursts also accompany the increase in the earlier negative FP component (compare bottom left and right near 20 msec). Y-axis calibration refers to FP only. A more complete description of this penetration appears in Kulics and Cauller (1986).

positive-polarity source activity from the larger surrounding population of pyramidal cells.

Our CSD analysis confirms that the serrations are small localized sinks. The time course of depolarization and the absence of a fast hyperpolization in individual cortical pyramidal cell somatic action potentials (Connors et al., 1982; Li and Chou, 1962) suggest that they could summate to produce such extracellular potentials. The close association of MUA bursts and deep-current sinks during P1 and P1a intervals indicates that this explanation may also apply to the generation of these sinks. This interpretation contrasts with that of Mitzdorf (1985), who believes the deep-current sinks reflect EPSPs in the basilar dendrites of the pyramidal cells. Further study is required to resolve this.

LIMITS ON EXPRESSION OF FIELD POTENTIAL AND UNIT CORRELATIONS

Our results show that FPs and MUA in each of three temporal intervals are functionally related. The trial-by-trial depth profiles and correlations, together with the CSD, have considerably reduced the ambiguity in relating slow-wave and unit activity. However, even though the observed relationship is consistent, it is not obligatory, for reasons discussed in the following.

The development of the N1 sink, as indicated earlier, leads to unit discharges within the deeper cortical laminae. The amplitude of the N1 sink and MUA vigor show a significant positive correlation (Kulics and Cauller, 1986). The correlation is rather robust, requiring only 50 to 100 trials to achieve significance at the $p <$ .01 level, but is by no means perfect (i.e., $r = 0.4$ to 0.8). Several factors set upper limits on the expression of this relationship. One is the ongoing background electroencephalogram (EEG) in the awake monkey. This is superimposed on the FP and unpredictably enhances or diminishes, on a moment-by-moment basis, the stimulus-locked voltage changes seen over trials. Normally, signal averaging reduces this influence, but this technique obviously cannot be applied to single-trial measurements. A further source of variability that limits the expression of the correlation is intrinsic to the MUA. Although the average temporal profile of MUA is stable, the trial-by-trial MUA is expressed differently. It may be more or less clustered with differing central tendencies within the time window analyzed, which tends to increase the measurement uncertainty.

The MUA behavior further suggests a theoretical limit, as well as the preceding practical ones, governing the magnitude of the FP–MUA correlation. In addition to variation in the dispersion and central tendency of MUA, there appears to be a change in which units participate in the MUA pooled response. For example, a particularly large, identifiable single-unit response sometimes dominates an MUA recording. During a trial sequence, its activity may drop out and reappear in the profile independently of the dominant response tendency at the moment. The unit firing may further display temporal features that may or may not resemble the typical MUA response over trials. These observations suggest that a unit's membership within the responding pool of neurons reflected by the MUA can change over time, its participation being governed by selection processes as yet unknown.

Given the variability intrinsic to the FP and MUA over trials and the problems this causes for measurement, a single unit's response may not be expected to accurately reflect the FP associated with it. However, the average tendency of the neuronal pool of which the unit is a member would change with the FP amplitude, as we have shown. Each unit's activity should nevertheless demonstrate a small, but nonetheless real, correlation with the FP because of its membership within the pool. This correlation is vanishingly small in many cases, and requires many trials to reveal, as Fox and O'Brien (1965) have illustrated. They found that the seemingly random discharges of unit activity accurately reflect the associated field potential, but only after many samples of the activity are combined. These considerations suggest that the optimum conditions for expression of a significant correlation between the FP and MUA are yet unknown. A better understanding of the appropriate size cell population to sample and the means of adequately assessing this activity are required.

THE RELATIONSHIP OF CORTICAL NEURAL EVENTS TO BEHAVIOR

Our interest in N1 stems from its apparent relationship to the psychological dimensions of consciousness and subjective sensory experience. N1 is seen only in the awake monkey. In the sleeping (Cauller and Kulics, in preparation) or anesthetized monkey, as indicated earlier, the N1 component is abolished, but P1 activity is not. The relationship of N1 to sensation has been suggested by earlier work (see Figure 6-8 and Kulics, 1982) showing that N1 amplitude predicts the choice performance of monkeys engaged in a cutaneous intensity discrimination. In contrast, P1 amplitude does not correlate with sensory choices (Kulics et al., 1977); it appears more related to the process of stimulus representation or registration. In addition, the study of N1 temporal dynamics—such as its apparent propagation to adjacent cortical areas—may reveal additional clues to the neuronal mechanisms underlying the aforementioned psychological processes.

The interrelationship between FPs and MUA discussed earlier suggests the possibility of direct correlations between sensory performance measures and cellular population discharges. In general, combining neurophysiologic and behavioral investigations like those outlined here for studying the neural correlates of sensation and perception appears to be well justified (Uttal, 1973). A similar need exists for an integrative approach that uses multielectrodes to study the cortical basis of goal-directed behavior, as suggested by Evarts and colleagues (1984).

THE SIGNIFICANCE OF FINDINGS IN THE MONKEY FOR HUMANS

The similarity of the sensory neural mechanisms between monkey and human suggests that our findings and generalizations in the form of the model may apply directly to questions concerning the neural basis of the FP and EEG in humans. For example, recordings from postcentral gyrus in the awake human (Goff et al., 1980) have revealed the presence of a large negative potential, called the somatic late potential (SLP), following the primary evoked response. The SLP is similar

E28R

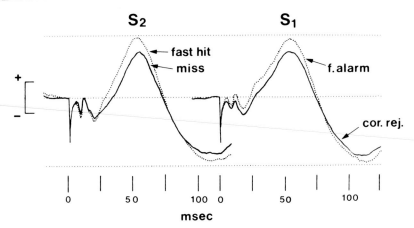

Figure 6-8 Average FPs from a recording tip within cortex near the thumb representation in a performing monkey trained to respond to the higher (S₂) of two touch stimulus intensities. FPs were sorted on the basis of correct (hits or correct rejections) or incorrect behaviors (misses or false alarms), and averaged. The average amplitude of N1 (and P2) was greater for fast hits ($N = 29$) than for misses ($N = 25$) and greater for false alarms ($N = 15$) than for correct rejections ($N = 30$). P1 amplitude showed larger changes with stimulus intensity (S₂, S₁) than with psychophysical performance. The bracket indicates ± 1 SD of prestimulus baseline amplitude of the response population ($N = 200$) from which the samples were drawn (1SD = 30 μV). This type analysis is described in detail in Kulics, et al. (1977) and Kulics (1982).

to N1; however, it remains to be determined whether the SLP amplitude, like that of N1, predicts sensory performance or attentiveness. If N1 in the monkey and the SLP in the human prove to be operationally identical, the case for a similar neural process underlying both, such as the one described here, is strengthened.

Another potentially fruitful line of inquiry concerns the neural basis of the *Bereitschaftspotential* (Kornhuber and Deecke, 1965), or readiness potential (Vaughan et al., 1968)—a ramplike negativity recorded from the scalp before preprogrammed voluntary movements are initiated. In the monkey, a similar negativity, which occurs before trained self-initiated movements, has been observed in motor cortical recordings (Arezzo and Vaughan, 1975). We are in the process of studying the negativity, with the multielectrode methodology, in monkeys trained on a task operationally identical to that described for humans. The aim is to describe the "readiness" potential in a way similar to the sensory FP model, to better appreciate how cooperative cellular activity leads to goal-directed behavior.

The development of positron-emission tomography (PET) methodology in humans permits localization of functionally active cortical regions to a degree not possible with scalp-recording techniques. Currently, the biggest problem with PET scans is their limited spatial and temporal resolution. Multielectrode studies within or between cortical areas in behaving monkeys could serve as a useful adjunct to the PET studies and help clarify the temporal dynamics of cortical processing and the neural mechanisms underlying cognitive function.

CONCLUSIONS—A MATTER OF PERSPECTIVE

Our studies indicate that the multielectrode approach in awake or performing monkeys occupies an important niche between single-unit methods and the evoked potential or PET scan techniques.

First, in the awake monkey, recordings from many sites and from surface-to-depth in cortex have revealed dynamic cooperative neural phenomena that were unnoticed by other techniques. The best example of this is the N1 evoked component generated within postcentral gyrus and propagated over a wide area. Its study may lead to a better understanding of the role of neural activity in the production of sensory experience and the mechanism by which sensory information is transmitted throughout the cortex.

Second, multielectrode recordings have led to a better appreciation of the fundamental interrelatedness among field potentials, cellular discharges and extracellular currents elicited by touch stimuli. Other techniques ignore one or another of these information sources, which can blind one to clues about the underlying neural processes. The simultaneous repeated observation of these phenomena over trials has revealed the operation of a single underlying sensory process expressed in the form of a working model. It is hoped that, through this model, the implications of results obtained by any specific technique will be appreciated in the broader context of which they are a part.

Third, multielectrode recordings in performing monkeys permit assessment of those features of the cortical process that might have functional significance by allowing direct correlation of the neural and behavioral response dimensions on a trial-by-trial basis. This is much more difficult (if not impossible) with single-unit or scalp-recorded evoked response techniques.

Fourth, multielectrode recordings may be important for studies of the neural determinants of psychological processes that were previously difficult to formulate. Two examples of FP activity amenable to such an analysis in humans—the SLP and readiness potentials—were discussed.

In general, the advantage of the multielectrode technique lies in the perspective it offers, which is neither too fine nor too coarse. It has revealed an unexpected, but coherent richness in the neural processes governing somatosensory cortical function.

ACKNOWLEDGMENTS

This work was supported by grants from the National Institute of Mental Health (MH38157) and the Ohio Board of Regents Research Challenge Fund.

REFERENCES

Amassian, V. E., Waller, H. J., and Macy, J. Jr. (1964). Neural mechanism of the primary somatosensory evoked potential. *Ann. NY Acad. Sci.* 112:5–32.

Andersen, P., Bliss, T. V. P., and Skrede, K. K. (1971). Unit analysis of hippocampal population spikes. *Exp. Brain Res.* 13:208–221.

Arezzo, J., and Vaughan, H. G. Jr. (1975). Cortical potentials associated with voluntary movements in the monkey. *Brain Res.* 88:99–104.

Connors, B. W., Gutnick, M. J., and Prince, D. A. (1982). Electrophysiologic properties of neocortical neurons in vitro. *J. Neurophysiol.* 48:1302–1320.

Eccles, J. (1984). The cerebral neocortex: a theory of its operation. In: E. Jones and A. Peters (eds.). *Cerebral Cortex: Functional Properties of Cortical Cells* Vol. 2, pp. 1–36. New York: Plenum.

Edelman, G. M., and Finkel, L. H. (1984). Neuronal group selection in the cerebral cortex. In: G. M. Edelman, W. E. Gall, and W. M. Cowan (eds.). *Dynamic Aspects of Neocortical Function* pp. 653–695. New York: John Wiley & Sons.

Evarts, E. V., Shinoda, Y., and Wise, S. P. (1984). *Neurophysiologic Approaches to Higher Brain Functions,* p. 147. New York: John Wiley & Sons.

Feldman, M. L. (1984). Morphology of the neocortical pyramidal neuron. In: A. Peters and E. G. Jones (eds.). *Cerebral Cortex: Cellular Components of the Cerebral Cortex,* Vol. 1, pp. 123–200. New York: Plenum.

Fox, S. S., and O'Brien, J. H. (1965). Duplication of evoked potential waveform by curve of probability of firing of a single cell. *Science* 147:888–890.

Galambos, R., and Hillyard, S. A. (1981). Electrophysiologic approaches to human cognitive processing. *Neurosci. Res. Prog. Bull.* 20:143–265.

Goff, W. R., Williamson, P. D., VanGilder, J. C., Allison, T., and Fisher, T. C. (1980). Neural origins of long latency evoked potentials recorded from the depth and from the cortical surface of the brain in man. In: J. Desmedt (ed.). *Progress in Clinical Neurophysiology* vol. 7: Clinical uses of cerebral brainstem and spinal somatosensory evoked potentials, pp. 126–145. Basel: Karger.

Kornhuber, H. H., and Deecke, L. (1965). Hirnpotentialanderungen bei Willkurbewegungen und passiven Bewegungen des Menschen: Bereitschaftspotential und reafferente Potentiale. *Pflugers Arch. ges Physiol.* 284:1–17.

Kulics, A. T. (1982). Cortical neural evoked correlates of somatosensory stimulus detection in the rhesus monkey. *Electroenceph. Clin. Neurophysiol.* 53:78–93.

Kulics, A. T., and Cauller, L. J. (1986). Cerebral cortical somatosensory evoked responses, multiple unit activity and current source-densities: their interrelationships and significance to somatic sensation as revealed by stimulation of the awake monkey's hand. *Exp. Brain Res.* 62:46–60.

Kulics, A. T., Lineberry, C. G., and Roppolo, J. R. (1977). Neurophysiological correlates of cutaneous discrimination performance in rhesus monkey. *Brain Res.* 136:360–365.

Li, C., and Chou, S. N. (1962). Cortical intracellular synaptic potentials and direct cortical stimulation. *J. Cell. Comp. Physiol.* 60:1–16.

Mitzdorf, U. (1985). Current source-density method and application in cat cerebral cortex: investigation of evoked potentials and EEG phenomena. *Physiol. Rev.* 65:37–100.

Mountcastle, V. B. (1984). Introduction. In: G. M. Edelman, W. E. Gall, and W. M. Cowan (eds.). *Dynamic Aspects of Neocortical Function,* pp. 1–3. New York: John Wiley & Sons.

Roland, P. E. (1982). Cortical regulation of selective attention in man. *J. Neurophysiol.* 48:1059–1078.

Roland, P. E. (1983). Intensity and localization of cortical activity in man during sensory discrimination, directed attention and thinking. *Proc. Int. Union. Physiol. Sci.* 15:217.

Roland, P. E., Larsen, B., Lassen, N. A., and Skinhoj, E. (1980). Supplementary motor area and other cortical areas in organization of voluntary movements in man. *J. Neurophysiol.* 43:118–136.

Szentagothai, J. (1975). The module-concept in cerebral cortex architecture. *Brain Res.* 95:475–496.

Szentagothai, J. (1978). The Ferrier Lecture. The neuron network of the cerebral cortex: a functional interpretation. *Proc. R. Soc. Lond. B* 201:219–248.

Uttal, W. R. (1973). *The Psychobiology of Sensory Coding.* New York: Harper & Row.

Vaughan, H. G. Jr., Costa, L. D., and Ritter, W. (1968). Topography of the human motor potential. *Electroenceph. Clin. Neurophysiol.* 25:1–10.

Woolsey, C. N., Marshall, L. L., and Bard, W. H. (1942). Representation of cutaneous tactile sensibility in the cerebral cortex of the monkey as indicated by evoked potentials. *Bull. Johns Hopkins Hosp.* 70:399–441.

7

Neural Mechanisms
of Pain Discrimination

WILLIAM D. WILLIS, JR.

Pain involves a complex set of sensory and behavioral responses. Under ordinary circumstances, pain is the consequence of stimuli that either threaten or cause injury. The responses to such stimuli include both the perception of pain, which may have any of several sensory qualities (e.g., pricking, burning, or aching sensation), and a series of motivational-affective responses (e.g., somatic and autonomic reflexes, endocrine changes, arousal, suffering). Another type of pain can result from injury to nervous tissue. It has some of the attributes of pain caused by stimulation, but the mechanisms are less well defined and the clinical management is different.

Stimuli perceived as painful activate a sequence of neurons that include nociceptors, ascending somatosensory tract cells, neurons of the somatosensory thalamus and cerebral cortex, and cells of higher processing centers such as the parietal association cortex (Willis, 1985). Operating in parallel with this perceptual system for pain is another, partially overlapping system responsible for triggering motivational-affective responses (Melzack and Casey, 1968). This chapter emphasizes the perceptual system for cutaneous pain sensation. This system resembles those responsible for other qualities of somatic sensation, such as touch-pressure, flutter vibration, and position sense.

One problem for experimentalists interested in pain sensation is stimulus control. Unlike stimuli used for the study of other sensory systems, such as the visual system, which can be well defined and repeatedly applied, painful stimuli are hard to control, and their repeated application can dramatically alter the behavior of the sensory receptors. For cutaneous pain, commonly used stimuli include mechanical stimuli that compress a fold of the skin and thermal stimuli that heat (or cool) the skin surface to levels that human subjects report as painful.

Another area of practical difficulty in studying pain is use of anesthesia. Most investigators use well-anesthetized animals, which by definition do not experience pain. However, it appears that the processing of nociceptive information at the levels of the sensory receptors and ascending tract cells is relatively little altered by anesthesia. For studies of the processing of information related to pain at the cortical level, it may be necessary to use awake, trained subjects who tolerate mildly painful stimuli but can terminate stimuli they judge to be too strong. Such an approach precludes the use of very intense stimuli.

NOCICEPTORS

Nociceptors are defined by Sherrington as sensory receptors that signal stimuli that threaten or produce damage (Sherrington, 1906). Two main classes of cutaneous nociceptors have been recognized. These are termed *Aδ mechanical nociceptors* and *C polymodal nociceptors,* based on the type of primary afferent nerve fiber and their responsiveness to different forms of noxious stimuli. Aδ mechanical nociceptors are supplied by finely myelinated nerve fibers and respond to mechanical but not thermal or chemical noxious stimuli, unless they were previously sensitized (Burgess and Perl, 1967; Fitzgerald and Lynn, 1977). C polymodal nociceptors, on the other hand, are supplied by unmyelinated nerve fibers and can be activated by noxious mechanical, thermal, and chemical stimuli (Bessou and Perl, 1969; Kumazawa and Perl, 1977). There appear to be other classes of cutaneous nociceptors, as well, including Aδ nociceptors that respond to both mechanical and thermal noxious stimuli (Fitzgerald and Lynn, 1977); Georgopoulos, 1976; LaMotte et al., 1982).

The receptive fields of cutaneous nociceptors are restricted in size. The receptive fields of Aδ mechanical nociceptors consist of 3 to 20 spots less than 1 mm^2 (Bessou and Perl, 1969). Thus, one or a few nociceptors could provide the information needed for localizing a discrete noxious stimulus, such as a pinprick. There is evidence that such spatial coding does occur. In experiments on human subjects using stimulation through microneurography electrodes to stimulate C fibers, Torebjork and Ochoa (1983) found that the receptive fields the subject projected were very small, and congruent with those mapped while the discharges of the afferent units were being recorded.

The thresholds of Aδ mechanical nociceptors and of C polymodal nociceptors to mechanical stimuli are similar and considerably higher than those of sensitive mechanoreceptors (Georgopoulos, 1976). Responses to graded intensities of mechanical stimuli increase well into the range that causes damage; the responses have phasic and static components. Strong or repeated stimuli may cause receptor fatigue (Perl, 1968; Kumazawa and Perl, 1977).

C polymodal nociceptors often have thresholds to noxious heat stimuli in the range of the threshold for human heat pain, about 45° C (Bessou and Perl, 1969; LaMotte and Campbell, 1978). The responses of these receptors increase in either a linear or a positively accelerating fashion as the intensity of the noxious heat stimuli increases (Beck et al., 1974; LaMotte and Campbell, 1978). Most Aδ mechanical nociceptors fail to respond to noxious heat unless the skin has been sensitized by repeated noxious heat stimuli (Fitzgerald and Lynn, 1977). Those Aδ nociceptors that are heat sensitive have thresholds higher than those of C polymodal nociceptors; most require stimuli exceeding 51° C to be activated in the absence of skin sensitization (LaMotte et al., 1982).

The response properties of Aδ mechanical nociceptors and C polymodal nociceptors seem well suited to account for the two major qualities of cutaneous pain: pricking and burning sensation (Lewis, 1942). Pricking pain appears to result from activation of Aδ nociceptors and burning pain from excitation of C polymodal nociceptors (Torebjork and Ochoa, 1983).

Activity in cutaneous nociceptors is transmitted to neurons of the dorsal horn of the spinal cord (or of the medulla, in the case of the trigeminal nerve). Aδ mechanical nociceptors have been shown by Light and Perl (1979), using intracellular labeling of individual electrophysiologically identified axons, to project chiefly to laminae I and V (with some endings in laminae X) of Rexed (1952). C polymodal nociceptors have been thought to end in laminae I and II, based on somewhat indirect evidence (reviewed in Willis, 1985). Recent evidence by Perl and his associates using anterograde labeling with a sensitive technique involving the *Phaseolus vulgaris* lectin (Sugiura et al., 1986) indicates that these receptors do, indeed, project to laminae I and II (with a few endings in laminae III and IV).

PROCESSING OF INFORMATION FROM NOCICEPTORS BY ASCENDING SOMATOSENSORY TRACT CELLS

Activity in primary afferent fibers supplying nociceptors is known to activate interneurons in the same laminae (I, II, V, X) that receive direct nociceptive projections (e.g., Bennett et al., 1980; Cervero et al., 1976; Christensen and Perl, 1970; Honda, 1985; Menetrey et al., 1977; Nahin et al., 1983; Price and Mayer, 1974; Wall, 1967). The meaning of the activity of a particular interneuron, in terms of transmission of nociceptive signals by ascending tract cells, is generally unclear. Interneurons undoubtedly play a role in exciting ascending tract cells, but some are involved in inhibitory interactions. Others participate in reflex activity, and may not be involved at all in sensory processing. The present account emphasizes the responses of ascending tract cells rather than those of interneurons.

The spinothalamic tract (STT) is generally regarded as the more important ascending tract for transmitting discriminative information concerning pain (Willis, 1985). This opinion is based on the following evidence: (1) anterolateral cordotomy produces analgesia in human subjects on the contralateral side below the segmental level of the lesion (Spiller and Martin, 1912; Foerster and Gagel, 1932; White and Sweet, 1955) and a contralateral reduction in reactivity to painful stimuli in animal subjects (Vierck and Luck, 1979); (2) pain sensation is unimpaired after the spinal cord is transected, except for one anterolateral quadrant (Noordenbos and Wall, 1976); (3) pain can be produced by stimulating the anterolateral quadrant with the appropriate stimulus parameters (Mayer et al., 1975); and (4) recordings from STT cells in animal subjects indicate that most of these neurons are nociceptive and have contralateral thalamic projections (Willis et al., 1974; Price et al., 1978; Chung et al., 1979; Kenshalo et al., 1979). However, other nociceptive pathways are present in the anterolateral quadrant of the monkey spinal cord, including the spinoreticular and spinomesencephalic tracts (Haber et al., 1982; Yezierski et al., 1987). These are likely to play an important role in pain, although whether they are crucial for pain discrimination is debatable. Nociceptive neurons also project in ascending tracts located in the posterolateral and posterior funiculi (Willis, 1985). However, the success of anterolateral cordotomies in relieving pain in humans argues against a substantial role for these pathways in human pain.

Since most of the experimental work on the spinothalamic tract has been done on monkeys, this account is limited to the primate STT. There is good evidence that Aδ and C polymodal nociceptors can activate primate STT cells. Volleys in either Aδ and C fibers evoked by electrical stimulation of the peripheral nerve supplying the receptive field cause discharges in most STT cells tested (Foreman et al., 1975; Beall et al., 1977; Chung et al., 1979). STT cells also respond to noxious mechanical and thermal stimuli that would activate Aδ mechanical and thermal nociceptors and C polymodal nociceptors (Chung et al., 1979; 1986a; Surmeier et al., 1986a,b; Ferrington et al., 1987). Responses to noxious heat stimuli are diminished more than those to noxious mechanical stimuli after blockade of C fibers by local application of capsaicin to the peripheral nerve innervating the receptive field (Chung et al., 1984). STT cells, for example small ones in lamina I, not included in the sample investigated may respond exclusively to activity in Aδ or to C polymodal nociceptors. However, based on the evidence available to date, the typical primate STT cell appears to receive a convergent input from both classes of nociceptor. One implication is that the same neurons may contribute to the central processing of both pricking and burning pain. The distinction between these may relate to temporal factors in the discharges (see Lewis, 1942). Alternatively, different classes of STT cells activated differentially by one or the other type of nociceptor may be responsible for these different qualities of pain sensation.

The receptive fields of STT cells are much larger than those of nociceptors (Figure 7-1), which argues for considerable convergence of information from different primary afferent fibers on a given STT cell (Willis et al., 1974; Applebaum et al., 1975; Giesler et al., 1981; Ferrington et al., 1987). The receptive fields of some STT cells may, however, be small enough to provide useful information for stimulus localization (Figure 7-1). This is especially true for STT cells in lamina I, which on the whole, have smaller receptive fields than do STT cells in deeper laminae (Table 7-1). Only 2 percent of a sample of 212 STT neurons in laminae IV through VI had receptive fields that were confined to an area equivalent to or less than that of a digit, but this proportion increased to 18 percent for a sample of 57 STT cells in lamina I. Thalamic projection neurons in the monkey nucleus gracilis (NG) examined under similar experimental conditions (Willis et al., 1986) had much smaller receptive fields than did STT cells of laminae IV through VI (Table 7-1); the correspondence between the receptive fields of NG neurons and STT cells of lamina I was much closer. Although information useful for localization may be provided by STT neurons, this does not argue against a role for coactivated mechanosensitive neurons in pain localization (Poggio and Mountcastle, 1960).

Most STT cells respond in a graded fashion to graded noxious mechanical stimuli applied to their receptive fields. However, the threshold mechanical stimulus may be so weak that only sensitive mechanoreceptors are activated (e.g., movement of single hair; see Willis et al., 1974; 1975). Other STT cells have thresholds for mechanical stimuli that are in the noxious range. In an effort to

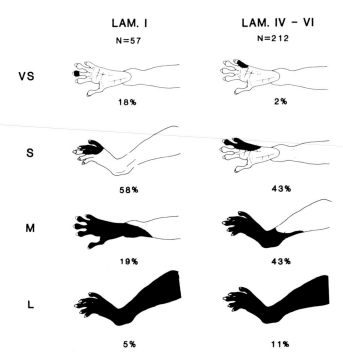

Figure 7-1 Receptive fields (RFs) of STT cells in lamina I and laminae IV through VI. Very small (VS) RFs had an area equivalent to or less than a digit. Small (S) RFs occupied an area corresponding to that of the foot or less. Medium (M) RFs included an area larger than the foot but smaller than the foot and leg. Large RFs covered an area greater than the foot and leg.

Table 7-1 Sizes of Receptive Fields of STT Cells Compared with Cells of the Nucleus Gracilis

	Lamina I		Laminae IV–VI		Nucleus Gracilis	
	No.	%	No.	%	No.	%
VS[a]	10	17.5	4	1.9	15	32.6
S	33	57.9	91	42.9	19	41.3
M	11	19.3	92	43.4	2	4.3
L	3	5.3	24	11.3	0	0
CX	0	0	1	0.5	0	0
Total	57		212		46	

[a]Key to abbreviations.
 VS = ≤ area of digit
  S = ≤ area of foot
 M = > area of foot, < area below knee
  L = > area below knee
 CX = bilateral

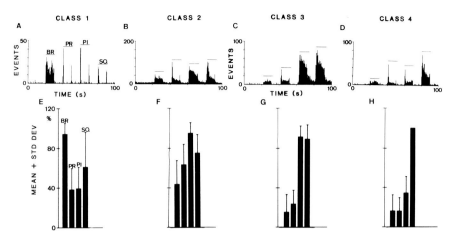

Figure 7-2 Responses of different classes of STT cells to graded intensities of mechanical stimuli. The single-pass peristimulus histograms in A through D show the responses of typical examples of STT cells belonging to response classes 1 through 4 to brushing of the skin (with a camelhair brush), pressure (application of a large arterial clip to the skin), pinching (application of a small arterial clip to the skin), and squeezing of the skin (with forceps). Abbreviations: BR, brush; PR, pressure; PI, pinch; and SQ, squeeze. The bar graphs in E through H show the mean responses (+1 SD) for the populations of STT cells in the different response classes to the same four stimuli. The proportion of STT cells in each class is as follows: 1, 12%; 2, 14%; 3, 45% and 4, 29%.

classify the types of STT cells based on mechanical stimuli, we have recently examined, in anesthetized monkeys, the response profiles of a large series of STT cells to four standard stimuli: brushing the skin repeatedly (to activate a variety of sensitive mechanoreceptors) and a series of three compression stimuli of different intensities in the noxious range (application of a large arterial clip to a fold of skin; application of a small arterial clip to the skin, and squeezing the skin with serrated forceps). The large arterial clip is just above the pain threshold for human skin. The small arterial clip is distinctly painful, and the stimulus from forceps would not be tolerated by an unanesthetized subject. Responses to each of these stimuli were quantitated over a 10-second period by means of a peristimulus time histogram. Background activity was subtracted, and the responses were normalized by taking the largest response as 100 percent.

The initial sample of 128 STT cells could be subdivided into three response classes using a k means cluster analysis (Chung et al., 1986a). Recently, analysis of the responses of a population of 318 STT cells revealed at least four response classes. Histograms showing response profiles typical of each of the four classes are shown in Figure 7-2A–D. The bar graphs in Figure 7-2E–H are the mean (+1 standard deviation [SD]) of the responses of all the STT cells in each class for a population of 318 neurons. Some of the STT cells (12%) responded best to brushing (Figure 7-2A and E). Others (14%) were activated effectively by the large arterial clip, although the little clip often produced a stronger response (Figure 7-2B and F). However, most of the STT cells (74%) were excited only weakly by brushing or by the large arterial clip and strongly by the intensely noxious stimuli. These nociceptive STT cells belonged to response category 3 (Figure 7-2C–G) or 4 (Figure 7-D and H). Class 3 cells (45%) were maximally activated by the small arterial clip, whereas the most vigorous discharges of class 4 cells (29%) were elicited by squeezing the skin with the forceps. Evidently, class 3 cells have steeper stimulus-response curves than class 4 cells, and the responses saturate at lower stimulus intensities.

The response properties of the population of STT cells could be displayed on a principal components plot. Principal components analysis provides a weighted response value for each cell based on the responses to the four stimuli. The principal components are calculated by computer and assigned coordinates in a four-dimensional space. The first principal component is the coordinate passing through the data space along the axis of greatest variance. The second principal component is orthogonal to the first and passes through the region of the next greatest variance. Higher-order principal components are successfully orthogonal and account for the remaining variance. A plot of the data in the plane of the first two principal components is a satisfactory representation of the distribution, since such a plot accounts for most of the variance in the data. Figure 7-3 shows the distribution of the responses of 318 STT cells to the four standard mechanical stimuli in the plane of the first two principal components. Figure 7-3A shows the response locations of cells belonging to the four classes, and Figure 7-3B shows the locations of the responses of STT cells in lamina I (the symbol used for such cells is I), laminae IV through VI (V), or an undetermined location (dot). The location of a point depends on the weighting assigned to the different responses for the two principal components. For positions along the ordinate (principal component 1), a large positive weighting was assigned to the response to brushing and a large negative weighting to that to squeezing the skin with forceps. For the abscissa (principal component 2), large positive weightings were given for responses to the two arterial clips and small weightings to the responses to brushing and squeezing of the skin. Responses of STT cells belonging to class 1 are plotted in the upper left quadrant (good responses to brushing, poor responses to the other stimuli); those for cells in class 2 are in the upper right quadrant (good responses to brushing and to the arterial clips); the responses of most of the clearly nociceptive cells of classes 3 and 4 are plotted below the abscissa. Class 3 cells are in the lower right quadrant (good response to the small arterial clip and squeezing the skin with forceps and poor response to brushing), and class 4 cells are in the lower left quadrant (good response to squeezing, but weak responses to the other stimuli).

It is worth noting that the responses of STT cells in lamina I are distributed somewhat differently from those of STT cells in laminae IV through VI. Only one STT cell of lamina I belonged to class 1, and only seven to class 2; thus, only a few of these neurons are shown in the upper part of the principal component plot. By contrast, the responses of STT cells of laminae IV through VI are distributed in all four quadrants, although most are in the lower half of the plot.

Primate STT cells usually respond well to noxious heat stimuli as well as to noxious mechanical stimuli (Kenshalo et al., 1979; Surmeier et al., 1986a,b). The threshold for activating STT cells may be as low as 43 or 45° C, suggesting an input from C polymodal nociceptors, or it may be 49° C or more, presumably reflecting an input from heat-sensitive Aδ nociceptors (LaMotte et al., 1982) or

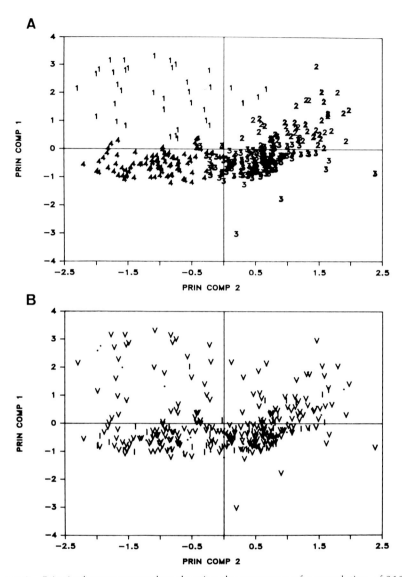

Figure 7-3 Principal component plots showing the responses of a population of 318 STT cells to the four standard mechanical stimuli. The weightings given to the different responses for principal components (PRIN COMP) 1 and 2 are described in the text. (A) The distribution of the four classes of STT cells in the plane of the first two principal components. (B) The distribution of the responses of STT cells in lamina I (I), laminae IV through VI (V), or undetermined locations (dot).

sensitized Aδ mechanical nociceptors. The mean stimulus–response function for noxious heat stimuli is a positively accelerating curve over the range of heat stimuli from 45 to 52° C (Kenshalo et al., 1979; Surmeier et al., 1986a). The responses of STT cells to noxious heat stimuli show several temporal components that may reflect input from different types of nociceptors (Surmeier et al., 1986b).

When the thermal responses of STT cells belonging to the clearly nociceptive classes based on mechanical stimuli were compared, some of the STT cells were found to have steeply rising stimulus–response curves for mechanical stimuli and thermal responses that may have resulted from an input from both C polymodal and Aδ nociceptors. It was suggested that these STT cells might be particularly suited to signal potentially damaging stimuli (Surmeier et al., 1986b). Another group of STT cells responded best to the most intense mechanical and thermal stimuli. Their dominant input appeared to be from Aδ nociceptors. It was proposed that these might signal frankly damaging stimuli (Surmeier et al., 1986b).

PROCESSING OF INFORMATION FROM NOCICEPTORS IN THE VENTRAL POSTERIOR LATERAL NUCLEUS OF THE THALAMUS

Investigations of nociception at higher levels of the nervous system are difficult because of the potential of anesthetics to alter responses. Most anesthetics appear to have little influence on the activity of nociceptors or to affect their sensitivity. Indeed, the responses of primate STT cells to volleys in C fibers are actually enhanced by small doses of barbiturates (Hori et al., 1984). However, the effects of anesthetics are likely to be predominantly suppressive at the levels of the thalamus and cerebral cortex, and so the responses examined at these levels in the presence of anesthetic may not reflect the activity that would be recorded in the absence of anesthetic. In addition, it is a challenge to examine nociceptive responses in unanesthetized animals in an ethical fashion. This has been accomplished by using awake, behaving animals trained to accept mildly painful stimuli and provided with the means to avoid stimuli they find intolerable (Casey and Morrow, 1983; Maixner et al., 1986). However, only a limited amount of information concerning nociceptive responses in the ventrobasal thalamus is available from such studies (Casey and Morrow, 1983).

It is well known that the spinothalamic tract in primates projects to the ventral posterior lateral (VPL) nucleus of the thalamus (reviewed in Willis, 1985). The VPL nucleus serves as the main somatosensory relay in the thalamus and is believed to play a major role in discriminative sensory functions (Jones, 1985). Before direct evidence was available, one might have thought the component of the STT that terminates in the VPL nucleus serves a tactile function, since the STT plays a tactile role as well as nociceptive and thermoreceptive roles (Foerster and Gagel, 1932). However, experiments have been reported in which the axons of individual nociceptive STT cells were traced into the VPL nucleus, by means of antidromic activation (Applebaum et al., 1979).

Recordings from neurons in the VPL nucleus of the monkey thalamus in anesthetized preparations reveal neurons in this nucleus that respond to noxious mechanical and thermal stimuli (Kenshalo et al., 1980; Chung et al., 1986a,b). These neurons can be excited by volleys in Aδ and C fibers evoked by electrical stimulation of the peripheral nerve supplying the receptive field (Chung et al., 1986b). Although the receptive fields of these cells are often restricted in extent and located on the contralateral side of the body, some of the receptive fields are quite large and may be bilateral (Chung et al., 1986b). Nociceptive neurons in the VPL nucleus can have stimulus–response functions to graded noxious mechanical stimuli similar to those of STT cells of classes 3 or 4 (Chung et al., 1986a). Thresholds for noxious heat stimuli may be as low as 45° C, and the mean stimulus–response function for graded noxious heat stimuli is a positively accelerating curve in the range of 45 to 55°C (Kenshalo et al., 1980).

Nociceptive neurons in the monkey VPL nucleus are located in the somato-topically appropriate part of the nucleus: cells that have receptive fields on the hindlimb are in the lateral part of the nucleus, and those with forelimb receptive fields are in the medial part (Kenshalo et al., 1980). Almost all of the nociceptive cells in the VPL nucleus so far tested were found to project to the SI region of the cerebral cortex by antidromic activation with microstimulation (Kenshalo et al., 1980). The best location for successful antidromic activation of these cells was a point in area 1 near its junction with area 3b.

It is not necessarily obvious which ascending pathway provides nociceptive input to a given VPL neuron. In a number of cases, it was shown that the no-ciceptive input depended on the integrity of the lateral quadrant ipsilateral to the thalamic neuron (contralateral to the receptive field) by sectioning part of the spinal cord white matter rostral to the segmental level of the input (Kenshalo et al., 1980; Chung et al., 1986b). In at least some cases, however, the nociceptive input was by way of the dorsal half of the spinal cord (Chung et al., 1986b). The nociceptive information reaching the VPL nucleus could have been transmitted by the spinocervicothalamic pathway (Downie et al., 1986), or perhaps by the postsynaptic dorsal column pathway (Willis et al., 1986).

Nociceptive neurons in the monkey VPL nucleus have also been reported in anesthetized animals by Gaze and Gordon (1954) and Perl and Whitlock (1961) and in unanesthetized, behaving monkeys by Pollin and Albe-Fessard (1979) and Casey and Morrow (1983).

PROCESSING INFORMATION FROM NOCICEPTORS BY NEURONS IN THE SOMATOSENSORY CEREBRAL CORTEX

Mountcastle and Powell (1959) reported nociceptive responses for a few neurons in the SI region of the monkey cerebral cortex. However, the receptive fields were large and often included ipsilateral as well as contralateral regions of the body.

Kenshalo and Isensee (1983) have recorded nociceptive responses from the SI cerebral cortex, near the boundary between areas 3b and 1, in anesthetized mon-keys. These cells had either restricted, contralateral receptive fields or large bi-lateral receptive fields. They responded in a graded fashion to graded mechanical and thermal stimuli in the noxious range. The mean stimulus–response function for noxious heat stimuli was a positively accelerating curve in the range of 45 to 55° C.

Several investigations have been made of neurons in the SII region of the monkey cerebral cortex in unanesthetized preparations. Whitsel et al. (1969) found nociceptive neurons with large, often bilateral, receptive fields in the caudal part of the SII cortex. Convergent input from the auditory and visual systems was common. Robinson and Burton (1980) found that only 3 percent of their sample of neurons in the SII cortex of unanesthetized monkeys responded to noxious stimuli. They found a higher percentage of nociceptive neurons in area 7b than in the SII cortex, and a few nociceptive cells were also found in the retroinsular and granular insular cortex.

CONCLUSION

It is evident that there is a neural system that can account for the discriminative aspects of cutaneous pain sensation. This system includes primary afferent fibers supplying nociceptors, ascending somatosensory tracts, such as the spinothalamic tract, neurons in the somatosensory thalamus and cerebral cortex, and neurons in the parietal association cortex. It will be important for the evidence from anesthetized animals to be confirmed and extended in experiments using unanesthetized animals (within the limits imposed by ethical constraints). Unfortunately, full understanding of the neural system responsible for the discriminative aspects of pain sensation is likely to be insufficient to explain many clinical pain phenomena. The latter will require further investigations of the neural basis of the motivational–affective aspects of pain.

ACKNOWLEDGMENTS

The author thanks Helen Willcockson and Griselda Gonzales for their expert technical assistance in the experiments described and Phyllis Waldrop for typing the manuscript. Support for work in the author's laboratory included NIH Research Grants NS 09743 and NS 11255.

REFERENCES

Applebaum, A. E., Beall, J. E., Foreman, R. D., and Willis, W. D. (1975). Organization and receptive fields of primate spinothalamic tract neurons. *J. Neurophysiol.* 38:572–586.

Applebaum, A. E., Leonard, R. B., Kenshalo, D. R., Jr., Martin, R. F., and Willis, W. D. (1979). Nuclei in which functionally identified spinothalamic tract neurons terminate. *J. Comp. Neurol.* 188:575–586.

Beall, J. E., Applebaum, A. E., Foreman, R. D., and Willis, W. D. (1977). Spinal cord potentials evoked by cutaneous afferents in the monkey. *J. Neurophysiol.* 40:199–211.

Beck, P. W., Handwerker, H. O., and Zimmermann, M. (1974). Nervous outflow from the cat's foot during noxious radiant heat stimulation. *Brain Res.* 67:373–386.

Bennett, G. J., Abdelmoumene, M., Hayashi, H., and Dubner, R. (1980). Physiology and morphology of substantia gelatinosa neurons intracellularly stained with horseradish peroxidase. *J. Comp. Neurol.* 194:809–827.

Bessou, P., and Perl, E. R. (1969). Response of cutaneous sensory units with unmyelinated fibers to noxious stimuli. *J. Neurophysiol.* 32:1025–1043.

Burgess, P. R., and Perl, E. R. (1967). Myelinated afferent fibres responding specifically to noxious stimulation of the skin. *J. Physiol.* 190:541–562.

Casey, K. L., and Morrow, T. J. (1983). Ventral posterior thalamic neurons differentially responsive to noxious stimulation of the awake monkey. *Science* 221:675–677.

Cervero, F., Iggo, A., and Ogawa, H. (1976). Nociceptor-driven dorsal horn neurones in the lumbar spinal cord of the cat. *Pain* 2:5–24.

Christensen, B. N., and Perl, E. R. (1970). Spinal neurons specifically excited by noxious or thermal stimuli: marginal zone of the dorsal horn. *J. Neurophysiol* 33:293–307.

Chung, J. M., Kenshalo, D. R., Jr., Gerhart, K. D., and Willis, W. D. (1979). Excitation of primate spinothalamic neurons by cutaneous C-fiber volleys. *J. Neurophysiol.* 42:1354–1369.

Chung, J. M., Lee, K. H., Hori, Y., and Willis, W. D. (1984). Effects of capsaicin applied to a peripheral nerve on the responses of primate spinothalamic tract cells. *Brain Res.* 329:27–38.

Chung, J. M., Surmeier, D. J., Lee K. H., Sorkin, L. S., Honda, C. N., Tsong, Y., and Willis, W. D. (1986a). Classification of primate spinothalamic and somatosensory thalamic neurons based on cluster analysis. *J. Neurophysiol.* 56:308–327.

Chung, J. M., Lee, K. H., Surmeier, D. J., Sorkin, L. S., Kim, J., and Willis, W. D. (1986b). Response characteristics of neurons in the ventral posterior lateral nucleus of the monkey thalamus. *J. Neurophysiol.* 56:370–390.

Downie, J. W., Ferrington, D. G., and Willis, W. D. (1986). Response properties of neurons of the primate lateral cervical nucleus (LCN). *Soc. Neurosci. Abstr.* 12:227.

Ferrington, D. G., Sorkin, L. S., and Willis, W. D. (1987). Responses of spinothalamic tract cells in the superficial dorsal horn of the primate lumbar spinal cord. *J. Physiol.* 388:681–703.

Fitzgerald, M., and Lynn, B. (1977). The sensitization of high threshold mechanoreceptors with myelinated axons by repeated heating. *J. Physiol.* 265:549–563.

Foerster, O., and Gagel, O. (1932). Die Vorderseitenstrangdurchschneidung beim Menschen. Eine klinisch-patho-physiologisch-anatomische Studie. *Z. ges. Neurol. Psychiat.* 138:1–92.

Foreman, R. D., Applebaum, A. E., Beall, J. E., Trevino, D. L., and Willis, W. D. (1975). Responses of primate spinothalamic tract neurons to electrical stimulation of hindlimb peripheral nerves. *J. Neurophysiol.* 38:132–145.

Gaze, R. M., and Gordon, G. (1954). The representation of cutaneous sense in the thalamus of the cat and monkey. *J. Exp. Physiol.* 39:279–304.

Georgopoulos, A. P. (1976). Functional properties of primary afferent units probably related to pain mechanisms in primate glabrous skin. *J. Neurophysiol.* 39:71–83.

Giesler, G. J., Yezierski, R. P., Gerhart, K. D., and Willis, W. D. (1981). Spinothalamic tract neurons that project to medial and/or lateral thalamic nuclei: evidence for a physiologically novel population of spinal cord neurons. *J. Neurophysiol.* 46:1285–1308.

Haber, L. H., Moore, B. D., and Willis, W. D. (1982). Electrophysiological response properties of spinoreticular neurons in the monkey. *J. Comp. Neurol.* 207:75–84.

Honda, C. N. (1985). Visceral and somatic afferent convergence onto neurons near the central canal in the sacral spinal cord of the cat. *J. Neurophysiol.* 53:1059–1078.

Hori, Y., Lee, K. H., Chung, J. M., Endo, K., and Willis, W. D. (1984). The effects of small doses of barbiturate on the activity of primate nociceptive tract cells. *Brain Res.* 307:9–15.

Jones, E. G. (1985). *The Thalamus*. New York: Plenum Press.

Kenshalo, D. R., Jr., Giesler, G. J., Leonard, R. B., and Willis, W. D. (1980). Responses of neurons in primate ventral posterior lateral nucleus to noxious stimuli. *J. Neurophysiol.* 43:1594–1614.

Kenshalo, D. R., Jr., and Isensee, O. (1983). Responses of primate SI cortical neurons to noxious stimuli. *J. Neurophysiol.* 50:1479–1496.

Kenshalo, D. R., Jr., Leonard, R. B., Chung, J. M., and Willis, W. D. (1979). Responses of primate spinothalamic neurons to graded and to repeated noxious heat stimuli. *J. Neurophysiol.* 42:1370–1389.

Kumazawa, T., and Perl, E. R. (1977). Primate cutaneous sensory units with unmyelinated (C) afferent fibers. *J. Neurophysiol.* 40:1325–1338.

LaMotte, R. H., and Campbell, J. N. (1978). Comparison of responses of warm and nociceptive C-fiber afferents in monkey with human judgments of thermal pain. *J. Neurophysiol.* 41:509–528.

LaMotte, R. H., Thalhammer, J. G., Torebjork, H. E., and Robinson, C. J. (1982). Peripheral neural mechanisms of cutaneous hyperalgesia following mild injury by heat. *J. Neurosci.* 2:765–781.

Lewis, T. (1942). *Pain*. London: Macmillan.

Light, A. R., and Perl, E. R. (1979). Spinal termination of functionally identified primary afferent neurons with slowly conducting myelinated fibers. *J. Comp. Neurol.* 186:133–150.

Maixner, W., Dubner, R., Bushnell, M. C., Kenshalo, D. R., Jr., and Oliveras, J. L. (1986). Wide-dynamic-range dorsal horn neurons participate in the encoding process by which monkeys perceive the intensity of noxious heat stimuli. *Brain Res.* 374:385–388.

Mayer, D. J., Price, D. D., and Becker, D. P. (1975). Neurophysiological characterization of the anterolateral spinal cord neurons contributing to pain perception in man. *Pain* 1:51–58.

Melzack, R., and Casey, K. L. (1968). Sensory, motivational, and central control determinants of pain. In: D. R. Kenshalo (ed.), *The Skin Senses*, pp. 423–439. Springfield, IL: Charles C. Thomas.

Menetrey, D., Giesler, G. J., and Besson, J. M. (1977). An analysis of response properties of spinal cord dorsal horn neurones to nonnoxious and noxious stimuli in the spinal rat. *Exp. Brain Res.* 27:15–33.

Mountcastle, V. B., and Powell, T. P. S. (1959). Neural mechanisms subserving cutaneous sensibility, with special reference to the role of afferent inhibition in sensory perception and discrimination. *Bull. Johns Hopkins Hosp.* 105:201–232.

Nahin, R. L., Madsen, A. M., and Giesler, G. J. (1983). Anatomical and physiological studies of the gray matter surrounding the spinal cord central canal. *J. Comp. Neurol.* 220:321–335.

Noordenbos, W., and Wall, P. D. (1976). Diverse sensory functions with an almost totally divided spinal cord. A case of spinal cord transection with preservation of part of one anterolateral quadrant. *Pain* 2:185–195.

Perl, E. R. (1968). Myelinated afferent fibres innervating the primate skin and their response to noxious stimuli. *J. Physiol.* 197:593–615.

Perl, E. R., and Whitlock, D. G. (1961). Somatic stimuli exciting spinothalamic projections to thalamic neurons in cat and monkey. *Exp. Neurol.* 3:356–296.

Poggio, G. F., and Mountcastle, V. B. (1960). A study of the functional contributions of the lemniscal and spinothalamic systems to somatic sensibility. *Bull. Johns Hopkins Hosp.* 106:266–316.

Pollin, G., and Albe-Fessard, D. (1979). Organization of somatic thalamus in monkeys with and without section of dorsal spinal tracts. *Brain Res.* 173:431–449.

Price, D. D., Hayes, R. L., Ruda, M. A., and Dubner, R. (1978). Spatial and temporal transformations of input to spinothalamic tract neurons and their relation to somatic sensations. *J. Neurophysiol.* 41:933–947.

Price, D. D., and Mayer, D. J. (1974). Physiological laminar organization of the dorsal horn of *M. mulatta*. *Brain Res.* 79:321–325.

Rexed, B. (1952). The cytoarchitectonic organization of the spinal cord in the cat. *J. Comp. Neurol.* 96:415–494.

Robinson, C. J., and Burton, H. (1980). Somatic submodality distribution within the second somatosensory (SII), 7b, retroinsular, postauditory, and granular insular cortical areas of *M. fascicularis*. *J. Comp. Neurol.* 192:93–108.

Sherrington, C. S. (1906). *The Integrative Action of the Nervous System*, New Haven, CT: Yale University Press. 2nd ed. (1947; Yale paperbound, 1961).

Spiller, W. G., and Martin, E. (1912). The treatment of persistent pain of organic origin in the lower part of the body by division of the anterolateral column of the spinal cord. *JAMA* 58:1489–1490.

Sugiura, Y., Lee, C. L., and Perl, E. R. (1986). Central projections of identified, unmyelinated (C) afferent fibers innervating mammalian skin. *Science* 234:358–361.

Surmeier, D. J., Honda, C. N., and Willis, W. D. (1986a). Responses of primate spinothalamic neurons to noxious thermal stimulation of glabrous and hairy skin. *J. Neurophysiol.* 56:328–350.

Surmeier, D. J., Honda, C. N., and Willis, W. D. (1986b). Temporal features of the responses of primate spinothalamic neurons to noxious thermal stimulation of hairy and glabrous skin. *J. Neurophysiol.* 56:351–369.

Torebjork, E., and Ochoa, J. (1983). Selective stimulation of sensory units in man. *Adv. Pain Res. Ther.* 5:99–104.

Vierck, C. J., and Luck, M. M. (1979). Loss and recovery of reactivity to noxious stimuli in monkeys with primary spinothalamic cordotomies, followed by secondary and tertiary lesions of other cord sectors. *Brain* 102:233–248.

Wall, P. D. (1967). The laminar organization of dorsal horn and effects of descending impulses. *J. Physiol.* 188:403–423.

White, J. C., and Sweet, W. H. (1955). *Pain: Its Mechanisms and Neurosurgical Control.* Springfield, IL: Charles C. Thomas.

Whitsel, B. L., Petrucelli, L. M., and Werner, G. (1969). Symmetry and connectivity in the map of the body surface in somatosensory area II of primates. *J. Neurophysiol.* 32:170–183.

Willis, W. D. (1985). *The Pain System. The Neural Basis of Nociceptive Transmission in the Mammalian Nervous System*. Basel: Karger.

Willis, W. D., Ferrington, D. G., and Downie, J. W. (1986). Response properties of neurons of the primate nucleus gracilis. *Soc. Neurosci. Abstr.* 12:227.

Willis, W. D., Maunz, R. A., Foreman, R. D., and Coulter, J. D. (1975). Static and dynamic responses of spinothalamic tract neurons to mechanical stimuli. *J. Neurophysiol.* 38:587–600.

Willis, W. D., Trevino, D. L., Coulter, J. D., and Maunz, R. A. (1974). Responses of primate spinothalamic tract neurons to natural stimulation of hindlimb. *J. Neurophysiol.* 37:358–372.

Yezierski, R. P., Sorkin, L. S., and Willis, W. D. (1987). Response and receptive field properties of primate spinomesencephalic tract (SMT) cells. Fifth World Congress on Pain, International Association for the Study of Pain, Hamburg, Germany.

III

HEARING

The auditory system presents as many features of research interest as the visual system, although in humans auditory input has been overshadowed by vision. The sensory life of bats, however, is experienced almost entirely in auditory terms, and we have learned more about certain aspects of sensory processing from studies of the auditory system in bats and birds than from studies in any other species. In humans, the auditory system may be subservient to vision in locating objects in the environment, but think of the richness of the musical world and the intricacies of spoken language. This brings up the same problem that was discussed earlier in terms of the taste and olfactory systems, that of the need to appreciate the particular analytic strengths of the system under study, because birds, bats, and people put their auditory pathways to quite different uses in their everyday lives. Are these different behaviors a result of differences in receptors or in the structure and analytic possibilities of the central relays of the brain? Or are all species capable of similar performance but simply do not use their capabilities in the same fashion? Each of these factors is probably partly true.

The next two chapters introduce us to auditory system studies. In Chapter 8 Tsuchitani and Johnson discuss an issue that concerns all sensory system physiologists: how to determine what is the significant signal feature of a neuron's pattern of activity. What is the significant "message" that the neural system is extracting from the activity of its components? No longer is our concern a Beethoven symphony, or even a single note, but rather the ionic changes in nerve cells into which all sensory messages are converted. It is indeed amazing that the brain eventually assigns different qualities to the incoming sensation that make the world so delightful and interesting! Tsuchitani and Johnson bring the sensory investigator's task sharply into focus; they introduce the question and the experimental means of analyzing the initial message in the flow of activity in auditory pathways.

The auditory cortex is the subject of Chapter 9, in which Phillips reviews the current status of investigations of cortical auditory function in primates, cats, and other animals. We again find stripe arrays, puzzles over the meaning of different topographies, and discussion of the relationship of physiological data and behavior patterns. This chapter brings to our attention the interrelations between subcortical and cortical pathways and the existence and interrelationship of multiple cortical areas devoted to the same sensory modality. The different qualities of the sensory messages are discussed, including the auditory location of sounds in space, tonal quality, and even human language. The questions that concerned Tsuchitani and

Johnson about the nature of the coding of auditory input in the neuron's activity patterns are seen to be equally important at a cortical level, where the experimenter struggles to understand the nature of the analyses being conducted by the cortical neuropil. Temporal factors in stimulus coding and binaural interaction are shown to be of crucial importance, and the need to know the range of auditory signals normally relevant to the particular animal under study is discussed.

8

Statistical Modeling of
Auditory Neuron Discharges

CHIYEKO TSUCHITANI AND DON H. JOHNSON

The nervous system has been described as "a communication machine" that processes information (Perkel and Bullock, 1968). In the case of sensory systems, the sensory receptors convert the information (e.g., light, sound, chemicals in the air or food) into a form that can be conveyed and processed by the nervous system. The sensory information, once transformed by the receptors, is then transmitted and further transformed by the nervous system, which represents, interprets, stores, and, if necessary, responds to the sensory information. Much research in sensory neurophysiology is concerned with how the nervous system represents sensory information. Psychophysical studies of sensory systems usually characterize the critical aspects of sensory information that must be represented by the nervous system; the physical/chemical attributes of the stimulus are related to the basic sensations elicited by common, naturally occurring stimuli. Incomplete or inadequate characterization can seriously hamper sensory neurophysiological research. For example, the flavor of a steak cannot be described by a combination of the so-called four basic tastes (sweet, sour, salty, and bitter; see Chapter 2). When knowledge of the sensory information to be processed is inadequate, the neurophysiological investigator must catalog stimulus–response relationships without guidance as to what the salient stimulus or response features may be. Both the information to be processed and how it is processed must be described. Fortunately, most sensory systems have been amenable to psychophysical analysis, and the basic sensory attributes as well as their physical correlates have been fairly well defined. In the case of the auditory system, the waveform (frequency and phase relationships), amplitude, and temporal variations of the acoustic stimulus form the basic physical attributes to be processed. Binaural information (i.e., information unique to stimulating the two ears simultaneously) can also be defined in terms of interaural frequency, phase, amplitude, and time differences.

Within a sensory receptor organ, the stimulus is delivered to a set of spatially distributed receptors. Within this set of receptors, the stimulus may alter the activity of a limited subset of receptors (e.g., a red light on a 0.03-mm^2 area of the fovea affects only the activity of "red" cone receptors in the light field). These receptors are connected to a specific group of afferent neurons, which function as labeled lines that encode information about the location (the area in the fovea) and quality (red light) of the stimulus. Within the auditory system, the spatial distribution of the stimulus on the receptor organ is related to the frequency of the stimulus. Hence, the *location* of a nerve fiber within a nerve/tract or of a neuron soma within a ganglion, nucleus, or cortex expresses the *frequency* of the acoustic stimulus to which it is maximally sensitive; this principle is known as *tonotopic* organization. The perceived spatial location of the acoustic stimulus source is related to binaural stimulus cues, which are not directly represented spatially within the auditory system. Within most sensory systems, the neurons that form or synapse with the receptors—the first-order or *primary* afferents—generate trains of spike potentials propagated to neurons within the central nervous system. The spatial distribution of the activated single afferent neurons (i.e., the activated labeled lines) and their spike-trains provide the representation of the physical/chemical stimulus to the central nervous system. According to the all-or-none principal of the nerve impulse, each spike is identical in waveform to all others in a spike-train. Hence, the information-bearing aspect of the spike train (discharge) is considered to be contained in the sequence of the times of occurrences of these spikes. Because the time periods between spikes—the interspike intervals—of afferent neural discharges vary, they are best characterized by statistical models. Consequently, in most models of auditory neuron discharges it is assumed that the processes underlying the neuronal responses are inherently probabilistic.

According to this view, the time points at which a neuron produces a spike potential can be described mathematically as isolated events generated by a point process. Because the timing of individual spikes generally differs in response to identical stimuli, and is thus to a certain degree unpredictable, the spike-train is considered to be equivalent to a *stochastic* point process. If the statistical properties of the spike-train do not change with time or repetition of the stimulus, the discharge can be represented as a realization of a *stationary* stochastic point process; statistical measures of a selected portion within the spike train will not differ significantly from those selected from another portion, although the timing of discharges will. Models were initially developed from statistical analyses of spontaneous discharges (discharges occurring in the absence of an experimenter-controlled stimulus) or from the maintained discharges to continuous stimuli, which are the most likely to meet the stationarity requirement. Whether or not a spike-train is the realization of a stationary point process is an unsolved problem, complicated by uncertainty about the quantities needed to define the process (Perkel et al., 1967).

Assuming that the stationarity issue is resolved, the next issue to be addressed is the statistical relationships between interspike intervals. If the intervals between successive spike potentials are independent (i.e., no significant correlations exist between adjacent interspike intervals), the point process can be characterized as a *renewal* point process. The assumption that the underlying process is renewal simplifies the modeling task enormously. For example, the probability distribution of interspike intervals completely characterizes the statistical properties of a stationary renewal point process. This critical quantity is easily measured by computing the interspike inteval histogram (Perkel et al., 1967; Johnson, 1978). The simplest renewal point process is the Poisson process, in which "the probability of an event in any small interval of time is proportional to the length of that interval and is independent of the occurrence of any previous events generated by the process" (Glaser and Ruchkin, 1976). This process is often used in modeling auditory neuron discharges because of its mathematical tractability. The probability density of the interspike intervals is an exponential, with the form $\lambda e^{-\lambda \tau}$ where λ is the average rate at which events occur and τ denotes an interspike interval duration.

The regularity of a stationary discharge pattern is often measured by the coefficient of variation (CV). This quantity is defined as the ratio of the standard deviation and the mean of the interspike intervals, and is roughly interpreted as the relative variation of the interspike interval durations about their average value. In a Poisson process, the coefficient of variation equals unity, regardless of the average rate's value. The Poisson process is considered to be irregular (100% variation about the mean); a CV value of about 0.4 to 0.5 is usually considered as the value demarcating regular and irregular discharges. Theoretically, a renewal process can have any CV value.

The problem of determining whether the data meet the renewal requirement (i.e., whether there is any serial dependence between interspike intervals) is often ignored in practice. Detecting and describing serial dependencies are difficult, again because the underlying distribution is unknown and many measures assume that the distribution is of a certain type. For example, the first-order serial correlation coefficient, which provides a scalar quantity indicating the type and degree of serial dependence between successive interevent intervals (Perkel et al., 1967), suffices *only* when the intervals are drawn from a Gaussian distribution and the serial dependence is linear. One measure that is relatively distribution-free and does not assume linear relationships is the conditional mean function, which is derived from the joint interval plot (Rodieck et al., 1962; Johnson et al., 1986). The first-order joint interval plot is a scatter diagram of the length of an interspike interval versus the length of its succeeding interval. The conditional mean function describes the relationship between the mean of the intervals immediately following (i.e., the forward conditional mean) or preceding (i.e., the backward conditional mean) a "conditioning" interval of a specified duration. Deviation of the conditional mean function from a constant would be indicative of serial dependence. For example, a negatively sloped forward conditional mean function indicates that short intervals tend to precede long intervals and long intervals tend to precede short intervals. Evidence of serial dependence would rule out renewal processes and could be suggestive of a kth-order Markov process "in which the dependency of the interval upon the past extends only over the previous several [k] intervals." (Glaser and Ruchkin, 1976). Assuming the serial dependence is well understood, a point process model can be derived to characterize a given set of discharges.

STATISTICAL PROPERTIES OF AUDITORY NEURON DISCHARGES

The Auditory Nerve

The neurons forming the auditory nerve are the primary afferents of the auditory nervous system. Their responses to sound stimuli provide the initial representation of the acoustic environment to central auditory structures. Paradoxically, the most studied aspect of auditory nerve fiber discharges has been the spontaneous discharges generated in the absence of experimentally produced stimulation (Kiang et al., 1965; Walsh et al., 1972). The temporal pattern of these discharges has been described as irregular and random. The shapes of histograms of successive interspike intervals are similar for all fibers regardless of mean discharge rate. They are asymmetrical (skewed to the right) with a mode between 3 and 7 msec and a long-interval tail that drops exponentially from the modal value; the CVs are greater than 0.8. The intervals between successive spikes were also described as independent of one another. It has been suggested that the statistical properties of auditory nerve fiber interspike intervals resemble those of the interevent intervals generated by a Poisson process (Rodieck et al., 1962; Kiang et al., 1965). The major difference between the interval statistics of the auditory nerve fiber discharges and those of a Poisson process is the lack of small intervals in the former; the auditory nerve fiber discharges exhibit a minimum nonzero interval period—the dead time—during which the fiber produces no discharges and those of the Poisson process do. The critical issue of stationarity remains. Although the preceding analyses are based on the assumption of stationarity, a recent report (Johnson and Kumar, 1985) suggests that auditory nerve fiber spontaneous activity is not well-described as a stationary renewal process, because the rate of discharges apparently undergoes unexplained random variations. While this preliminary result does not completely obviate the previous results, it certainly complicates an exact statistical analysis.

The interspike intervals of auditory nerve fiber—maintained discharges, such as those elicited by continuous tones of frequency above 5 kHz, have statistical properties similar to those of their spontaneous discharges (Kiang et al., 1965). The spacing of the spike discharges are irregular, resulting in similarly asymmetric interval histograms. When a lower-frequency sine wave serves as the stimulus, periodicity can be measured in the elicited discharges related to the period of the sinusoid (Tasaki, 1954). The folded or period histogram, which indicates the occurrence of a discharge relative to the positive zero crossing time of the sine wave, is used to examine these synchronized discharges (Rose et al., 1967). The synchronization index, originally called the vector strength (Greenwood and Durand, 1955), is a measure derived from the period histogram, which indicates the degree to which discharges are synchronized to the tonal waveform (Anderson, 1973; Johnson, 1980). With this measure, it was determined that the auditory nerve fiber discharges to tones with frequency up to 6 kHz were synchronized to a particular phase of the sinusoidal stimulus (Johnson, 1980). The preferred phase at which a fiber discharged is determined by the stimulus frequency and the frequency to which the fiber was most sensitive (the fiber's characteristic frequency or CF) (Pfeiffer and Molnar, 1970; Anderson et al., 1971). The synchronized discharges are probabilistic, however, in that they do not occur with every cycle or every n^{th} cycle of the tone; they can occur at any time throughout most of a half-cycle of the sinusoid. Thus, the statistical characteristics governing the probability of the occurrence of a spike are periodic. These discharge patterns are nonstationary, having a rate of discharge that varies in synchrony with the low-frequency sinusoidal stimulus.

The time course of auditory nerve fiber discharges to transient stimuli have been examined with the poststimulus time (PST) histogram, which indicates the time of the occurrence of a discharge relative to the time of stimulus onset (Johnson, 1978). In response to a short stimulus burst (e.g., a tone of 300 msec) the discharges of auditory nerve fibers reach a maximum instantaneous rate within a few milliseconds of the stimulus onset and rapidly drop to lower rates as a function of time (Kiang et al., 1965). Most typically the initial onset response decays rapidly with a time constant of 5 msec and is followed by a slower, exponential decay with a time constant of 40 msec (Smith and Zwislocki, 1975). Within 150 msec of the onset of a stimulus, a steady rate of discharge is reached and maintained for the duration of the stimulus. Following termination of the stimulus, a transient decrease in discharge occurs to rates below spontaneous levels. When the frequency of the tone burst is sufficiently low (e.g., less than 200 Hz), a periodicity in the discharges related to the period of the sinusoid is also observed. Increasing the amplitude of an acoustic stimulus generally results in increases in the mean discharge rate (Kiang et al., 1965; Sachs and Abbas, 1974), and in the synchronization to low-frequency sinusoids (Anderson et al., 1971; Johnson, 1980); each measure eventually approaches a limiting value. Many of these features of the auditory nerve fiber responses to acoustic stimuli are believed to be related specifically to the input to the nerve fibers—the stimulus drive function of the receptor organ. For example, the response properties related to fiber CF are thought to be determined by the mechanical properties of the receptor organ, and therefore are specified in the input to the statistical model of auditory nerve fiber discharges.

The Cochlear Nuclear Complex

The auditory nerve fibers terminate within the cochlear nuclear complex, forming different types of terminal endings on the cells of the complex. Within the rostral anteroventral cochlear nucleus (aAVCN), an auditory nerve fiber forms a large terminal ending, the endbulb of Held, that envelopes the cell body of a spherical bushy cell (Ramon y Cajal, 1909; Osen, 1969, 1970; Brawer and Morest, 1975). Each ending contains a number of synaptic contact areas with the soma of a spherical bushy cell (Lenn and Reese, 1966; Cant and Morest, 1979). This large terminal ending is believed to generate an extracellular prepotential—the presynaptic potential—that precedes the spherical bushy cell spike potential by 0.5 msec (Pfeiffer, 1966a). In more caudal areas of the AVCN (pAVCN), two to four club-shaped endbulbs of Held terminate on the cell body of a globular bushy cell (Lorente de No, 1933; Feldman and Harrison, 1969) and generate small extracellular prepotentials that may require signal averaging to be detected (Bourk, 1976). Each of these endings forms multiple synaptic contacts on the cell soma (Tolbert and Morest, 1982). The discharges of cochlear nuclear units generating spikes preceded by presynaptic potentials have statistical properties similar to those of auditory nerve fiber discharges, and thus have been designated *primarylike* units (Pfeiffer, 1966b; Bourk, 1976). Their spontaneous and maintained discharges generate positively skewed interval histograms with CVs greater than 0.7, and their tone burst–elicited discharges produce primarylike PST histograms (Pfeiffer and Kiang, 1965; Bourk, 1976; Goldberg and Brownell, 1973). Other neurons within the AVCN are multipolar in shape and receive numerous small bouton-type terminals from auditory nerve fibers like neurons in the rostral posteroventral cochlear nucleus (PVCN), (Osen, 1969, 1970; Feldman and Harrison, 1969; Brawer and Morest, 1975). These neurons generate spike potentials with no prepotentials and a discharge pattern termed *chopper-type* (Pfeiffer, 1966b). This cell-type/response-type relationship has been confirmed by intracellular marking tehniques (Rhode et al., 1983; Rouiller and Ryugo, 1984). Chopper patterns are characterized by initial, transient, but regularly spaced discharges with interspike intervals unrelated to stimulus frequency (Pfeiffer, 1966b) and are elicited by high-frequency tone bursts. The interval histograms of their later maintained discharges are more symmetrically shaped than those of primarylike units, resembling Gaussian-type distributions (Bourk, 1976). Hence, the temporal pattern of discharges of these units is considered more regular than those of primarylike units. The spike-trains of regularly discharging ventral cochlear nucleus units exhibit a negative serial dependence between successive interspike intervals, indicating that a renewal model would be inappropriate for describing them (Rodieck et al., 1962; Goldberg and Greenwood, 1966). The chopper-type units of the AVCN differed from the primarylike units in other ways (Bourk, 1976; Goldberg and Brownell, 1973; Rhode and Smith, 1986). They are less likely to be spontaneously active than primarylike units. They generally have longer latencies, but can also discharge at higher maximum rates to tonal stimuli. The primarylike units are often capable of synchronizing their discharges to tonal stimuli with frequencies as high as 5 to 7 kHz, but the chopper-type units often fail to synchronize their discharges at frequencies as low as 2 kHz. Although other morphological and physiological types of units are found in the cochlear nuclear complex (see, e.g., Cant and Morest, 1984; Young, 1984), only the statistical characteristics of the primary and chopper-type discharges of the cochlear nuclear complex units have been studied in any great detail.

The Superior Olivary Complex

The statistical properties of the discharges of only two cell groups of the superior olivary complex—the medial nucleus of the trapezoid body (MNTB) and the lateral superior olive (LSO)—have been studied in any detail. The MNTB neurons are responsive only to stimulation of the contralateral ear (Guinan et al., 1972). A large calycelike terminal envelopes the cell soma of the principal cell of the MNTB (Ramon y Cajal, 1909). The calyce terminal forms multiple synaptic contacts on the cell soma (Lenn and Reese, 1966) and is the ending of a thick myelinated axon that originates from a globular bushy cell of the contralateral pAVCN (Tolbert et al., 1982). Hence, this input to the MNTB principal cell is primarylike in temporal pattern. The discharges of most MNTB units are characterized by a prepotential that precedes each unit action potential by 0.5 msec (Li and Guinan, 1971). The calyce terminal is believed to generate the prepotential. The MNTB unit discharges with prepotentials have statistical properties similar to those of the auditory nerve fibers and the primarylike units of the AVCN (Guinan et al., 1972). LSO units respond to sound presented binaurally as well as monaurally; most of these units are sensitive to high-frequency stimuli (Tsuchitani and Boudreau, 1966).

The principle LSO neuron is spindle-shaped, with two polar dendrites forming disk-shaped fields that extend parallel to the sagittal plane (Ramon y Cajal, 1909; Scheibel and Scheibel, 1974). The neuron makes numerous synaptic contacts with small terminal boutons distributed over the surface of its dendrites and cell soma (Stotler, 1953; Cant, 1984). Neurons in the LSO do not generate discharges characterized by prepotentials (personal observations). The input from the ipsilateral ear is excitatory (Tsuchitani and Boudreau, 1966) and is believed to be primarylike in temporal pattern, originating from the spherical bushy cells of the aAVCN (Warr, 1966, 1982). The input from the contralateral ear is inhibitory (Tsuchitani and Boudreau, 1966) and is also believed to be primarylike in temporal pattern, originating from the principal cells of the MNTB (Glendenning et al., 1985; Spangler et al., 1985). The discharges of LSO units to stimulation of the ipsilateral ear are chopper-like, like the multipolar cells of the AVCN and PVCN (Boudreau and Tsuchitani, 1970). If a high-frequency tone is presented for an extended time, stationary portions of the response of LSO units can be isolated (Tsuchitani, 1982; Johnson et al., 1986). These discharges tend to be regular, having interspike interval CVs of less than about 0.3 to 0.5. The LSO unit discharges to ipsilateral stimulation also exhibit a negative serial dependence between successive interspike intervals (Tsuchitani, 1982). Although stimulation of the contralateral ear results in a decreased rate of discharge elicited by ipsilateral stimulation, negative serial dependence is still present in the resulting discharges (Tsuchitani, 1988b).

Other Auditory Brainstem Structures

The statistical properties of other auditory brainstem neuron discharges have not been studied in any great detail, in some cases because the discharges exhibited nonstationarities that precluded statistical analyses.

STATISTICAL MODELS

Two general approaches have been taken in the statistical modeling of auditory neural representation of acoustic information: the physiological-based *threshold* models and the more abstract *point process* models.

Threshold Models

The threshold models assume that a random component in the spike production process is responsible for the statistical properties of the generated spike-trains. Basic to these models are the assumptions that the neuron membrane is electrochemically sensitive and is permeable to ions. According to these models, selective membrane permeability to ions maintains a transmembrane potential in the unstimulated neuron called the *resting potential*. The application of a stimulus— physical or chemical (e.g., a neurotransmitter)—affects the neuron's membrane permeability to ions and results in a shift in the membrane potential. The stimulus may induce hyperpolarization or depolarization of the membrane potential, depending on the nature of the stimulus and the properties of the receiving neuron.

This postsynaptic potential change is considered excitatory with depolarization and inhibitory with hyperpolarization. Although the amplitude of the postsynaptic potential is related to the magnitude of the stimulus, the duration of the postsynaptic potential is related to both the duration of the stimulus and the electrochemical properties of the neural membrane. When the stimulus is terminated, the membrane potential decays toward the resting potential. A passive electronic spread of the postsynaptic potential occurs throughout the neuron, and the potential degrades as it spreads from its site of origin; its amplitude decreases and its rise-time increases with distance traveled. When individual postsynaptic potentials occur sufficiently close in time, they summate or integrate. When the membrane potential reaches a critical level—the threshold membrane potential—by means of the superposition of an adequate number of excitatory postsynaptic potentials at the site of the spike generator, the neuron discharges and produces a spike potential. Following the spike discharge, the neuron becomes unresponsive to further input for a short, fixed period of time—the absolute refractory period. The absolute refractory period may be followed by a period of increased threshold—the relative refractory period. Thus, the maximum rate or shortest interspike interval at which a unit could discharge would be determined by the rate at which the unit "recovers" from a spike discharge and by stimulus intensity. At high stimulus levels the neuron can be driven to discharge during its relative refractory period, but not during its absolute refractory period.

Moore et al. (1966) subdivided the threshold models into two categories: those in which the random component of the model is *intrinsic* to the neuron (e.g., in the random fluctuations of the membrane potential; see (Viernstein and Grossman, 1961; Weiss, 1964, 1966; Goldberg et al., 1964; Geisler and Goldberg, 1966; Geisler, 1968), and those in which it is *extrinsic* to the neuron (e.g., in the random nature of the synaptic *input* to the neuron, see Gerstein and Mandelbrot, 1964; Fetz and Gerstein, 1963; Molnar and Pfeiffer, 1968; Geisler, 1981).

Intrinsic Noise Models

In the first class of threshold models, the randomness of the discharges is determined by intrinsic neuron noise, in the form of probabilistically distributed fluctuations of the membrane potential. In most cases, the random component was assumed to have a Gaussian distribution with zero mean and standard deviation σ_N. The neuron discharged when the value of the membrane potential, determined by the Gaussian noise and the stimulus intensity, exceeded the membrane threshold. Spontaneous discharges were the result of the random fluctuation of the membrane potential to above-threshold levels in the absence of an experimentally controlled stimulus. The stimulus was described by a driving function that was deterministic and generated an output related to the waveform of the acoustic stimulus (Weiss, 1964, 1966; Geisler and Goldberg, 1966; Geisler, 1968). In these models, the time course of the relative refractory period determined the recovery of the threshold or membrane potential over time to the level preceding the spike discharge. It was described by an exponential function with a recovery time constant τ_R. The bandwidth $(1/\tau_N)$ and standard deviation of the Gaussian noise, the duration of the absolute refractory period, and the recovery time constant determined the statistical properties of the model unit discharges. The basic models are so complex that analytical solutions are extremely difficult to obtain; in the statistical literature, such models are termed *level-crossing* models, and only the simplest have analytic solutions. Simulation techniques are thus required to determine whether the data can be reproduced by the proposed models. If the membrane potential was not reset to or below the resting level following a spike discharge, a serial dependency between interevent intervals occurred when the bandwidth of the Gaussian noise was too narrow—when τ_N was long compared to τ_R (Weiss, 1966). Weiss (1966) reports that setting the bandwidth of the noise between 2 and 5 kHz and the time constant of threshold recovery between 0.3 and 1 msec resulted in simulations with irregular discharge patterns that generated interval histograms similar in shape to those of auditory nerve fiber spontaneous discharges. The spontaneous discharge rate was determined by the ratio of the standard deviation of the noise to the threshold potential; the greater the ratio, the higher the probability that the threshold would be exceeded per unit of time. The ability of this type of model to simulate faithfully the auditory nerve fiber discharges to click and low-frequency sinusoidal stimuli depended on the selection of the input-driving function used to describe the transducer function of the receptor organ (Weiss, 1966; Gray, 1966).

In other intrinsic noise threshold models, the membrane potential was reset to or below resting levels following a spike discharge; the model developed by Goldberg and co-workers for ventral cochlear nucleus and superior olivary complex units is one example (Goldberg et al., 1964; Goldberg and Greenwood, 1966; Geisler and Goldberg, 1966). In this model, the regularity of the discharges was related to the magnitudes of the standard deviation of the Gaussian noise and the recovery time constant (Geisler and Goldberg, 1966). The greater the standard deviation of the noise or the shorter the recovery time constant, the more irregular the discharge pattern became. Consequently, when the standard deviation of the Gaussian noise was held constant, short recovery times produced irregular discharge patterns—asymmetrically distributed interval histograms with CV > 0.8— and longer recovery times produced more regular discharge patterns—symmetrically distributed interval histograms with CV < 0.2 (Goldberg et al., 1964). Computer simulations with this model produced events that generated interval histograms resembling those generated by the spontaneous and maintained discharges of high CF primarylike and regularly discharging units of the ventral cochlear nucleus and superior olivary complex (Geisler and Goldberg, 1966). The ability of this model to simulate the discharges to transient acoustic stimuli was not tested. In a study simulating auditory nerve fiber discharges to low-frequency sinusoids with an adaptation of the Geisler and Goldberg model, Geisler (1968) reported that the selection of the driving function describing the input to the model was critical in determining the ability of the simulation to match fiber discharges.

In most of these threshold models, there was assumed to be no statistical relationship between the interspike intervals in the spike trains being modeled. To account for observed serial dependencies, Geisler and Goldberg (1966) hypothesized that the membrane potential was hyperpolarized immediately following spike generation to a level beyond that before spike generation, and it decayed exponentially from hyperpolarization to a level determined by the stimulus intensity and Gaussian noise. Their model neuron also had a "memory": if the neuron discharged after a "short" interspike interval, what remained of the hyperpolarization produced by the first spike was added to that produced by the second spike. Consequently, it would take longer to recover from this increased hyperpolarization, and the next spike would occur after a long interspike interval. Thus, a discharge with a short interspike interval would be followed by a discharge with a longer interspike interval. Conversely, if a unit discharged after a long interspike interval, the hyperpolarization from the first spike would have dissipated and the unit would be required to recover only from the hyperpolarization of the second spike. Hence, a discharge with a long interspike interval would be followed by a discharge with a shorter interspike interval. It is interesting to note that intracellular recordings from cochlear nuclear units producing regular chopper discharge patterns are depolarized for the duration of a tone burst and are not hyperpolarized following a spike discharge (Romand, 1979; Rhode et al., 1983; Rouiller and Ryugo, 1984). Unfortunately, the serial dependence characteristics of these discharge patterns have not been reported.

Extrinsic Noise Models

Although random fluctuations in the membrane potential may have some effect on the production of auditory neuron discharges, studies of the activity of denervated auditory nerve fibers and ventral cochlear nuclear units have indicated that the spontaneous discharges of these neurons depend on the presence of an input source. For example, auditory nerve fibers that produce no spontaneous discharge and no response to acoustic stimuli following ototoxic drug or noise exposure can discharge to electrical stimulation of the cochlea (Kiang et al., 1970; Liberman and Kiang, 1978). In addition, units within the ventral cochlear nuclei cease to produce spontaneous discharges following section of the auditory nerve (Koerber et al., 1966). On the basis of these and other findings, the occurrence and statistical properties of the spontaneous discharges of auditory nerve fibers and ventral cochlear units may depend on their synaptic inputs (Furukawa et al., 1978; Geisler, 1981; Molnar and Pfeiffer, 1968). Furthermore, the processes that produce spontaneous discharges may well be identical to those that produce discharges to acoustic stimuli. Thus, the probabilistic properties of the acoustically driven discharges can also be determined by the synaptic input. In the extrinsic noise threshold models, the temporal properties of the neuron's inputs determine the statistical properties of its discharges. A number of mutually independent, randomly timed inputs to a neuron (sites at which synaptic vesicles release quantized packets of neurotransmitter) are hypothesized, which may yield postsynaptic potentials differing in amplitude and waveform (Gerstein and Mandelbrot, 1964; Fetz and Gerstein, 1963). Spontaneous discharges were considered to result from the spontaneous release of neurotransmitter at the synapses provided by the input receptors/neurons. The effect of normal acoustic stimulation was the increase in the rate of release of the neurotransmitter. The inputs, which might have to occur within a set period of time to summate, combined to drive the membrane potential toward or away from the threshold potential. When the membrane potential reached the threshold level, the unit discharged and the membrane potential was reset to resting level.

In early models (Gerstein and Mandelbrot, 1964; Fetz and Gerstein, 1963), the membrane threshold remained constant (i.e., no refractory period was specified) and the membrane potential was reset to resting levels immediately following a spike discharge. This type of model was first applied to cochlear nuclear unit discharges. In a random walk model, Gerstein and Mandelbrot (1964) assumed that a large number of excitatory and inhibitory inputs occurred at regular time intervals. Each input produced a discrete, small-step change in the membrane potential that did not decay with time. An absolute refractory period results by virtue of the necessity for the membrane potential to build up to threshold by integrating a number of small input potentials. The movement of the membrane potential toward the threshold level was viewed as analogous to the "random walk" or diffusion process. This model, borrowed from physics, describes the probability density function of interspike intervals in terms of the distribution of first passage times of a Gaussian-distributed diffusion current toward an absorbing barrier. To resolve some differences between the model-generated and the observed data, the arrival rate of the excitatory inputs was assumed to be greater than that of the inhibitory inputs. The resulting drift velocity of the diffusion process describes the difference between the probability of the membrane potential moving toward threshold and that of the membrane potential moving away from threshold. The model was described by a simple analytical expression with two parameters— the number of increments between the resting potential and the threshold and the ratio of the excitatory to inhibitory rates—and one variable—the membrane potential. Unfortunately, the problem of applying the diffusion equation to discharges elicited by time-varying acoustic stimuli, especially transient stimuli, is not easily solved. Computer simulations using Monte Carlo methods are required, and they generate threshold crossings with statistical properties that match those of cochlear nuclear unit discharges to click stimuli. A similar model was proposed by Nilsson (1975) with the input specified by a nonlinear differential equation. The random-walk model has been criticized because it does not incorporate certain neurophysiological observations (e.g., the irregularity of auditory nerve fiber discharges that serve as input to ventral cochlear nuclear neurons, the decay of postsynaptic potentials, and relative refractory effects; (see MacGregor and Lewis, 1977). In the Fetz and Gerstein model (1963), each input had independent Poisson arrival rates and generated postsynaptic potentials of equal amplitudes that were subject to exponential decay. Because this modification of the random-walk model proved to be mathematically unwieldy, a resistor-capacitor network simulation of the postsynaptic mechanisms with discrete input processes was employed to study the interval statistics of the events generated by the model. The spontaneous discharges of many cochlear nuclear units were then simulated by varying the values of the network equivalents to the two model parameters. Simulation of the discharges to time-varying acoustic stimuli was not attempted with this version of the model.

In the model proposed by Molnar and Pfeiffer (1968), the input to a ventral cochlear nuclear neuron was described as the superposition of discharges from a number of auditory nerve fibers. The arrival time at each terminal was specified by a renewal stochastic process. Assuming the process is renewal, it is then characterized entirely by the probability density function of the interspike intervals. The criteria for the analytic form of the interspike interval's probability density function were a mathematical convenience as well as a reasonable fit to the interval histograms of auditory nerve fiber spontaneous discharges. The discharges in the different auditory nerve fibers were also assumed to be statistically independent of one another. Simultaneous recordings from pairs of auditory nerve fibers with similar CFs indicate that the discharges of different auditory nerve fibers can be described as statistically independent (Johnson and Kiang, 1976). The AVCN neurons producing spikes with prepotentials were assumed to produce a spike discharge each time a spike arrived at a large endbulb terminal of an auditory nerve fiber. This presumption of one-for-one mimicking of the input by the output was based partly on the observation that this type of ending contains numerous areas of synaptic specialization (Lenn and Reese, 1966), which presumably produce a synchronized release of neurotransmitter with the arrival of a spike potential in the terminal. Hence these anatomical constraints suggest that the primarylike units discharge to each of a small number of superpositioned inputs, that is, to a small number of auditory nerve fibers that are each represented by an independent renewal process.

One complication of the superposition model is that interevent intervals occur at time intervals much shorter (< 0.7 msec) than those observed in the discharge of auditory neurons. Another complication is that the model produces a sequence of events with interevent intervals that are not statistically independent: the superposition of two to five of the selected renewal processes results in a sequence of events that exhibit a first-order negative serial dependence. Although the mathematics of the superposition of a small number of renewal processes has been described (Cox, 1962), including the refractory properties of neurons complicates the modeling procedure sufficiently to require simulation techniques. The major effect of including the absolute and relative refractory periods in the simulations was to shift the interspike interval distribution away from the origin, thereby resulting in interval histograms similar to those of AVCN units producing primarylike discharge patterns and action potentials preceded by prepotentials. Linebarger and Johnson (1986) report that imposing a dead time (an absolute refractory period) in simulations of the superposition of two renewal processes removes much of the serial dependence present in simulations that have no imposed dead time.

Molnar and Pfeiffer (1968) attributed the PVCN extracellularly recorded action potentials unassociated with prepotentials to neurons that receive numerous small, distributed synaptic inputs. The activation of any one synaptic input was assumed to result in the generation of a small, subthreshold postsynaptic potential. These units are believed to correspond to the multipolar cells of the PVCN, which are covered by numerous small terminal bouton endings of auditory nerve fibers (Lorente de No, 1933; Feldman and Harrison, 1969; Osen, 1969, 1970). In the model, it was assumed that a number of auditory nerve fibers provided independent inputs to a PVCN unit through these small, distributed synapses. Since an individual auditory nerve fiber branches extensively to form a "nest" of terminal branches around an individual PVCN neuron (Lorente de No, 1933), this assumption does not appear to be quite correct. A PVCN neuron may receive input from a small number of parent auditory nerve fibers, and the arrival time of a spike at any one bouton would have a fixed-time relationship with those provided by the same parent axon at other boutons. In any case, the result of the superposition of numerous independent renewal processes is a point process that has statistics approaching those of a Poisson process (Cox and Smith, 1954). Based on the assumption that a Poisson process model describes the times at which input events occurred, a Markov process can be used to model the cumulative effects of the individual input events. Molnar and Pfeiffer also assumed that the threshold remained constant, that each small incremental change in the postsynaptic potential decayed exponentially, and that following a spike discharge the membrane potential returned immediately to resting level. The value of three parameters—the number of increments from resting to threshold level, the mean rate of the input Poisson process, and the decay rate of the postsynaptic potential—determined the value of the state variable of the Markov random process, considered to be analogous to the membrane potential. Whenever the membrane potential exceeded threshold, the unit was considered to have produced a spike discharge. As in the random-walk models, the refractory properties of the discharges were accounted for by the time required for small incremental changes in the membrane potential to build up to threshold level. Simulations of the Markov process model could generate interval histograms similar to three of the six types of unimodal interval histograms generated by the spontaneous and maintained discharges of cochlear nuclear units. When the threshold was low (3.2 to 4.5 increments) and the ratio of the mean rate of the Poisson process to the decay rate was low (2), the interval histograms resembled those of primarylike units. When the threshold was high (10.67 increments) and the ratio of the mean rate of the Poisson process to the decay rate was high (16), the interval histograms resembled those of chopper-type units. Serial dependence between successive interevent intervals was not mentioned. A more generalized model was suggested to describe all interval histogram forms, but was not developed because it would have been "difficult and expensive" to implement the solution of the resulting model. No attempt was made to model the cochlear nucleus discharges to time-varying acoustic stimuli.

Van Gisbergen et al. (1975) have concluded that this superposition model predicts the tightly time-locked first and second spike discharges of the tone burst–elicited responses of cochlear nuclear chopper units better than the recovery model of Geisler and Goldberg (1966) would. They argued that the low CVs (<0.3) of the first spike latency and the first interspike interval in the tone burst discharges were indicative of regularly discharging units whose discharges were governed by the temporal integration required for small input events to summate to threshold rather than by a long recovery process. The inability of chopper-type units to synchronize to sinusoids at frequencies above 2 kHz was also interpreted as evidence that the discharge statistics of these units were governed by the temporal integration of numerous small inputs. They postulated that certain chopper-type units could discharge at higher maximum rates than primarylike units because there was a "strong convergence of excitatory inputs" to these units. The Molnar–Pfeiffer superposition model has also been used to determine whether it could describe and predict the behavior of LSO units to binaural stimulation (Colburn and Moss, 1981). Stimulation of the contralateral (inhibitory) ear was considered to produce an input of polarity opposite to that produced by stimulation of the ipsilateral (excitatory) ear. Each input produced a step change in membrane potential that decayed exponentially to the membrane resting potential. The arrival times of the ipsilateral and contralateral inputs were assumed to be described by Poisson processes. The rate-level functions generated by the simulated responses to "binaural stimulation" were similar to those generated by LSO units under similar stimulus conditions. However, the interval histograms generated by the simulated discharges to binaural stimulation differed from that generated by LSO unit discharges to binaural stimulation: while the simulation interval histograms were positively skewed and unimodal, many of the LSO unit interval histograms are positively skewed but bimodal (Tsuchitani, 1988b).

The discharges of auditory nerve fibers have also been simulated by a threshold model with random input from auditory receptor cells (Geisler, 1981). Geisler's model assumed that an auditory receptor cell released quantal packets of neurotransmitter that produced postsynaptic potentials in the dendritic terminal of an auditory nerve fiber. The arrival time of each input, equivalent to the release of a quantum of neurotransmitter, was governed by Poisson statistics. Each quantum release was assumed to produce a unitary increase in the membrane potential, which had a waveform that was exponential with a time constant of 1 msec. When the rate of neurotransmitter release was sufficiently high, the unitary changes in membrane potential would summate linearly. When the membrane potential reached threshold level, the simulated auditory nerve fiber generated a spike potential and the membrane potential was reset to resting level following an absolute refractory period of 1 msec. No relative refractory period was specified. The simulation technique of Molnar and Pfeiffer (1968) was used to study the behavior of this model. With the proper selection of the values for the two model parameters—the number of increments from resting level to threshold and the rate of the Poisson process—the discharge rate and interval histograms of auditory nerve fiber spontaneous discharges could be well described. Low thresholds (< 1 quantum) were associated with high discharge rates and high thresholds (>1 quantum) with low discharge rates for both spontaneous and driven activity. Simulation of the maintained discharges to continuous tones indicated that the model should be expanded to include relative refractory effects.

Other extrinsic noise models attribute most of the statistical properties of auditory nerve fiber discharges to the processes involved in neurotransmitter release from the receptor cell (Schroeder and Hall, 1974; Oono and Sujaku, 1974; Geisler et al., 1979; Schwid and Geisler, 1982; Smith and Brachman, 1982). The input to the hair cell was described by a function that controlled the hair cell's permeability to the neurotransmitter or its receptor potential. The probability of neurotransmitter release determined the probability of a spike from an auditory nerve fiber. The depletion of neurotransmitter was assumed to be responsible for setting the maximum discharge rate, determining the discharge rate/stimulus level functions, and transiently decreasing the discharge rate with time following stimulus onset (i.e., adaptation). According to the Schroeder–Hall analytical model, the nerve fiber threshold remained constant, no refractory period was involved, and the release of a quantum of neurotransmitter from the transmitter pool of a single receptor cell could produce a spike discharge from the auditory nerve fiber. The probability of a nerve fiber discharge was a function of the amount of neurotransmitter available in the receptor pool and the permeability of the receptor cell membrane to the neurotransmitter. Three parameters were required: the average rate of neurotransmitter production, the rate at which the neurotransmitter was removed with no resulting discharge, and the rate at which the neurotransmitter was released with a resulting discharge when there was no input signal (i.e., the mean spontaneous discharge rate). The soft half-wave rectifier law was used to describe the permeability of the hair cell membrane to the neurotransmitter as a function of the spontaneous discharge rate and the stimulus-driving function. Because the amount of neurotransmitter available to the hair cell was determined by the rate of production and the amount released, the analytical model reproduced the sigmoid shape of auditory nerve fiber rate-level functions. Simulations with an equivalent RC circuit were also used to examine the temporal patterns of events elicited by acoustic stimuli at different intensities. The period histograms of events elicited with 1-kHz tones and the PST histograms of events generated by tone burst stimuli were similar, but not identical, to those measured from auditory nerve fiber discharges. Incorporation of nerve fiber refractory properties in computer simulations improved the match only slightly. Geisler et al. (1979) attempted to improve the match by limiting the rate at which the probability of discharge increased over time with stimulus level. These authors suggested that imposing an upper limit to the permeability function should improve the fit of the modified model to the tone-burst PST histograms, but would not produce a better match of the 1-kHz period histograms.

Schwid and Geisler (1982) amended the Schroeder–Hall hair cell model by adopting the Oono–Sujaku (1974) model of the hair cell. Instead of releasing a single quantum of neurotransmitter per unit of time, the Oono–Sujaku hair cell releases a variable amount of neurotransmitter per unit of time that is a fraction of the amount available. Because the release of any amount of neurotransmitter per unit of time results in a nerve fiber discharge, the amount released in excess of a single quantum is wasted, that is, similar in effect to the neurotransmitter depletion rate in the Schroeder–Hall model. The effect of depleting the stored neurotransmitter in this manner resulted in better simulations of low-frequency period histograms and tone-burst PST histograms than the amended Schroeder–Hall model produced (Geisler et al., 1979). Schwid and Geisler (1982) extended the hair cell model further, proposing a model with multiple pools of neurotransmitter, each with independent permeability functions limited in the amount of neurotransmitter released per unit of time to a fraction of the amount available. Each pool was also characterized by different thresholds for release and replacement of neurotransmitter—the higher the threshold for release, the lower the threshold for replacement. The maximum amount of neurotransmitter replaced per unit of time was arbitrarily limited. Computer simulations that included neuron refractory properties reportedly produced rate-level functions, low-frequency period histograms, tone-burst PST histograms, and "lifelike" interval histograms. A similar model with a more complex replacement mechanism was proposed by Smith and Brachman (1982).

Abstract Point Process Models

The abstract point process models generally have been developed to examine the information-processing aspects of neuronal discharges. These models attempt to describe the flow of stimulus information through the nervous system without ascribing model parameters to underlying chemical/physiological mechanisms. They seek to understand the fundamental relationships between the stimulus/signal input and neural discharge output.

Recovery Models

Modern developments in point process theory have provided a basis for modeling the discharges of auditory neurons elicited by time-varying acoustic stimuli. Recovery models are based on a renewal point process characterized by independent, identically distributed intervals. The *intensity* of this process is described to be the product of two functions: a stimulus drive function related to the input signal and the neuron recovery function that describes how the probability of a discharge varies with time following the preceding discharge. Because the underlying process is assumed to be a renewal point process, the recovery function is proportional to the hazard function, a quantity directly related to the interval histogram of a spike-train with a constant discharge rate (Gray, 1967; Johnson et al., 1986). This modeling approach has been used to determine the relative contributions of the auditory receptor organ (i.e., the stimulus-drive function) and the auditory nerve fibers (i.e., the unit recovery function) to the statistical properties of auditory nerve fiber discharges. A major goal of this type of modeling effort is the specification of the signal being processed as it is transformed at the different stages of the auditory system. Basically, the recovery models of auditory nerve fibers assume that the timing of the input to the auditory nerve fiber is probabilistic, that each input event is capable of discharging the nerve fiber, and that the nerve fiber refractory properties determine whether or not the fiber discharges to an input event. Consequently, the problem of determining the nerve fiber contribution to its response characteristics is reduced to describing the nerve fiber refractory process. By determining the effect of auditory nerve fiber refractory properties on the statistics of the nerve fiber discharges, one can eliminate refractory effects to reveal the contributions of the receptor organ—the receptor transducer function—to the nerve fiber discharges (Weiss, 1966; Gray, 1966, 1967) or, in the context of communications theory, to reveal the information or signal being conveyed by the discharges (Johnson and Swami, 1983). At higher levels of the auditory system, the specification and removal of the neuron refractory effects on the neuron discharges reveal the representation of the stimulus signal in the neural discharges (Johnson et al., 1986).

Gray (1967) was one of the earliest to use this approach. Based on statistical measures of auditory nerve fiber discharges (Kiang et al., 1965), he assumed that the point process underlying the discharges could be described by a renewal process. The probability of a discharge at a given point in time was proportional to the values of the stimulus-drive function and the unit recovery function at that point in time. The increase in the probability of a spike with time since the previous spike was considered to reflect the "recovery" of the nerve fiber from the immediately preceding discharge. These recovery functions are characterized by an initial period of zero probability (i.e., the dead time) that is 0.7 to 1 msec long, followed by a fairly rapid increase in discharge probability within the next 2 to 4 msec, gradually increasing to a plateau within an additional 15 to 20 msec.

According to Gray (1967), as the probability of a discharge reaches maximum value at an interspike interval of 25 msec, the nerve fibers are fully recovered from a discharge once this interval passes without an intervening discharge. Implicit in this argument is the assumption that the dead time represents the absolute refractory period of the nerve fiber, and the remainder of the recovery time represents the relative refractory period. Recovery effects were eliminated from acoustically driven discharges by selecting discharges that occurred with interspike intervals greater than 25 msec. Using this approach, Gray demonstrated that the period and PST histograms generated from the "recovered" discharges differed from those generated from all discharges, and presumably provide a more direct measure of the stimulus input signal generated by the auditory receptor organ.

More recent studies attempting to remove neuron recovery effects from auditory nerve fiber discharges have involved the selection of different renewal processes or of different methods for determining the neuron recovery function and for eliminating the recovery effects (Gaumond et al., 1982, 1983; Johnson and Swami, 1983; Miller, 1985; Jones et al., 1985; Teich and Khanna, 1985). For example, Johnson and Swami (1983) modeled auditory nerve fiber discharges by passing the output of a Poisson process through a system that had a recovery function with an absolute dead time (a delayed step function) or a relative dead time (a clipped ramp function), or both. When simulating responses to high-frequency tone bursts, the stimulus-drive function was specified as a step function with amplitude equal to mean discharge rate. A nonlinear transformation of a sinusoid represented the stimulus-drive function for the responses to low-frequency tonal stimuli. The simulated events generated interval, PST, and period histograms that were similar to those generated by auditory nerve fibers. The simulation PST histograms illustrated that the time course (transient, adaptive features) of tone burst–elicited discharges could result from the inclusion of either or both dead times in the recovery function. The period histograms of the simulated discharges to tones with frequency below 1 to 1.3 kHz were skewed, just as those of auditory nerve fiber discharges. However, simulations with compensation for recovery effects indicated that the skew observed in the period histograms to these low-frequency tones result primarily from nonlinear mechanisms in the auditory receptor organ.

This modeling approach has also been applied to the discharges of LSO units elicited by monaural stimulation (Johnson et al., 1986). The hazard functions of the sustained discharges to ipsilaterally presented tone bursts and continuous tones were used to derive estimates of the unit recovery function. Examination of the discharges using conditional statistics (the forward conditional mean and the conditional hazard function) indicated that the negative serial dependence observed in LSO unit discharges is first-order (i.e., between successive interspike intervals) and that the dependence was well described as a shift in the interspike interval distribution, the degree of shift depending on the duration of the previous interval. The discharges to ipsilateral tone-burst stimuli were modeled by passing the output of a Poisson process through a system characterized by a recovery function with an absolute dead time and a relative dead time estimated from the discharge hazard function and by a shifting function estimated from the measured conditional mean function. The parameters of the model were derived from the hazard and conditional mean functions generated by the later (>30 msec past the stimulus onset time), more stationary, sustained discharges in a tone burst response. The stimulus-drive function was a step function with duration equal to that of the tone burst used to generate the simulated response and with an amplitude equal to the mean discharge rate of the simulated response. The simulations reproduced (within 30 msec of the stimulus onset time) the transient chopper pattern in the initial portion of the tone burst response. Thus, contrary to the view of van Gisbergen and associates (1975), the tightly time-locked discharges of chopper-type responses could be described as the result of longer recovery (refractory) processes. For example, in these simulations of LSO unit tone burst discharges, long recovery times were associated with lower chopping rate, more regular peaks in the chopper PST histograms, and smaller-interval CV. Short recovery times (e.g., the auditory nerve fibers with absolute dead times of 0.7 to 1 msec) were associated with primarylike PST histograms. The negative serial dependence also had a role in maintaining the regularity of discharges, the net effect being a correction for extremes in interval times—long intervals were followed by short intervals, and vice versa.

This model was also used to examine a hypothesis first proposed by Goldberg et al., (1964) to account for the LSO unit discharges resulting from binaural stimulation (Tsuchitani, 1988a). According to their hypothesis, stimulation of the contralateral (inhibitory) ear produces a net reduction in the stimulus drive function of the ipsilateral (excitatory) ear. As a consequence, the statistical properties of the LSO unit discharges should not differ under monaural (i.e., ipsilateral) and binaural stimulus conditions—the statistics of a monaural response at a given discharge rate should be identical to those of a binaural response of similar discharge rate. However, because the hazard functions of the binaurally elicited, sustained discharges often differ from those of monaurally elicited discharges with similar mean discharge rate, it was concluded that the hypothesis that binaural stimulation results in the simple reduction of the ipsilateral stimulus-drive function was incorrect (Tsuchitani, 1988a). The observation that the LSO unit recovery functions differs under monaural and binaural stimulus conditions while estimates of the unit shifting functions are matched suggests that different mechanisms may underlie these two functions (Tsuchitani, 1988a).

Psychoacoustic Models

Statistical models of auditory neural discharges have been developed in an attempt to relate neurophysiological phenomena to behavioral/psychophysical measures of auditory perception. For example, the works of Siebert and others were concerned with using statistical models to examine the limits the probabilistic behavior of auditory nerve fiber discharges places on an "ideal central processor" involved in auditory perception (Siebert, 1965, 1968, 1970, 1973; Siebert and Gray, 1963; Goldstein, 1973; Colburn, 1973, 1977; Colburn and Latimer, 1978; Srulovicz and Goldstein, 1983; Miller et al., 1987). The neural noise evident in the stochastic behavior of auditory nerve fiber discharges was interpreted as limiting auditory discrimination performance, since acoustic information can reach the central nervous system only by way of the auditory nerve. Based on this view, the performance of an ideal central processor provided with an input generated by the model auditory neuron can be compared with that of human observers. If the ideal processor performs better than human observers, central auditory structures can be considered to place greater limitations on auditory discrimination performance than the "noisy" auditory nerve discharges. Thus, certain stimulus information available in the neural discharges would not be used fully by central neural structures. For example, an optimum sinusoidal frequency discrimination system operating with simulated auditory nerve fiber phase-locked discharges was determined to perform significantly better than the human observer in a frequency discrimination task (Siebert, 1970). Thus, the representations of tonal stimuli by synchronized discharges of auditory nerve fibers were not considered to be a critical factor in frequency discrimination performance. They could, however, play an important role in the processing of temporal acoustic information (Colburn and Latimer, 1978). An ideal processor performance inferior to that of a human observer would imply that the model is missing important features required to match the performance of the human observer. Matched performance is interpreted as indicating that the model specified the important features of neural responses that determine discrimination performance and that the peripheral auditory system sets the limits for the performance capabilities of the human observer.

The auditory nerve fiber discharges were modeled as renewal point processes. The probability of discharge at any given point in time was a continuous function of the magnitude of the stimulus-drive function—the receptor transducer function. An important feature of these types of models is that the ideal central processor used signal information carried by the entire population of active auditory nerve fibers. Hence, a major task in these modeling efforts is to describe the functional properties of the mechanical and transducer components of the peripheral auditory system that determine which auditory nerve fibers are activated and provide the stimulus-drive function for each of them. In most cases, the nonlinear behavior of the peripheral auditory system (Kim and Molnar, 1975) was ignored to simplify the modeling of the stimulus-drive functions. For example, the peripheral auditory system could be assumed to operate like a bank of linear, minimum-phase filters with frequency response characteristics described by the tuning characteristics of auditory nerve fibers. Using such a model of the peripheral auditory system, the latency of response, the discharge pattern to click stimuli and the preferred phase of firing to low-frequency sinusoids were predicted on the basis of linear system theory (Goldstein et al., 1971). However, none of the models using linear stimulus drive functions could simulate the responses of auditory nerve fibers to high-intensity or complex stimuli (i.e., two tones or speech).

CONCLUSIONS

Information processing by the auditory nervous system is modeled at the subcellular level, invoking hypothetical electrochemical mechanisms, and at more abstract levels, involving the mathematics of statistical communication theory. Many of the early models were developed to provide mathematical analogs of the processes involved in generating spike discharges. Analytical models are generally preferred because statistical methods can be developed for objectively validating these types of models, predicting the behavior of the model under different stimulus conditions, and determining other properties of the model (e.g., Fienberg, 1974; Yang and Chen, 1978; de Kwaadsteniet, 1982). The threshold models include a minimum set of parameters selected to represent the essential physiological processes involved in spike generation. Because little is known about the synaptic processes involved in auditory neuron spike generation, the selection of these model parameters is necessarily arbitrary and often based on observations of nonauditory neurons. Most minimum-parameter models reproduce only certain aspects of the modeled spontaneous discharges and must be expanded to include other parameters (e.g., the absolute and relative refractory periods) to fit the data better. As more parameters are added to these models, however, they become so complex as to be intractable for analytic techniques. Consequently, digital computer simulations or equivalent electrical networks of the model are necessary to study their operation. Abstract models suffer from similar problems: although the basic model (e.g., the Poisson process) is fairly simple and amenable to analysis, as additional features are added, it becomes increasingly complex and more difficult to use. The modeling of stimulus-driven spike-trains produces even more complex solutions. Often, simulations of acoustically driven neural discharges indicate that additional parameters and/or complex driving functions are required to match the spike-train data (e.g., Geisler et al., 1979). Because these expanded forms of the models are also unrealizable, simulation techniques are required to validate the accuracy of the model's ability to replicate the acoustically elicited discharges. In most studies, qualitative judgments as to how well the simulations fit the data are made. In some cases, important features of the discharges that could invalidate the model under study (e.g., discharges with negative serial dependence modeled by a renewal process) are ignored.

No single model has successfully simulated the spontaneous and acoustically driven discharges of auditory nerve fibers and lower auditory brainstem units. They have been most useful in developing a coherent view of the physical, chemical, and anatomical features of auditory neurons and of the statistical characteristics of their discharges that are important to consider in modeling information processing by auditory neurons. They have also demonstrated that the task of modeling need not be limited to producing an exact replica of the thing being modeled. The threshold models have evolved over the years, incorporating new findings and providing new insights into other properties or constraints that should be considered in future models. The abstract models have provided a means of detecting and specifying the signal encoding features of the neural discharges. Both types of modeling efforts contribute to our understanding of the information-processing function of the auditory nervous system and offer a framework for developing and testing hypotheses in neurophysiological and anatomical studies of the auditory system.

ACKNOWLEDGMENTS

This work was supported in part by Grants NS20994 and NS20964 from the National Institutes of Health.

REFERENCES

Anderson, D. J. (1973). Quantitative model for the effects of stimulus frequency upon synchronization of auditory nerve discharges. *J. Acoust. Soc. Am.* 54:361–364.

Anderson, D. J., Rose, J. E., Hind, J. E., and Brugge, J. F. (1971). Temporal position of discharges in single auditory nerve fibers within the cycle of a sine-wave stimulus: frequency and intensity effects. *J. Acoust. Soc. Am.* 49:1131–1139.

Boudreau, J. C., and Tsuchitani, C. (1970). Cat superior olive S-segment cell discharge to tonal stimuli. In: W. D. Neff (ed.). *Contributions to Sensory Physiology*, pp. 143–213. New York: Academic Press.

Bourk, T. R. (1976). Electrical responses of neural units in the anteroventral cochlear nucleus of the cat. Ph.D. Dissertation, Department of Electrical Engineering, M.I.T., Cambridge.

Brawer, J. R., and Morest, D. K. (1975). Relations between auditory nerve endings and cell types in the cat's anteroventral cochlear nucleus seen with the Golgi method and Nomarksi optics. *J. Comp. Neurol.* 160:491–506.

Cant, N. B. (1984). The fine structure of the lateral superior olivary nucleus of the cat. *J. Comp. Neurol.* 227:63–77.

Cant, N. B., and Morest, D. K. (1979). The bushy cells in the anteroventral cochlear nucleus of the cat. A study with the electron microscope. *Neurosci.* 4:1925–1945.

Cant, N. B., and Morest, D. K. (1984). The structural basis for stimulus coding in the cochlear nucleus of the cat. In: C. I. Berlin (ed.). *Hearing Science: Recent Advances*, pp. 371–421. San Diego: College Hill Press.

Colburn, H. S. (1973). Theory of binaural interaction based on auditory-nerve data. I. General strategy and preliminary results on interaural discrimination. *J. Acoust. Soc. Am.* 54:1458–1470.

Colburn, H. S. (1977). Theory of binaural interaction based on auditory-nerve data. II. Detection of tones in noise. *J. Acoust. Soc. Am.* 61:525–533.

Colburn, H. S., and Latimer, J. S. (1978). Theory of binaural interaction based on auditory-nerve data. III. Joint dependence on interaural time and amplitude differences of discrimination and detection. *J. Acoust. Soc. Am.* 64:95–106.

Colburn, H. S., and Moss, P. J. (1981). Binaural interaction models and mechanisms. In: J. Syka and L. Aitkin (eds.). *Neuronal Mechanisms of Hearing*, pp. 283–288. New York: Plenum Press.

Cox, D. R. (1962). *Renewal Theory*. London: Methuen.

Cox, D.R., and Smith, W. L. (1954). On the superposition of renewal processes. *Biometrika* 41:91–99.

Feldman, M. L., and Harrison, J. M. (1969). Projections of acoustic nerve to ventral cochlear nucleus of rat. A Golgi study. *J. Comp. Neurol.* 137:267–295.

Fetz, E. E., and Gerstein, G. L. (1963). An RC model for spontaneous activity of single neuron. *Q. Prog. Rept., M.I.T. Res. Lab. Elect.* 71:249–257.

Fienberg, S. E. (1974). Stochastic models for single neuron firing trains: a survey. *Biometrics* 30:399–427.

Furukawa, T., Hayashida, Y., and Matsuura, S. (1978) Quantal analysis of the size of excitatory post-synaptic potentials at synapses between hair cells and afferent nerve fibers in goldfish. *J. Physiol.* 276:211–226.

Gaumond, R. P., Kim, D. O., and Molnar, C. E. (1983). Response of cochlear nerve fibers to brief acoustic stimuli: role of discharge-history effects. *J. Acoust. Soc. Am.* 74:1392–1398.

Gaumond, R. P., Molnar, C. E., and Kim, D. O. (1982). Stimulus and recovery dependence of cat cochlear nerve fiber spike discharge probability. *J. Neurophysiol.* 48:856–873.

Geisler, C. D. (1968). A model of peripheral auditory system responding to low-frequency tones. *Biophys. J.* 8:1–15.

Geisler, C. D. (1981). A model for discharge patterns of primary auditory-nerve fibers. *Brain Res.* 212:198–201.

Geisler, C. D., and Goldberg, J. M. (1966). A stochastic model of the repetitive activity of neurons. *Biophys. J.* 6:53–69.

Geisler, C. D., Le, S., and Schwid, H. (1979). Further studies on the Schroeder-Hall hair cell model. *J. Acoust. Soc. Am.* 65:985–990.

Gerstein, G. L., and Mandelbrot, B. (1964). Random walk models for the spike activity of a single neuron. *Biophys. J.* 4:41–68.

van Gisbergen, J. A. M., Grashuis, J. L., Johannesma, P. I. M., and Vendrik, A. J. H. (1975). Statistical analysis and interpretation of the intitial response of cochlear nucleus neurons to tone bursts. *Exp. Brain Res.* 23:407–423.

Glaser, E. M., and Ruchkin, D. S. (1976). *Principles of Neurobiological Signal Analysis.* New York: Academic Press.

Glendenning, K. K., Hutson, K. A., Nudo, R. J., and Masterton, R. B. (1985). Acoustic chiasm. II: Anatomical basis of binaurality in lateral superior olive of cat. *J. Comp. Neurol.* 232:261–285.

Goldberg, J. M., Adrian, H. O., and Smith, F. D. (1964). Response of neurons of the superior olivary complex of the cat to acoustic stimuli of long duration. *J. Neurophysiol.* 27:706–749.

Goldberg, J. M., and Brownell, W. E. (1973). Discharge characteristics of neurons in anteroventral and dorsal cochlear nuclei of cat. *Brain Res.* 64:35–54.

Goldberg, J. M., and Greenwood, D. D. (1966). Response of neurons of the dorsal and posteroventral cochlear nuclei of the cat to acoustic stimuli of long duration. *J. Neurophysiol.* 29:72–93.

Goldstein, J. L. (1973). An optimum processor theory for the central formation of the pitch of complex tones. *J. Acoust. Soc. Am.* 54:1496–1516.

Goldstein, J. L., Baer, T., and Kiang, N. Y. S. (1971). A theoretical treatment of latency, group-delay and tuning characteristics for auditory nerve responses to clicks and tones. In: M. B. Sachs (ed.). *Physiology of the Auditory System* pp. 133–141. Baltimore: National Ed. Consultants.

Gray, P. R. (1966). A statistical analysis of electrophysiological data from auditory nerve fibers in cat. Tech. Rept. 451, Res. Lab. Elec., M.I.T., Cambridge.

Gray, P. R. (1967). Conditional probability analyses of the spike activity of single neurons. *Biophys. J.* 7:759–777.

Greenwood, J. A., and Durand, D. (1955). The distribution of length and components of the sum of n random unit vectors. *Ann. Math. Sci.* 26:233–246.

Guinan, J. J., Norris, B. E., and Guinan, S. S. (1972). Single auditory units in the superior olivary complex. II. Locations of unit categories and tonotopic organization. *Int. J. Neurosci.* 4:147–166.

Johnson, D. H. (1978). The relationship of PST and interval histograms to the timing characteristics of spike trains. *Biophys. J.* 22:413–430.

Johnson, D. H. (1980). The relationship between spike rate and synchrony in responses of auditory-nerve fibers to single tones. *J. Acoust. Soc. Am.* 68:1115–1122.

Johnson, D. H., and Kiang, N. Y. S. (1976). Analysis of discharges recorded simultneously from pairs of auditory-nerve fibers. *Biophys. J.* 16:719–734.

Johnson, D. H., and Kumar, A. (1985). Analysis of the stationarity of models of auditory-nerve fiber discharge patterns. *J. Acoust. Soc. Am.* 77:S93.

Johnson, D. H., and Swami, A. (1983). The transmission of signals by auditory-nerve fiber discharge patterns. *J. Acoust. Soc. Am.* 74:493–501.

Johnson, D. H., Tsuchitani, C., Linebarger, D. A., and Johnson, M. J. (1986). Application of a point process model to responses of cat lateral superior olive units to ipsilateral tones. *Hearing Res.* 21:135–159.

Jones, K., Tubis, A., and Burns, E. M. (1985). On the extraction of the signal-excitation function from a non-Poisson cochlear neural spike train. *J. Acoust. Soc. Am.* 78:90–94.

Kiang, N. Y. S., Moxon, E. C., and Levine, R. A. (1970). Auditory nerve activity in cats with normal and abnormal cochleas. In: G. E. W. Wolstenholme and J. Knight (eds.). *Sensorineural Hearing Loss*, pp. 241–273. London: Churchill.

Kiang, N. Y. S., Watanabe, T., Thomas, E. C., and Clark, L. F. (1965). *Discharge Patterns of Single Fibers in the Cat's Auditory Nerve.* M.I.T. Res. Mono. 35. Cambridge, M.I.T. Press.

Kim, D. O., and Molnar, C. E. (1975). Cochlear mechanics: measurements and models. In: D. B. Towers (ed.). *The Nervous Sytem. Vol. 3: Human Communication and Its Discorders,* pp. 57–68. New York: Raven Press.

Koerber, K. C., Pfeiffer, R. R., Warr, W. B., and Kiang, N. Y. S. (1966). Spontaneous spike discharges from single units in the cochlear nucleus after destruction of the cochlea. *Exp. Neurol.* 16:119–130.

de Kwaadsteniet, J. W. (1982). Statistical analysis and stochastic modeling of neuronal spike-train activity. *Math. Biosci.* 60:17–71.

Lenn, N. J., and Reese, T. S. (1966). The fine structure of nerve endings in the nucleus of the trapezoid body and the ventral cochlear nucleus. *Am. J. Anat.* 118:375–389.

Li, R. Y.-S., and Guinan, J. J. (1971). Antidromic and orthodromic stimulation of neurons receiving calyces of Held. *Q. Prog. Rept., M.I.T. Res. Lab. Elect.* 100:227–234.

Liberman, M. C., and Kiang, N. Y. S. (1978). Acoustic trauma in cats: cochlear pathology and auditory-nerve activity. *Acta Otolaryngol. Suppl.* 358:1–63.

Linebarger, D. A., and Johnson, D. H. (1986). Superposition models of the discharge patterns of units in the lower auditory system. *Hearing Res.* 23:185–198.

Lorente de No, R. (1933). Anatomy of the eighth nerve. The central projection of the nerve endings of the internal ear. *Laryngoscope* 43:1–38.

MacGregor, R. J., and Lewis, E. R. (1977). *Neural Modeling.* New York: Plenum Press.

Miller, M. I. (1985). Algorithms for removing recovery-related distortion from auditory-nerve discharge patterns. *J. Acoust. Soc. Am.* 77:1452–1464.

Miller, M. I., Barta, P. E., and Sachs, M. B. (1987). Strategies for the representation of a tone in background noise in the temporal aspects of the discharge patterns of auditory-nerve fibers. *J. Acoust. Soc. Am.* 81:665–678.

Molnar, C. E., and Pfeiffer, R. R. (1968). Interpretation of spontaneous spike discharge patterns of neurons in the cochlear nucleus. *Proc. I.E.E.E.* 56:993–1004.

Moore, G. P., Perkel, D. H., and Segundo, J. P. (1966). Statistical analysis and functional interpretation of neuronal spike data. *Annu. Rev. Physiol.* 28:493–522.

Nilsson, H. G. (1975). Model of discharge patterns of units in the cochlear nucleus in response to steady state and time-varying sounds. *Biol. Cybern.* 20:113–119.

Oono, Y., and Sujaku, Y. (1974). A probablistic model for discharge patterns of auditory nerve fibers. *Sys. Comp. Controls* 5:35–44.

Osen, K. K. (1969). Cytoarchitecture of the cochlear nuclei in the cat. *J. Comp. Neurol.* 136:453–484.

Osen, K. K. (1970). Course and termination of the primary afferents in the cochlear nuclei of the cat. An experimental anatomical study. *Arch. Ital. Biol.* 108:21–51.

Perkel, D. H., and Bullock, T. H. (1968). Neural coding. A report based on a NRP work session. *Neurosci. Res. Prog. Bull.* 6:221–348.

Perkel, D. H., Gerstein, G. L., and Moore, G. P. (1967). Neuronal spike trains and stochastic point processes. I. The single spike train. *Biophys. J.* 7:391–418.

Pfeiffer, R. R. (1966a). Anteroventral cochlear nucleus: wave forms of extracellularly recorded spike potentials. *Science* 134:667–668.

Pfeiffer, R. R. (1966b). Classification of response patterns of spike discharges for units in the cochlear nucleus: tone-burst stimulation *Exp. Brain Res.* 1:220–235.

Pfeiffer, R. R., and Kiang, N. Y. S. (1965). Spike discharge patterns of spontaneous and continuously stimulated activity in the cochlear nucleus of anesthetized cats. *Biophys. J.* 5:301–316.

Pfeiffer, R. R., and Molnar, C. E. (1970). Cochlear nerve fiber discharge patterns: relationship to cochlear microphonic. *Science* 167:1614–1616.

Ramon y Cajal, S. (1909). *Histologie du système nerveux de l'homme et des vertébrés.* Paris: Maloine.

Rhode, W. S., Oertel, D., and Smith, P. H. (1983). Physiological response properties of cells labeled intracellularly with horseradish peroxidase in the cat ventral cochlear nucleus. *J. Comp. Neurol.* 213:448–463.

Rhode, W. S., and Smith, P. H. (1986). Encoding time and intensity in the ventral cochlear nucleus of the cat. *J. Neurophysiol.* 56:261–286.

Rodieck, R. W., Kiang, N. Y. S., and Gerstein, G. L. (1962). Some quantitative methods for the study of spontaneous activity of single neurons. *Biophys. J.* 2:351–368.

Romand, R. (1979). Intracellular recording of 'chopper responses' in the cochlear nucleus of the cat. *Hearing Res.* 1:95–99.

Rose, J. E., Brugge, J. L., Anderson, D. J., and Hind, J. E. (1967). Phase-locked response to low-frequency tones in single auditory-nerve fibers of the squirrel monkey. *J. Neurophysiol.* 30:769–793.

Rouiller, E. M., and Ryugo, D. K. (1984). Intracellular marking of physiologically characterized cells in the ventral cochlear nucleus of the cat. *J. Comp. Neurol.* 225:167–186.

Sachs, M. B., and Abbas, P. J. (1974). Rate versus level functions for auditory-nerve fibers in cats: tone-burst stimulation. *J. Acoust. Soc. Am.* 56:1835–1847.

Scheibel, M. E., and Scheibel, A. B. (1974). Neuropil organization in the superior olive of the cat. *Exp. Neurol.* 43:339–348.

Schroeder, M. R., and Hall, J. L. (1974). A model for mechanical to neural transduction in the auditory receptor. J. Acoust. Soc. Am. 55:1055–1060.

Schwid, H. A., and Geisler, C. D. (1982). Multiple reservoir model of neurotransmitter release by a cochlear inner hair cell. *J. Acoust. Soc. Am.* 72:1435–1440.

Siebert, W. M. (1965). Some implications of the stochastic behavior of primary auditory neurons. *Kybernetik* 2:206–215.

Siebert, W. M. (1968). Stimulus transformation in the peripheral auditory system. In: P. A. Kohlers and M. Eden (eds.). *Recognizing Patterns*, pp. 104–133. Cambridge: M.I.T. Press.

Siebert, W. M. (1970). Frequency discrimination in the auditory system: place or periodicity mechanisms? *Proc. I.E.E.E.* 58:723–730.

Siebert, W. M. (1973). Hearing and the ear. In: *Engineering Principles in Physiology, Vol. I.*, pp. 139–183. New York: Academic Press.

Siebert, W. M., and Gray, P. R. (1963). Random process models for the firing pattern of single auditory neurons. *Q. Prog. Rept., Res. Lab. Elect., M.I.T.* 71:241–245.

Smith, R. L., and Brachman, M. L. (1982). Adaptation in auditory-nerve fibers: a revised model. *Biol. Cybern.* 44:107–120.

Smith, R. L., and Zwislocki, J. J. (1975). Short-term adaptation and incremental responses in single auditory-nerve fibers. *Biol. Cybern.* 17:169–182.

Spangler, K. M., Warr, W. B., and Henkel, C. K. (1985). The projections of principal cells of the medial nucleus of the trapezoid body in the cat. *J. Comp. Neurol.* 238:249–262.

Srulovicz, P., and Goldstein, J. L. (1983). A central spectrum model: a synthesis of auditory-nerve timing and place cues in monaural communication of frequency spectrum. *J. Acoust. Soc. Am.* 73:1266–1276.

Stotler, W. A. (1953). An experimental study of the cells and connections of the superior olivary complex of the cat. *J. Comp. Neurol.* 98:401–432.

Tasaki, I. (1954). Nerve impulses in individual auditory nerve fibers of guinea pig. *J. Neurophysiol.* 17:97–122.

Teich, M. C., and Khanna, S. W. (1985). Pulse-number distribution for the neural spike train in the cat's auditory nerve. *J. Acoust. Soc. Am.* 77:1110–1128.

Tolbert, L. P., and Morest, D. K. (1982). The neuronal architecture of the anteroventral cochlear nucleus of the cat in the region of the cochlear nerve root: Golgi and Nissl methods. *Neurosci.* 7:3013–3030.

Tolbert, L. P., Morest, D. K., and Yurgelun-Todd, D. A. (1982). The neuronal architecture of the anteroventral cochlear nucleus of the cat in the region of the cochlear nerve root: horseradish peroxidase labeling of identified cell types. *Neurosci.* 7:3031–3052.

Tsuchitani, C. (1982). Discharge patterns of cat lateral superior olivary units to ipsilateral tone-burst stimuli. *J. Neurophysiol.* 47:479–500.

Tsuchitani, C. (1988a). Discharge patterns of cat lateral superior olivary units to binaural tone-burst stimuli: I. The transient chopper response. *J. Neurophysiol.* 59:164–183.

Tsuchitani, C. (1988b). Discharge patterns of cat lateral superior olivary units to binaural tone-burst stimuli: II. The sustained discharges. *J. Neurophysiol.* 59:184–211.

Tsuchitani, C., and Boudreau, J. C. (1966). Single unit analysis of cat superior olive S-segment with tonal stimuli. *J. Neurophysiol.* 28:684–697.

Viernstein, L. J., and Grossman, R. G. (1961). Neural discharge patterns in the transmission of sensory information. In: C. Cherry (ed.). *Information Theory*, pp. 252–269. London: Butterworths.

Walsh, B. T., Miller, J. B., Gacek, R. R., and Kiang, N. Y. S. (1972). Spontaneous activity in the eighth cranial nerve of the cat. *Int. J. Neurosci.* 3:221–236.

Warr, W. B. (1966). Fiber degeneration following lesions of the anterior-ventral cochlear nucleus of the cat. *Exp. Neurol.* 14:453–474.

Warr, W. B. (1982). Parallel ascending pathways from the cochlear nucleus: neuroanatomical evidence of functional specialization. In: W. D. Neff (ed.). *Contributions to Sensory Physiology*, pp. 1–38. New York: Academic Press.

Weiss, T. F. (1964). A model for firing patterns of auditory nerve fibers. *Tech. Rept., Res. Lab. Elect., M.I.T.* 418. Cambridge: M.I.T. Press.

Weiss, T. F. (1966). A model of the peripheral auditory system. *Kybernetik* 4:153–177.

Yang, G. L., and Chen, T. C. (1978). On statistical methods in neuronal spike-train analysis. *Math. Biosci.* 38:1–34.

Young, E. D. (1984). Response characteristics of neurons in the cochlear nuclei. In: C. I. Berlin (ed.). *Hearing Science: Recent Advances*, pp. 423–460. San Diego: College Hill Press.

9

The Neural Coding of Simple and Complex Sounds in the Auditory Cortex

DENNIS P. PHILLIPS

The auditory cortex is the target of a strikingly complex, divergent and convergent ascending afferent pathway (see Aitkin et al., 1984; Phillips, 1988a). This structural organization implies a wealth of neural processing power, and it is perhaps no surprise that interest in the coding of complex sounds may be traced to the earliest single-unit studies of this cerebral region (Suga, 1965; Whitfield and Evans, 1965). Some of the most elegant expressions of this research interest are the reports of Nobuo Suga and his colleagues on the cortex of the mustached bat. Local cortical regions in that species, contain neurons that are functionally specialized for the analysis of various parameters of the acoustic echoes reflected from targets. In part, this is made possible by the relatively stereotyped acoustic structure of the emitted sonar pulses; comparing the emitted pulse and its echo in terms of amplitude, delay, and spectra may provide information about the targets distance, size, and velocity. Much of this work has been reviewed elsewhere (Suga, 1982; Suga et al., 1983), and the reader is referred to those sources for detailed accounts of how the study of complex sound coding can be guided productively by considering the acoustic behavior of the relevant species.

The aims of this chapter are to examine the stimulus selectivity of cortical auditory neurons in cats and primates and to explore the nature of the neural or other processes that may underlie those stimulus response functions. With the possible exception of their vocal repertoires (e.g., Winter et al., 1966), the acoustic behaviors of these species have been less precisely defined than those of bats, and the range of ethologically relevant auditory signals can only be guessed at. These factors may have prompted the more generic or eclectic approaches to the study of these species' auditory nervous systems. A fortuitous feature of the auditory behavioral capacities of these animals is that, in some respects, they parallel those of the human. It may be optimistic to expect these similarities to extend to the underlying neural mechanisms, but the examination of cortical neural responsiveness to simple and complex sounds has been instructive in suggesting the neural mechanisms that shape that responsiveness in cats and primates and, if only presumptively, therefore in humans.

OVERVIEW OF AUDITORY CORTEX FUNCTIONAL ORGANIZATION

Microelectrode mapping studies in the 1970s provided our current understanding of the topographic organization of the auditory sensory cortex in cats (Merzenich et al., 1975; Reale and Imig, 1980) and primates (Imig et al., 1977; Merzenich and Brugge, 1973). These studies explored broad regions of the temporal cortex with tonal stimuli and examined the spatial distribution of neural "characteristic frequencies" (CF, the tone frequency to which a neuron is most sensitive) across those cortical territories. These experiments confirmed and extended the much earlier evoked-potential studies of Woolsey and his colleagues (Woolsey, 1960) in revealing a multiplicity of cortical auditory fields. Many of these fields contain neurons that are narrowly tuned to tone frequency and have very clearly defined CFs. Within such fields, the neurons are spatially arrayed so that a single locus on the cochlear partition is represented cortically as an isofrequency strip, that is, a thin band of tissue containing cells of similar CF. The isofrequency strips are themselves arrayed tonotopically, in an orderly and relatively complete reflection of the frequency organization of the peripheral receptor and, hence, the audible frequency spectrum. Other cortical fields contain neurons that may be broadly tuned to tone frequency, and show rather less internal organization according to the frequency selectivity of those elements (Imig et al., 1977; Merzenich and Brugge, 1973; Reale and Imig, 1980; Schreiner and Cynader, 1984). In the cat, these observations have led to the proposal that the auditory cortex contains two systems of fields, cochleotopic and diffuse, a notion that is bolstered by the finding that the cochleotopic and diffuse fields might have quite different, and largely nonoverlapping, patterns of thalamocortical connectivity (Andersen et al., 1980; Imig and Morel, 1983).

The various tonotopic fields are distinguishable on a number of grounds. They take their names from their spatial location (anterior field, posterior field, etc.) relative to a primary field (AI) that has a distinctly koniocortex appearance in Nissl material (Brugge, 1975; Merzenich and Brugge, 1973; Rose, 1949; Winer, 1984) that is less obvious in adjacent fields. *Cytoarchitectural* borders so defined usually correspond well with *physiological* borders in the same individuals. These functional boundaries are typically manifested as a reversal in the tonotopic sequence of neural CFs at the shared low- or high-frequency border. The correspondence between cytoarchitectonic and physiological field borders has been presented for primates (Imig et al., 1977; Merzenich and Brugge, 1973) and for rodents (McMullen and Glaser, 1982; Merzenich et al., 1976). In the cat, Rose and Woolsey (Rose, 1949; Rose and Woolsey, 1949; Woolsey, 1960) have provided joint cytoarchitectonic and evoked-potential maps of cortical auditory fields; the physiological parcellation has since been revised (Reale and Imig, 1980), but no corresponding cytoarchitectural evidence has been presented in the same animals.

Third, the tonotopic fields have been distinguished by their ascending afferent connections (Andersen et al., 1980; Imig and Morel, 1983, 1984). The vast majority of this input is derived from the auditory thalamus, which consists of the medial geniculate body and a number of related cell groups. The ventral division of the medial geniculate body (MGv) has a strikingly laminated appearance in Golgi material (Morest, 1965; Winer, 1985), which derives in part from the arrangement of its neurons and their tufted dendritic processes in sheets that broadly follow the curvature of the MGv's protrusion from the diencephalon. Electrophysiological studies (Aitkin et al., 1984, Imig and Morel, 1985a) indicate that the MGv has a strict tonotopic organization, with neurons of similar CF being disposed in curved sheets most likely following the contours of those seen in the Golgi studies. It is this structure that receives the bulk of the tonotopically constrained input from the auditory midbrain and that might properly be regarded as the core thalamic relay nucleus (Aitkin et al., 1984; Phillips, 1988a). A number of other nuclear groups surround the MGv. Some of these groups, notably the lateral part of the posterior group, may contain neurons that are sharply tuned to tone frequency and tonotopically arrayed (Imig and Morel, 1985b); others appear to receive a highly convergent, often multimodal sensory input.

The primary auditory cortical field receives dense, topographically organized projections from the MGv. Terminating most densely in layer IV, this is a convergent, sheet-to-strip projection: isofrequency *laminae* in the MGv direct their axons on isofrequency *lines* in the cortex (Andersen et al., 1980; Imig and Morel, 1984). The convergence of this thalamic input is far from complete, and evidence indicates that the projection may be constrained not only by tonotopy, but also by the binaural interactions of the participating neurons (Middlebrooks and Zook, 1983). AI also derives input from some of the cell groups abutting the MGv, particularly the lateral posterior group (Imig and Morel, 1984). The other tonotopic cortical fields similarly derive input from the MGv and adjacent cell groups, but the weightings of the projections are less biased toward MGv than is the case for AI. The nontonotopic, diffuse cortical fields receive their thalamic input almost exclusively from cell groups other than the MGv (Andersen et al., 1980; Imig and Morel, 1983).

In cats (Imig and Brugge, 1978; Imig and Reale, 1980, 1981) and probably in primates (Fitzpatrick and Imig, 1980), loci and AI are reciprocally interconnected with tonotopically homotypic loci in both the AI of the contralateral hemisphere and the other tonotopic fields of the ipsilateral hemisphere. The populations of AI neurons contributing to the callosal projections may only weakly overlap those participating in the ipsilateral corticocortical connections. This finding originates in Imig and Adrian's (1977) physiological mapping studies of cat AI. They categorized AI cells as either summative (with responses to binaural CF tones stronger than those to monaural tones of equivalent amplitude) or suppressive (with binaural responses weaker than the stronger monaural response) in their binaural interactions, and examined the distributions of those cell types across AI. They found that overlaying the continuous frequency representation in AI were discontinuous patches or territories of one or another binaural cell type. There has been some disagreement over what the relevant binaural categories are and over the geometry of the binaural patches, but the general finding that AI contains zones of binaurally response-specific neurons has been confirmed by other workers (Middlebrooks et al., 1980; Reale and Kettner, 1986).

Imig and Brugge (1978) revealed, in a combined physiological mapping/anatomical tracing study, that AI regions containing predominantly summative binaural cell types were the usual sources and terminations of the callosal axons. Soon thereafter, Imig and Reale (1981) reported that AI territories dominated by suppressive cell types participated most strongly in the ipsilateral, corticocortical connections. As mentioned earlier, there is some evidence that the MGv might also contain local regions of summative and suppressive cell types (Middlebrooks and Zook, 1983). Thus, not only may the thalamocortical projection preserve that binaural segregation, but its cortical manifestations may provide a framework for the organization of corticocortical connectivity.

RESPONSES TO TONE- AND NOISE-BURST STIMULI OF NEURONS IN THE PRIMARY FIELD

Most AI neurons in the anesthetized animal respond briskly and transiently to the onset of a tone pulse of appropriate frequency and amplitude. The most common means of depicting the range of excitatory tone frequencies has been the frequency tuning curve, which plots the minimum tone-pulse amplitude required for an excitatory response as a function of the frequency of the tone. Throughout the core nuclei of the afferent auditory system (Aitkin et al., 1975, 1984; Evans, 1975b; Kiang et al., 1965; Tsuchitani, 1977) and within AI (Phillips and Irvine, 1981a), most neurons have V-shaped tuning curves with clearly defined CFs to which they are most sensitive. The general form of this threshold frequency selectivity is based on the mechanical properties of the cochlear partition (Phillips, 1988). Each of the cochlear nerve fibers, which link the relevant sensory elements (inner hair cells) to the central auditory system, is narrowly tuned to tone frequency and derives that property from the particular hair cell to which it is connected. The acoustic input to the central nervous system is thus an array of finely tuned neurons whose CFs span the audible frequency range. The precision of the tonotopic constraints on afferent connectivity in the auditory system is indicated in the fact that tuning to a CF is retained at the cortex. As we discuss later, however, this does not suggest that some of the more quantitative features of the frequency tuning of cochlear nerve fibers are unmodified in that afferent pathway.

The distribution of minimum CF thresholds across the represented audible frequency range in the cortex usually parallels the sensitivity of the auditory periphery, and therefore behavioral sensitivity (Phillips and Irvine, 1981a; Phillips and Orman, 1984). There is considerable (30 to 50 dB) spread of absolute sensitivities for cortical neurons of any given CF. This may not be the case at the level of cochlear output, at least in cats reared in low-noise chambers (Liberman, 1978). In these animals, the spread of auditory nerve fiber sensitivities is usually less than 20 dB. On the arguable assumption that the animals used in the acute cortical experiments have had equally atraumatic acoustic histories, the broader spread of absolute sensitivities in the cortex must be a consequence of central processing. In general, the tuning curves for high-CF cortical neurons are narrower than those for low-CF cells (Phillips and Irvine, 1981a). This is also partly a reflection of the mechanical tuning of the basilar membrane, and is thus a property of cochlear output (Evans, 1975b; Kiang et al., 1965).

The sensitivity of central auditory neurons to the amplitude of a CF tone pulse or other stimulus is usually examined by plotting spike rate (expressed in spikes per second or spikes per stimulus trial) versus the intensity of that stimulus. For AI neurons, the resulting rate-level function typically takes one of two forms (Brugge et al., 1969; Brugge and Merzenich, 1973; Funkenstein and Winter, 1973; Pfingst and O'Connor, 1981; Phillips and Cynader, 1985; Phillips et al., 1985a). In some cases, the rate-level function is sigmoidal (monotonic), with the spike rate increasing from threshold over a 10 to 40-dB dynamic range toward a saturated response rate, which is maintained with further increments in stimulus level. For other neurons, rate-level functions for CF tones are bell-shaped (nonmonotonic). For these cells, there is an optimal CF tone level, above or below which spike rates may decline, often to zero. Neither of these cell types is unique to the cortex. Nonmonotonic neurons have been described in the dorsal cochlear nucleus (DCN; see Evans, 1975b; Evans and Nelson, 1973; Young and Brownell, 1976) and in the inferior colliculus to which the DCN directly projects (Rose et al., 1963; Semple and Aitkin, 1980). Monotonic neurons appear to be more common in the ventral cochlear nucleus (Evans, 1975b) and the lateral superior olivary complex (Tsuchitani, 1977), which constitute an alternate route from the cochlea to the auditory midbrain.

A second consequence of variations in the amplitude of tone pulses concerns the latent period between stimulus onset and the occurrence of the first spike evoked by the stimulus (e.g., Brugge et al., 1969, 1970; Phillips et al., 1985a). For threshold excitatory (and presumably also inhibitory) stimuli, response latencies are long and responses are relatively dispersed in time. As the stimulus level is increased, the mean latent period to first spike shortens, and the standard deviation of first spike times around that mean narrows significantly. The latency-intensity function is markedly nonlinear; it is steep for stimulus amplitudes near threshold, and asymptotes toward minima, in the range from 10 to 40 msec, for high stimulus levels. Among the shortest latency cells, the precision of first spike times may be on the order of a few tens of microseconds.

All cochlear nerve fibers have monotonic rate-level functions for CF, or other, tones (Evans, 1975b; Kiang et al., 1965), and it has therefore been argued that the descending limb of a nonmonotonic rate-level function must be produced by central inhibitory processes (e.g., Phillips et al., 1985a; Rose et al., 1963; Voigt and Young, 1980; Young and Brownell, 1976). A high-threshold inhibitory input at a neuron's CF could be produced by at least two means (Shofner and Young, 1985). First, it is possible that two neurons of similar CF, but differing thresholds, could be interconnected in such a way that spike activity in the high-threshold element inhibited that in the more sensitive neuron. Second, since frequency tuning curves are often V-shaped, a high-threshold inhibitory input at the CF of one neuron is likely to be provided by inhibitory connections from adjacent neurons of higher or lower CF, whose tuning curve skirts extended to the CF of the neuron in question.

For both brainstem and cortical neurons, two consequences follow from the presence of the sideband inhibitory inputs. First, for those nonmonotonic elements whose spike rates fall to zero at high stimulus levels, the excitatory tone response area may be completely circumscribed (Phillips et al., 1985a; Young and Brownell, 1976). Tonal stimuli for these neurons evoke spike discharges over narrow frequency and intensity ranges. This situation contrasts markedly with that of monotonic cells, whose spike rates continue to increment or saturate at high tone amplitudes. These cells have the more common V-shaped response areas. Second, if sideband inhibition is more common among nonmonotonic neurons, the two groups should be differentiable on the basis of their responsiveness to wide-band noise stimuli. This is because a broad-spectrum signal can simultaneously invade the excitatory and the inhibitory response areas of nonmonotonic cells. In the cortex (Phillips and Cynader, 1985; Phillips et al., 1985a) and in the DCN (Greenwood and Maruyama, 1965; Young and Brownell, 1976), this is indeed often the case. Thus, most nonmonotonic cortical cells are not excited by broadband noise pulses of any intensity, whereas most monotonic cells are. This correlation is imperfect, however, indicating that sideband inhibition may be present independently of that at CF.

Some of these notions have been tested in cortical cells by examining the effects on the CF tone rate-level function of simultaneously gated wide-band noise pulses (Phillips and Cynader, 1985). For monotonic cells, which are usually excited by noise pulses, the extent to which either the tone burst or the noise burst dominates the neuron's discharges depends on the relative levels of the two stimulus components. In effect, the stimulus that provides the stronger excitatory drive "captures" the cell's responses (strong-signal capture, see Rhode et al., 1978) in terms of both spike rate and spike timing, since the two stimulus elements can evoke responses with different latent periods.

For most nonmonotonic cortical cells, noise pulses provide a net inhibitory input, whereas CF tone pulses are excitatory at low amplitudes and inhibitory at high levels. When these neurons are studied with simultaneously gated tone-noise combinations, the effect of the noise pulse is to suppress the discharges that would otherwise be evoked by the tone. The greater the amplitude of the noise, the stronger the suppression of the tone response is. For what follows, it is noteworthy that the tone-evoked discharges that survive the noise mask have latent periods appropriate to those evoked in the absence of the noise (Phillips and Cynader, 1985). This means that the responses evoked by the tone-noise combination reflect the balance of excitatory and inhibitory inputs—that in the presence of the noise pulse, the tone response occurs with normal spike *timing*, but with reduced spike *rate* because of the competing inhibitory input evoked by the noise.

Figure 9-1 partly summarizes the foregoing observations. The left panels (A,C,E) present data for a single monotonic cortical neuron, and the right panels (B,D,F) show data for a nonmonotonic cortical neuron. Panels A and B depict the form of these neurons' excitatory frequency-intensity response areas. The unshaded, stippled, and shaded areas represent the tone stimulus domains that evoked firing rates of 10 to 50, 50 to 80, and 80 to 100 percent of maximum, respectively. Note that the monotonic neuron has a V-shaped response area, regardless of which spike-rate criterion is used to define its boundary. It is open-ended at high tone levels; the arrows indicate that the neuron continued to discharge spikes at intensities exceeding those for which spike count data were collected. The nonmonotonic neuron, on the other hand, had a circumscribed response area: tonal stimuli were effective within narrow domains of both frequency and level. As can be inferred from the slanted disposition of the unshaded and stippled areas in panel

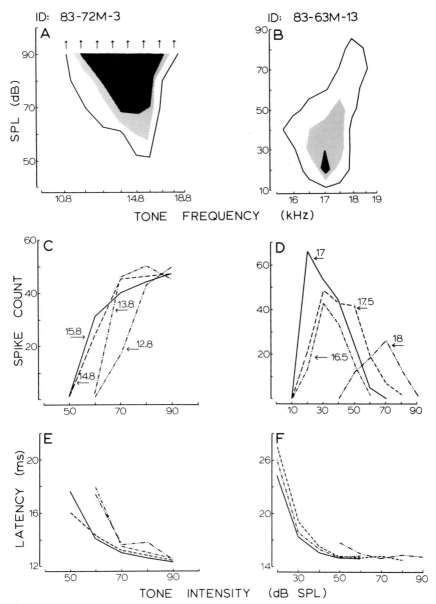

Figure 9-1 Some response properties of a monotonic (*A,C,E*) and a nonmonotonic (*B,D,F*) cortical neuron. (*A,B*) Pure-tone frequency-intensity response areas. Response rate (density of shading) is shown as a function of both the frequency (abscissa) and the intensity (ordinate) of a tonal stimulus. (*Unshaded areas*) Spike rates from 10 to 50 percent of maximum; (*stippled areas*) spike rates from 50 to 80 percent of maximum; (*black* areas) spike rates from 80 to 100 percent of maximum. Arrows in A indicate that the neuron continued to discharge spikes at intensities exceeding those for which quantitative data were collected. Abbreviation: SPL, sound pressure level (dB *re* 20 μPa). (*C,D*) Rate-level functions for tonal stimuli obtained from the same neurons. Numbers adjacent to each curve indicate the tone frequency (in kilohertz) used to obtain those data. (*E,F*) Latency-intensity functions for the responses depicted in panels *C* and *D*. The system of continuous and interrupted lines in *C* and *D* applies in *E* and *F*. All data are based on responses to 50 repetitions of every stimulus condition. [Data are from the study of Phillips et al (1985a).]

B, the neuron's optimal tone level varied with tone frequency. Only rarely is this not the case (Phillips and Orman, 1984).

A rate-level function represents a vertical slice through the tone response area. Four rate-level functions are shown for each of these neurons in panels C and D, and the tone frequencies used to obtain these functions are indicated. For the monotonic neuron (panel C), rate-level functions tended toward ceiling maxima, whereas for the nonmonotonic cell (panel D), the functions have clearly defined maxima and decline to zero at high stimulus levels. Recall that the descending limbs of these functions are shaped by inhibitory inputs at the test frequency. In the context of the response area from which these curves were derived, it is difficult to escape the conclusion that this neuron had a strong sideband inhibitory input on the low- and high-frequency sides of its excitatory response area.

Response latency data obtained from the responses depicted in panels C and D are shown in panels E and F, respectively. For both neurons, latent period to first spike declined toward limiting minima at high tone amplitudes. For the nonmonotonic neuron, recall that high-intensity tones are likely to activate both excitatory and inhibitory inputs. The timing of spike discharges follows the pattern for monotonic cells, while the number of spikes contributing to the responses declines.

BINAURAL INTERACTIONS AND SPATIAL PROPERTIES OF CORTICAL NEURONS

The vast majority of auditory cortex neurons receive input from both ears (Hall and Goldstein, 1968). Information from the two ears first converges in the auditory brainstem at the level of the superior olivary complex, and the neural sensitivity to interaural intensity and temporal parameters resulting from that convergence is largely preserved in the neurons of higher structures to which the superior olivary nuclei directly or indirectly project (Aitkin et al., 1984; Benson and Teas, 1976; Brugge et al., 1969, 1970; Phillips, 1988a; Yin and Kuwada, 1984).

The binaural interactions of central auditory neurons are of special interest because interaural disparities in the timing and intensity of stimuli at the ears constitute important cues for the azimuthal location of a sound source (Phillips and Brugge, 1985). The intensity cue (interaural intensity difference, IID) is relevant for sound sources with spectral energy at frequencies whose wavelengths are shorter than head diameter (or pinna height), because these cast a sound shadow for those signals. The temporal cue has two components. For any sound source located at nonzero azimuths (i.e., off the midsaggital plane), the greater path length to the further ear incurs an arrival time disparity at the two tympanic membranes. This arrival time difference imposes an ongoing interaural phase delay (IPD), which the central auditory system can resolve for spectral elements whose wavelengths are larger than the head diameter. Since the size of both the temporal and the intensity difference vary with stimulus frequency as well as with source eccentricity, spectrally complex sounds are likely to provide a wealth of cues for locating sound sources.

Cortical coding of interaural arrival time disparities of CF tone pulses has been described by Kitzes et al. (1980), and generally follows the form of that already described for the coding of IIDs. The relevant neurons are most likely the excitatory–inhibitory ones, and again, there is a strong contralateral bias in the range of time differences effective in evoking spike discharges. Neural coding of IPDs, on the other hand, probably uses quite different processes. The mechanisms underlying sensitivity to interaural differences in the phase of low-frequency stimuli have been described in greatest detail for neurons of the auditory brainstem (Brugge et al., 1970; Goldberg and Brown, 1969; Rose et al., 1966; Yin and Kuwada, 1984). Studies of cortical neurons (Benson and Teas, 1976; Brugge et al., 1969; Brugge and Merzenich, 1973; Orman and Phillips, 1984) have been less exhaustive in their treatment of this issue, but in large part they confirm the findings of the studies of pontine and mesencephalic auditory neurons.

The sensitivity of cortical neurons to IIDs is generally manifested in one of two forms. One group of neurons receives a net excitatory input at CF from the contralateral ear, and a net inhibitory input at the same frequency from the ipsilateral ear. For these cells, the spike discharge rate is a sigmoidal function of IID (Benson and Teas, 1976; Brugge et al., 1969; Brugge and Merzenich, 1973; Phillips and Irvine, 1981b; Orman and Phillips, 1984). Response rates are high when the IID favors the contralateral ear (i.e., when contralateral tone amplitude exceeds the ipsilateral tone amplitude), and are poor, or actively suppressed, when the IID significantly favors the ipsilateral ear. The steep, dynamic portion of the IID function usually occurs over IID ranges close to zero or slightly favoring the contralateral ear (Phillips and Irvine, 1981b). It is possible to extrapolate crudely, from these dichotic stimulus studies, the free-field azimuths of sources generating the IIDs responsible for different parts of the IID response function. The contralateral bias in the excitatory IID ranges suggests that such neurons would respond preferentially to sound sources in the contralateral acoustic hemifield. The steep part of the IID function suggests that medial edges of spatial receptive fields can be located relatively close to the midline. As is discussed later, these extrapolations are partly confirmed by direct examination of the spatial receptive fields of cortical neurons.

A second group of cortical (Kitzes et al., 1980; Orman and Phillips, 1984; Phillips and Irvine, 1981b) and subcortical (Wise and Irvine, 1984) neurons sensitive to IIDs has been described. These neurons appear to receive a sub-(spike)-threshold excitatory input from each ear, and, because of the apparent ineffectiveness of monaural stimuli, they have been termed predominantly binaural cells (Kitzes et al., 1980). Studied with dichotic stimuli, the discharge rates of these neurons are bell-shaped functions of both IID and interaural arrival time disparity. For both stimulus parameters, response rates are greatest for disparities close to zero. These data suggest that these neurons' spike rates reflect the temporal coincidence of excitatory afferent events from the two ears. A coincident arrival of these inputs at the binaural neuron is achieved when the stimuli reach the two ears simultaneously and with equal intensity. The coincidence would be disrupted, however, by either a difference in interaural stimulus arrival time or an IID—the latter by means of the latency–intensity relation.

Neurons sensitive to IPDs most likely receive bilateral inputs in which excitatory and inhibitory events are phase-locked to the stimulating waveforms at the tympanic membranes. This is made possible, in part, by the properties of the inner cochlear hair cells, and the structural organization of the projections of the relevant low-CF neurons transmitting phase information to central neurons (see Brugge and Geisler, 1978; Phillips, 1988a; Phillips and Brugge, 1985 for review). Briefly, the inner hair cells are functionally polarized and respond with depolarizations to unidirectional elevations of the basilar membrane. Transmitter release, and hence excitation of the cochlear nerve fibers linking the hair cell with the auditory brainstem, is tied to hair cell depolarization. Now, at low stimulus frequencies, the hair cell receptor potential is dominated by an alternating current (AC) component, which can thus synchronize transmitter release to the depolarizing half-period of the stimulus waveform. Structural and bioelectrical specializations of neurons in the brainstem auditory pathways (Aitkin et al., 1984; Brugge and Geisler, 1978; Phillips, 1988a) preserve this response timing, so that the spike discharges of those neurons show a temporal distribution resembling a partially rectified stimulus waveform. In the case of a low-frequency sinusoidal stimulus, spike discharges of auditory nerve fibers are locked in time to a preferred half-period of the stimulus waveform; in higher centers, an inhibitory response is often found to occupy the nonexcitatory half-period of the response (Brugge et al., 1969, 1970). Central neurons that receive such inputs from *both* ears can thus be thought of as coincidence detectors: when the excitatory afferent events from the two ears arrive synchronously, they summate and evoke a vigorous discharge rate. If the interaural phase is varied, the cadence of excitatory phase-locked events from one ear partially or totally overlaps in time with the cadence of inhibitory phase-locked events from the other ear, resulting in cancellation of the excitatory input to the neuron. Because the stimulus itself has a periodic waveform (i.e., a repeating sinusoid), further increments in IPD reestablish the coincident arrival of excitatory afferent events, leading to a spike rate that is a continuous, periodic function of IPD. The period of this response function equals that of the carrier stimulus; it is generated by the cycle-by-cycle comparison of the stimuli at the two ears.

Early studies of the auditory cortex with free-field stimuli revealed that stimuli with contralateral locations in auditory space evoked more vigorous responses than did the same stimuli located in the ipsilateral auditory hemifield (Benson et al., 1981; Evans, 1968). Only recently, however, parametric studies have characterized the spatial properties of cortical and subcortical auditory neurons and the factors contributing to that spatial selectivity (Calford and Pettigrew, 1984; Fuzessery and Pollak, 1985; Middlebrooks and Knudsen, 1984; Middlebrooks and Pettigrew, 1981; Moore et al., 1984; Palmer and King, 1985; Phillips et al., 1982; Semple et al., 1983). Middlebrooks and Pettigrew (1981) provided a major insight into these issues. These authors mapped the spatial receptive fields of cat cortical neurons, and described three major cell classes distinguished by their receptive field properties. One group of cells was termed *omnidirectional*: these neurons responded in a relatively undifferentiated fashion to a suprathreshold CF tone pulse, regardless of the sound source location. A second group of neurons, *hemifield* units, responded to CF tone pulses presented in the contralateral, but not the ipsilateral, acoustic hemifield. The medial edges of these cells' receptive fields were usually clearly defined and close to the midsaggital plane. The CFs of these cells were generally less than 12 kHz. The final group of cells were termed *axial* units. These neurons had CFs that were generally above 12 kHz, and their receptive fields were small, located in contralateral space, and in the apparent acoustical axis of the contralateral pinna.

These findings prompted direct examination of the nature of pinna directionality, to gain some appreciation for its contribution to the formation of central neuron receptive fields (Phillips et al., 1982). This study revealed that the cat pinna has a profound acoustical axis, manifested as a peak in tympanic sound pressure level for stimuli located on that axis. This directionality reflects the spatially selective amplification of signals on-axis, and the attenuation of signals presented off-axis. The pinna axis most likely has its basis in the joint effects of the resonance properties of the external auditory canal, and in the diffraction (or lack of it) of sound waves around the pinna itself (Calford and Pettigrew, 1984; Phillips et al., 1982). As might be predicted from physical acoustics, this directionality of sound-pressure transformations increases with stimulus frequency. In cats, pinna directionality may be negligible for frequencies less than 3 to 4 kHz (presumably depending on the size of the pinna), and becomes increasingly marked for frequencies from 5 to 12 kHz. It may asymptote at higher frequencies.

These data provide a partial understanding of the Middlebrooks and Pettigrew (1981) data on the spatial receptive fields of cortical neurons (see also Semple et al., 1983). It is likely that the axial receptive field shape of high-CF neurons reflects the directional sound-pressure transformations of the contralateral pinna. The hemifield units appear to have receptive field *boundaries* shaped more by binaural interactions than by pinna properties. To date, the firing rate of cortical neurons to tones *within* the receptive field has not been examined. Such evidence might provide more information on the relative contributions of binaural interactions and pinna directionality to the genesis of spatial receptive fields. At subcortical levels, it is becoming evident that pinna directionality can be superimposed on the spatial selectivity that would otherwise derive from binaural interactions alone (Fuzessery and Pollak, 1985; Moore et al., 1984; Palmer and King, 1985; Semple et al., 1983).

Together, the neurophysiological data on the binaural interactions and spatial properties of cortical (and subcortical) neurons have a number of further ramifications. First, there is a contralateral bias in the range of excitatory IIDs and IPDs in those cells, which is paralleled by the recent findings on the spatial receptive fields of central neurons. As far as the dichotic stimulus studies are concerned, it has been argued that the contralateral bias extends to the range of disparities encoded by the dynamic portions of the response rate–disparity functions (Phillips and Irvine, 1981b; Yin and Kuwada, 1984). These data suggest that each auditory cortex (and probably each side of the auditory brain rostral to the olivary nuclei; see Masterton and Imig, 1984; Phillips, 1988a; Phillips and Brugge, 1985; Phillips and Irvine, 1981b) encodes spatial information only for the contralateral acoustic hemifield and is capable of doing so independently. This conclusion has been reached independently in a number of behavioral lesion studies that have revealed that unilateral lesions of the primary auditory cortex result in deficits in sound localization performance only for stimuli in the sound field contralateral to the lesion (Kavanagh and Kelly, 1987; Jenkins and Masterton, 1982; Jenkins and

Merzenich, 1984; Strominger, 1969). Interestingly, animals with bilateral lesions may be unable to localize sound sources *within* either acoustic hemifield, but retain the ability to distinguish *between* sound source lateralities (Heffner, 1978; Heffner and Masterton, 1975; Kavanagh and Kelly, 1987). These data suggest that the neural mechanisms subserving discrimination of sound source *laterality* may be quite different from those subserving discrimination of sound source *location* within a hemifield.

SENSITIVITY OF CORTICAL NEURONS TO TONES PRESENTED AGAINST NOISE BACKGROUNDS

Cat cortical neurons have been studied for the effects of continuous wide-spectrum noise backgrounds on their sensitivity to tone pulses presented to the same ear (Phillips, 1985; Phillips and Cynader, 1985; Phillips and Hall, 1986). These studies parallel a history of similar studies of the cochlear nerve (Costalupes et al., 1984; Geisler and Sinex, 1980; Gibson et al., 1985; Kiang and Moxon, 1974; Rhode et al., 1978). A comparison of the effects of noise masking at these loci is instructive for three reasons. First, the comparison reveals some neural mechanisms available to cortical neurons that may figure significantly in the coding of temporally complex sounds. Second, it provides new evidence on the organization of afferent input to cortical cells. Finally, these studies have provided new insights into the neural correlates of stimulus level and perceived loudness.

All cochlear nerve fibers have monotonic rate-level functions for both tonal and wide-band noise stimuli (Evans, 1975b; Ruggero, 1973) and discharge spike action potentials for the duration of the stimulus. One effect of a continuous noise stimulus is therefore to set a base rate of spike discharges on which the responses to a simultaneously presented tone pulse must be superimposed (Costalupes et al., 1984; Geisler and Sinex, 1980). Since cochlear nerve fibers have finite maximum firing rates, the firing-rate baseline set by the noise background truncates the range of spike rates available for encoding the tone pulses. In addition, the part of the tone rate–level function that survives the baseline elevation is displaced toward higher-tone intensities. This dynamic range shift is based in the properties of basilar membrane motion—notably, the constraints on the mechanical response at the locus innervated by the fiber imposed by the more widespread mechanical response to the noise (Costalupes et al., 1984). This tone sensitivity shift very likely is general and equivalent across the entire range of frequencies to which the fiber is normally sensitive, so that the net effect of the noise background is to elevate thresholds for tonal stimuli equally across the tuning curve (see Geisler and Sinex, 1980; Kiang and Moxon, 1974). Further elevations in noise level increase the background response rate, further truncate the range of spike rates available to encode the tone level, and increase the tone dynamic range shift; the increases in dynamic range shift roughly match the increment in noise level (Costalupes et al., 1984; Gibson et al., 1985). These data are understandable in that a cochlear nerve fiber receives its input from a single cochlear hair cell whose own response is dominated largely by basilar membrane motion at its site (see Phillips, 1987). A subpopulation of cochlear nerve fibers is considerably more resistant to firing-rate saturation by noise, and this resistance is likely to be conferred on many more central neurons (see Costalupes et al., 1984; Gibson et al., 1985).

Cortical neurons typically do *not* respond continuously to maintained signals, particularly in anesthetized animals. Continuous noise, therefore, exerts little effect on the baseline rate of spike discharges. As a consequence, the whole range of the cortical neuron's spike rates is available for the encoding of tones presented against the noise. In practice, the net effect of a wide-band noise background is to displace the entire tone rate–level function toward higher-tone intensities. The magnitude of this sensitivity shift is linearly related to the level of the background noise, with a slope close to unity (Phillips and Cynader, 1985; Phillips and Hall, 1986). This suggests that cortical neurons dynamically adjust their sensitivity to a signal in proportion to the level of background noise, and therefore that the encoded signal amplitude parameter in the cortex may not be intensity per se, but *signal level re background*.

Now, recall that some cortical cells (mostly monotonic neurons) are excited by noise *pulses*, whereas others (mostly nonmonotonic cells) are inhibited by them. In contrast, the effects of *continuous* noise appear to be equivalent among cortical cells, suggesting that the factors shaping tone responsivity under the two masking conditions (simultaneous gating, and continuous noise) may be quite different. Studies of the timing of tone-evoked spikes (Phillips, 1985) have helped clarify the natures of the factors. Studied with simultaneously gated tones and noise, cortical neuron spike rates and latencies reflect those evoked by the tonal noise element of the combination which captures the cell's excitatory responses (see above; Phillips, 1985; Phillips and Cynader, 1985). When the noise mask is continuous, not only is the tone spike *rate*–level function displaced toward higher tone levels, but also the spike *latency*–level function, and by a comparable amount. The match between these effects on tone-evoked spike rates and spike timing suggests that the noise exerts a sensitivity shift for the tones, and not an obfuscation of the tone response. Independent evidence for this view comes from the finding that after a prolonged (800 msec) noise stimulus, recovery of tonal sensitivity, in terms of both spike rates and spike timing, proceeds exponentially with a specifiable time constant on the order of a few tens of milliseconds (Phillips, 1985). These data suggest that a maintained acoustic stimulus brings about a short-term adaptation of the cortical cell's sensitivity to that stimulus. The poststimulatory effects of such adaptation have been described in the responses to sequential tones in the awake cat's (Abeles and Goldstein, 1972) and monkey's cortex (Shamma and Symmes, 1985), indicating that it is not an anesthetic artifact. This adaptation has also been described in cochlear nerve fibers, in which it is less dramatic and has been ascribed to properties of the neurotransmitter reservoir at the hair cell–afferent fiber synapse (Smith and Brachman, 1982). It is perhaps likely that short-term adaptation in cortical cells is of the same form, but more severe in its effects on the time course of the excitatory or inhibitory responses evoked by the relevant stimulus.

In light of these findings, the differences in the effects of simultaneously gated and continuous noise on tone-pulse responses become understandable. If the onset of tone pulse is synchronous with that of the noise, then the cortical neuron's spike rate reflects the balance of excitatory and inhibitory events evoked by the two stimulus components, and its spike timing is dictated by the component that captures the cell's excitatory responses. The responses to noise (and tone) onset are transient, however, being shaped by short-term adaptation of the cell's sensitivity to the effective level of the mask. A tone pulse presented against a continuous mask evokes a response that must overcome, not the excitatory or inhibitory sequelae of noise onset, but rather the short-term adaptation incurred by the continued presence of the mask.

In cortical neurons, the tone sensitivity shifts brought about by continuous noise vary with the frequency of the test tone and are typically greatest at the neuron's CF (Phillips and Hall, 1986). This necessarily suggests that a cortical neuron can receive partially independent inputs across the frequency range making up its response area, and these inputs are differentially susceptible to noise masking. A further ramification of this conclusion is that the threshold tuning curve of a cortical neuron actually represents the envelope of the best sensitivities of a large number of afferent inputs (Phillips and Hall, 1986). In this respect, recall that a cochlear nerve fiber receives a unitary input—from the single cochlear hair cell it contacts—and such nonequivalence of wide-band noise masking effects across the tone response area is not seen (Kiang and Moxon, 1974). The conclusion for cortical cells is further justified by considering the structural organization of the afferent auditory pathway: it is highly divergent and convergent, but tonotopically constrained throughout (Aitkin et al., 1984; Phillips, 1988a). For this reason, cortical cells most likely have retained narrow frequency tuning while acquiring a host of other properties (e.g., binaural interactions, sideband inhibition) that initially result from processing at spatially separated brainstem nuclei.

A second, and related, feature of cortical responses to tones in continuous noise is that the tone rate–level functions that have been significantly displaced toward higher tone levels are also sometimes significantly steepened (Phillips and Hall, 1986). This effect is most marked among those cells whose unmasked rate–level functions have broad dynamic ranges. The effect meshes well with the hypothesis of a highly convergent input to AI neurons. It is not unreasonable to expect a cortical cell to receive multiple inputs at a test frequency, and that those inputs may have different absolute sensitivities, that is, dispersed thresholds. Now, continuous noise is likely to bring about sensitivity shifts in the responses of only those inputs whose thresholds are exceeded by the relevant noise spectrum levels. This brings into closer register the thresholds of the inputs at the test frequency. A supra-(masked)-threshold tone may therefore simultaneously activate more inputs than an unmasked tone of a comparable suprathreshold level. A steeper rate-level function for masked tones may reflect the smaller intensity increments needed to recruit new inputs. As is discussed in the following section, this view is supported by studies of cortical neural responses to amplitude modulated tones. For the present, however, note that this steepened rate-level function reflects the convergent nature of the input onto cortical cells, which is absent in that onto cochlear nerve fibers. In this respect, cochlear nerve fibers show no steepening of their rate-level functions for tones masked by continuous noise (Costalupes et al., 1984).

These data may be relevant to the neural basis of perceived loudness. There are two major hypotheses on this issue. One relies on the demonstrable spread of excitation along the basilar membrane, and hence, an increased number of discharging afferent fibers, that follows from increases in tone level; it suggests that perceived loudness is related to the number of active neural elements (see Evans, 1975a). This view has been questioned on both psychophysical and neurophysiological grounds (Phillips, 1987; Viemeister, 1983). An alternative hypothesis is that perceived loudness is related to the spike discharge rates of the relevant neurons tuned to the stimulus (Harrison, 1981; Harrison and Evans, 1977; Phillips, 1987; Viemeister, 1983). This view receives some support from a number of lines of investigation. The tone rate–level functions of cortical cells are displaced toward higher intensities and steepened by continuous wide-band noise masks, as are the loudness functions for noise-masked tones in normal listeners (Stevens and Guirao, 1967). Similarly, cochlear disease results in elevated behavioral thresholds for acoustic stimuli and an increased rate of loudness growth for suprathreshold tones. Recent evidence suggests that some cochlear nerve fibers innervating damaged organs of Corti have elevated tone thresholds and significantly steepened rate-level functions (Harrison and Evans, 1977; Phillips, 1987). If the slopes of rate-level functions do indeed contribute to the rate of loudness growth, these and the foregoing observations provide an important contrast: in cochlear disease, the steepened rate-level functions are expressed at the level of cochlear output, but not in noise masking; thus, they must represent the consequences of central processing.

NEURAL CODING OF AMPLITUDE AND FREQUENCY MODULATIONS

A longstanding interest has existed in the sensitivity of cortical neurons to time-varying stimuli, whether modulations of the amplitude or frequency of tonal stimuli (e.g., Suga, 1965; Whitfield and Evans, 1965) or the "natural" spectral and amplitude variations that occur in vocalizations (e.g., Wollberg and Newman, 1972). There is little doubt that some cortical neurons display marked selectivity to the direction of amplitude or frequency change, or that some cortical neurons are excited by only select members of extensive repertoires of behaviorally relevant vocalizations with which they have been tested. These observations are of considerable importance in themselves, because they may testify to a sophisticated level of central auditory processing. Of equal interest, however, is the question of the nature of the processing in which such selectivity has its genesis. This issue is only beginning to be investigated systematically, and it is the purpose of the remainder of this chapter to draw attention to some approaches to this question that have been particularly productive.

The sensitivity of cat cortical neurons to small (6-dB) linear amplitude modulations (AMs) of CF carrier tones has been studied by Phillips and Hall (1987). They examined the coding of the direction and the rate of AM, and the effects of the carrier level on which that AM was imposed. These data were compared with the responses of the same neurons to CF tone pulses over a 60-dB range (i.e., the rate-level function). Neurons that responded only to the onset of a CF tone pulse responded only to incremental AMs, regardless of carrier level (Figure 9-2B). This finding is explicable by the short-term adaptation of cortical neuron sensitivity that occurs in response to maintained stimulation (see preceding), that

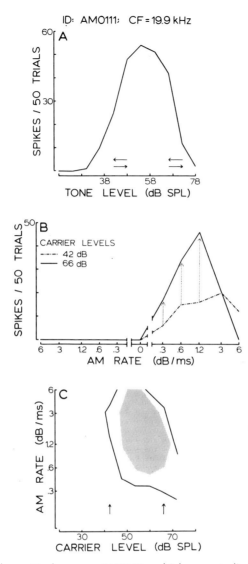

Figure 9-2 Data for cortical neuron AM0111, which was studied with both CF tone pulses (A) and 6-dB linear AMs of continuous CF carrier tones (B,C). (A) Rate-level function for CF tone pulses. Arrows indicate the intensity ranges covered by the AMs used to collect the data presented in B. (B) Sensitivity of the same neuron to the rate of amplitude change for intensity decrements (*left side*) and intensity increments (*right side*), at two specified carrier levels. Note that the neuron was unresponsive to intensity decrements, and for intensity increments, identical low-speed AMs were more effective when imposed on a high carrier level (dotted arrows show the comparison). (C) Spike rate of the same neuron evoked by the conjunctions of AM rate (ordinate) and carrier level (abscissa) of a continuous CF tone. Stippled area indicates AM rate/carrier level conjunctions that evoked spike rates within 33 percent of maximum. Bounded, unshaded area indicates AM rate/carrier level conjunctions evoking spike rates within 66 percent of maximum. Each data point is based on responses to 50 repetitions of every stimulus conjunction. [Data are from the study by Phillips and Hall (1987).]

is, the neuron's threshold is adjusted to the effective level of the carrier that precedes the AM. An intensity decrement may thus reduce the effectiveness of the signal from threshold to subthreshold levels, and therefore be ineffective in evoking spike discharges. In contrast, an amplitude increment most likely provides the neuron with a supra-(adapted)-threshold stimulus, and may therefore be efficacious in evoking spike responses. In general, the neurons found to respond to intensity *decrements* were also excited by tone-pulse *offset*.

For low carrier levels, high-speed AMs were typically more effective than were slow-rate AMs, and the converse was true for high carrier levels. Figure 9-2C illustrates this point by plotting response strength as a function of both AM rate (ordinate) and carrier level (abscissa). Note that the domain of the most effective AM rate/carrier level conjunctions, indicated by the stippled area, is somewhat slanted from upper left to lower right. This reflects the drift in preferred AM rate from high to low as the carrier level increases.

Consideration of the dynamic nature of cortical responses and the spectra of linear AMs may provide a partial understanding of the mechanisms responsible for this pattern of behavior (Phillips and Hall, 1987). Recall that, at threshold levels of stimulation, the timing of afferent input to cortical neurons is long in latency and asynchronous. It is reasonable to expect that the stimulus events (at CF) that are likely to maximize the synchrony of the afferent volleys, and therefore be effective in evoking spike discharges, are rapid amplitude increments (Goldstein and Kiang, 1958). Also recall that there may be multiple inputs at CF, that these may have dispersed thresholds, and that these thresholds may be brought into closer register by short-term adaptation of the more sensitive elements to a prolonged stimulus (Phillips and Hall, 1986). This means that when they are imposed on a high-level carrier, even low-speed AMs may simultaneously activate more inputs to the neuron than the same AM imposed on a low-amplitude carrier would. These factors might explain the relatively greater effectiveness of low-speed AMs at high carrier levels. Corroboration of this view comes from inspection of the absolute spike rates evoked by identical low-speed AMs imposed on different carrier levels (dotted arrows in Figure 9-2B). A small AM imposed on a low-level carrier whose amplitude is close to the cell's absolute threshold would be expected to evoke a response of comparable strength to that of a tone pulse of the same peak amplitude. When the same AM is imposed on a high carrier level, it would be expected to evoke a response stronger than that of a threshold tone burst, and this is commonly the case.

On the other hand, a *high*-speed AM may effectively be a broad-band stimulus. The peak of the stimulus short-term spectrum may be at the carrier frequency, but very significant, and not necessarily even, splatter of spectral energy may occur across the frequency domain. This splattered energy is likely to invade the inhibitory response areas of cortical cells and be particularly effective when the carrier level is high. The simultaneous activation of inhibitory and excitatory inputs by these stimuli may be responsible for reducing the effectivenness of high-speed AMs, and therefore for conferring band-pass tuning to AM velocity at high carrier levels (see Figure 9-2B and C). This account of the sensitivity of cortical cells to linear AMs thus draws on basic properties of cortical cells revealed earlier in studies with tones and with noise pulses: the timing of afferent events evoked by the stimulus, the dynamic character of those afferent events (or of cortical responses to them), and the overlap between the short-term stimulus spectrum and the neuron's excitatory and inhibitory frequency–intensity response areas. The same sensitivity to short-term stimulus spectra most likely engenders a marked sensitivity to the rise-time of tone pulses (Phillips, 1988b). This is because the faster the rise-time, the broader the short-term spectrum at tone onset. Neurons with a nonmonotonic rate-response for conventional rise-time tones, and which are sensitive to stimulus bandwidth, are increasingly suppressed by faster rise-time tone pulses, and become monotonic for long rise-time tones in which the onset spectrum is narrow.

Schreiner and Urbas (1986) have recently surveyed the cat's anterior auditory cortical field for sensitivity to the rate of sinusoidal modulation of CF tones. They found that most cells could be assignd a 'best modulation frequency,' which was typically less than 50 to 100 Hz. In many instances, the modulation transfer function (or tuning to the frequency of the AM) was relatively independent of AM depth. The factors responsible for this tuning are not completely clear. The high-frequency cut-off in the modulation transfer function might reflect the inability of cortical cell responses to follow high repetition rates of effective stimuli (e.g., Creutzfeldt et al., 1980; see also Rees and Møller, 1987). Because these modulation frequencies are very low, the sidebands in the stimulus spectra are close to the cell's CF, and therefore unlikely to invade inhibitory stimulus domains (Rees and Møller, 1987).

It is of equal interest that the best modulation frequencies of cortical cells are systematically lower than those of brainstem neurons (Creutzfeldt et al., 1980; Møller, 1976; Møller and Rees, 1986; Rees and Møller, 1983, 1987), and these, in turn, are lower than those of primary auditory nerve fibers (Møller, 1976). The lower best modulation frequencies of neurons in the auditory cortex may be paralleled by their relative inability to entrain to high-frequency, repetitive transient stimuli (cf. Godfrey et al., 1975; Møller, 1969; Ribaupierre et al., 1972; Rouiller and Ribaupierre, 1982). Neither the reasons for these properties, nor their further functional consequences, have been studied in detail.

Three forms of frequency-modulated (FM) stimuli have been applied in the study of cortical cells. These are the sinusoidal modulation of a continuous CF carrier tone, the linear FM of an ongoing stimulus, and the FM presented in the form of a tone pulse. The factors shaping responses to linear FMs, in the form of both tone pulses and linear modulations of otherwise invariant, ongoing signals, have been studied in some detail and, as might be expected, are probably quite different.

Suga (1965) studied bat cortical neurons with both static and FM tone pulse stimuli. The frequency excursions covered by the FMs were typically wider than the excitatory response areas of the cells studied, but Suga was careful to manipulate both the starting frequencies of the FMs and the stimulus amplitudes at which they were presented. Suga found that neurons varied in their sensitivity to the direction in which the FM passed through the excitatory response area (see also Suga, 1968): some cells responded preferentially to upward or downward FMs and others were nonselective, and in other cells static tone pulses were not excitatory but FMs were. Suga found that neural preference for the direction of frequency change in an FM pulse was understandable in terms of the temporal sequence with which the stimulus activated the inhibitory and excitatory response areas of the neurons (Suga, 1965, 1968). If the FM pulse first activated an inhibitory input, then, even though the stimulus thereafter traversed the excitatory response area, the neuron would not be excited by it. On the other hand, if the FM pulse first penetrated an excitatory tonal domain, the neuron would be excited by the FM pulse whether or not the stimulus later invaded an inhibitory stimulus domain. These observations were supported by the further observation that the sequential activation of inhibitory and excitatory response areas by successive tones produced like effects (Suga, 1965).

These data have two important implications. First, responses to brief FM pulses may be domained by the afferent events evoked by pulse onset. A plausible, if speculative, explanation is that an inhibitory response evoked by pulse onset outlasts the triggering stimulus event and therefore suppresses the excitatory response that would otherwise be evoked when the frequency of the stimulus invades the excitatory response area. In the case of excitatory FM pulses (i.e., those initiated in the excitatory response area), the neural circuitry leading to the generation of spike discharges has already been activated by the time the stimulus invades an inhibitory stimulus domain. A second ramification of these observations is that the mechanisms underlying sensitivity to the direction of frequency change, and therefore the direction of stimulus movement across the receptor surface (cochlear partition), are quite unlike those in the cortices of other sensory modalities. In the somatic (Gardner and Costanzo, 1980) and visual cortex (Goodwin et al., 1975), direction selectivity is a property manifested *within* an excitatory or inhibitory region of the receptive field and does not require the sequential activation of those regions.

Phillips and his colleagues (1985b) studied the sensitivity of cat cortical cells to narrow (2-kHz-wide) linear FMs of continuous carrier tones. They systematically varied the frequency range covered by the ramp in and around the excitatory response area of each neuron, and examined the sensitivity to the direction of FM for each excursion. Figure 9-3 presents data for one neuron. Panel A shows a frequency response function for the neuron (i.e., the spike rate is plotted as a function of the frequency of a tone pulse of fixed amplitude—solid line). The dashed line shows the cell's direction preference for 2-kHz-wide FMs centered at the frequency specified by the abscissa (quantified using the ratio of peak firing rates for [UP − DOWN]/[UP + DOWN] FMs).

Linear FMs whose excursions did not invade the excitatory response area failed to evoke spike discharges. For such FMs, direction preference quotients were zero (Figure 9-3A: the extremes of the direction preference curve). FMs that invaded the high- or low-frequency sides of the response area, but did not cross the CF, resulted in near-perfect direction preference for downward or upward FMs, respectively. FMs centered near CF evoked vigorous responses in both directions of frequency change.

These observations suggest that the direction preference for embedded FMs reflects the extent to which the FM provides the neuron with an increasingly effective stimulus. To explore this notion, Phillips et al. replotted the direction pref-

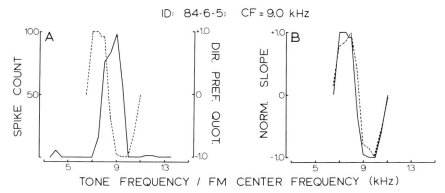

Figure 9-3 Data for neuron 84-6-5, which was studied with both tone pulses and 2-kHz wide, linear FMs. (A) Solid line is the spike-rate frequency function for constant-amplitude tone pulses. Dashed line indicates the neuron's direction preference for an FM stimulus centered at the frequency specified by the abscissa. Positive values of the direction preference quotient (DIR. PREF. QUOT.) indicate preference for ascending FMs; negative values indicate preference for descending FMs. (B) Solid line is a reproduction of the direction preference curve shown in A. The dashed line represents the normalized, instantaneous gradient of the frequency-response function measured over the same 2-kHz-wide frequency ranges traversed by FM ramps centered at the same frequency. [Data are from the study of Phillips et al. (1985b).]

erence data for each neuron, and superimposed it over a second curve representing the (normalized) instantaneous gradient of the frequency response function, measured over the same 2-kHz-wide excursions covered by the relevant FMs. These data, shown Figure 9-3B, are derived from the data in Figure 9-3A. The continuous line in Figure 9-3B reproduces the direction preference curve in panel A; the dashed line shows the normalized, instantaneous slope of the frequency response function. The correspondence between these curves was obvious in almost all cases.

In light of the foregoing discussions, the reasons for this pattern of responsiveness to linear FM ramps seem clear. It is likely that a cortical neuron shows short-term adaptation to the effective level of the maintained tone that precedes the FM. Only those FMs that provide the neuron with an inherently more effective input will therefore be able to evoke spike discharges from the neuron. The extent to which the FM can provide that input can reasonably be expected to reflect the extent to which the stimulus itself exceeds the adapted threshold, and hence to reflect the slope of the frequency response function over the range traversed by the FM ramp. It is of particular interest that, although neurons were relatively homogeneous in the organization of their responses to these embedded FMs, the same cells were strikingly heterogeneous in their direction selectivity for broad FM pulses that started and ended well beyond the limits of the pure tone response area (Phillips et al., 1985b). The latter data were concordant with those of Suga (1965) in that cat cortical cells were highly individual in their direction preferences for such FMs. The neuron whose data are presented in Figure 9-3 showed a complete preference for downward-directed broad FM pulses. These data confirm that the neural processes conferring direction selectivity for small, embedded FMs are quite different from those responsible for direction selectivity for very broad pulse FMs.

Whitfield and Evans (1965) were the first to describe the responses of cat cortical cells to FM stimuli. Their study was remarkable not only for its precedence, but also because it used unanesthetized animals. Their study revealed many cells whose preference for the direction of FM showed patterns described by Phillips et al. (1985b) two decades later. Whitfield and Evans also described two other forms of sensitivity to FM stimuli that warrant mention here. One group of cells responded to a single direction of frequency change, regardless of the absolute excursion covered by the FM within the excitatory response area. The nature of the mechanisms responsible for this behavior are not known. A second group of neurons were not excited by static tonal stimuli, but were sensitive to FM stimuli. Such cells have also been described by Suga (1965), who observed that these neurons were often actively inhibited by static tonal stimuli. Neurons excited by complex sounds, but inhibited by tone pulses, have been described in the cochlear nucleus (Voigt and Young, 1980; Young and Brownell, 1976), where some of the relevant neuronal circuitry has been worked out. In the cat's cortex, other authors have described cells that are responsive to wide-band noise but unresponsive to tones (Goldstein et al., 1968). The more recent quantitative studies, however, have employed tonal search stimuli, and therefore would have missed these cells.

SENSITIVITY OF CORTICAL NEURONS TO VOCALIZATIONS

In a pioneering study, Wollberg and Newman (1972) described the responses of auditory cortex neurons in alert primates to species-specific vocalizations. They found that some neurons were strikingly selective in the vocalizations to which they responded, whereas other neurons were considerably less vocalization-specific (see also Newman and Wollberg, 1973a). Since that study, a number of independent laboratories have confirmed the general finding of variable sensitivity of cortical cells to vocalizations, con-specific or otherwise (Creutzfeldt et al., 1980; Glass and Wollberg, 1983; Newman and Symmes, 1974, 1979; Newman and Wollberg, 1973b; Sovijarvi, 1975; Steinschneider et al., 1982; Winter and Funkenstein, 1973). These studies have also provided insights into the factors on which sensitivity to a given vocalization, or selectivity for that vocalization, might be based.

One approach to this question has focused on the extent to which a correspondence exists between the long-term spectral content of an effective vocalization, on the one hand, and the excitatory (or inhibitory) frequency response ranges of the relevant neurons, on the other. This approach has met with variable success. One study found little such correspondence (Newman and Wollberg, 1973a), but others reported a greater congruence (63%—Winter and Funkenstein, 1973; 68%—Sovijarvi, 1975; 76%—Newman and Wollberg, 1973b). At least three factors may contribute to these data. The first is methodological. In most of these studies, sensitivity to tones was examined at a single-stimulus amplitude, typically close to the peak (or rms) amplitude of the vocalizations used. This provides a limited assessment of a neuron's response area (see preceding). Moreover, the fact that the peak (or rms) levels of the vocalizations and tones may be matched does not imply that the relevant spectrum levels are also matched. These considerations temper the level of the conclusions to be derived from these studies. Second, an overlap of the frequency spectra of complex sounds and the effective frequency domains of cortical neurons has independently been shown to be insufficient to account for the responsiveness of cortical cells to those sounds. Recall that coextensive FMs (or AMs) that vary only in the direction of frequency (or amplitude) change are not necessarily equally effective in evoking spike discharges from cortical cells (see preceding section). Such neurons are not passive acoustic filters. Finally, the more recent studies of responses to AM stimuli have emphasized the sensitivity of cortical cells to short-term stimulus events and their spectra. These observations suggest that it might be more profitable to examine the correspondence between the spectral content of individual vocalization components and response area organization of a single neuron. Sensitivity to the sequencing of a vocalization's components may well be mediated by quite different processes.

A second approach to understanding the factors shaping responsiveness to vocalizations has explored the temporal organization of spike discharges evoked by the stimuli, and the effects of deletions of various acoustic elements from the vocalizations and additions to them. Wollberg and Newman (1972) observed that many neurons displayed complex spike patterns in response to a given vocalization, and these temporal patterns most likely followed the short-term acoustic structure of the stimulus. They also observed that the deletion of a vocalization component could significantly alter the responses to vocalization components following the deleted element. More recent studies have revealed that responsiveness to an isolated vocalization fragment is often greater than that to the same acoustic element when it is part of the complete vocalization from which it was drawn (Newman and Symmes, 1979). In the same vein, Steinschneider et al. (1982) found that the responses of cortical neurons in the awake primate to a synthetic human vowel sound were significantly modified by the acoustic identity of a preceding consonant. These observations confirm that responsiveness to a given element in a temporally complex signal may be significantly conditioned or biased by the responses to a preceding stimulus component.

Glass and Wollberg (1983) have provided further evidence on this issue. They studied the responses of cortical cells in alert primates to normal and reversed species-specific vocalizations. This stimulus manipulation maintains the spectral content of the stimulus, but reverses the order of its acoustic components. For some neurons, reversing the vocalization eliminated responsiveness to the stimulus. In other cells, it reversed the temporal pattern of spike discharges evoked, and in still other cells the reversal altered the cadence of spike discharges unpredictably.

Perhaps three conclusions might reasonably be drawn from studies of the sensitivity of cortical neurons to vocalizations. First, an overlap between the long-term stimulus frequency spectra and the excitatory (or inhibitory) response areas of the neuron's inputs may be a necessary, but not a sufficient, condition for the effectiveness of that vocalization in eliciting (or suppressing) spike discharges. Second, to some extent, the time structure of the responses to a vocalization may follow the short-term acoustic structure of the stimulus. Third, responses to successive stimulus elements in the vocalization need not be independent. These conclusions are in keeping with findings from the studies of cortical cells with combined tone-noise and FM/AM stimuli.

SPECULATIONS ON SOME EFFECTS OF AUDITORY CORTEX LESIONS

It remains to examine the extent to which the foregoing data may be concordant with the available evidence on the behavioral consequences of auditory cortex lesions. This issue is complicated by the fact that, in some instances, visualization of apparent behavioral deficits may depend as much on the nature of the motor response required in the testing procedure as on the specifically sensory processing required by the relevant discrimination (Heffner, 1978; Heffner and Masterton, 1975; Ravizza and Masterton, 1972; Thompson, 1960). Equally pertinent, ascribing behavioral deficits to lesion of a given cortical field (including AI) must be justified by factors other than the locus of the lesion with respect to the cerebral fissural pattern, because the location of these fields may vary between individuals (Merzenich et al., 1975; Merzenich and Brugge, 1973; Rose, 1949). In the case of human lesions, investigators' conclusions are constrained not only by the individuality of lesion locus, but also by the paucity of evidence on the identity of cortical fields within the damaged zone.

Neither animals nor human beings with bilateral lesions of the primary auditory cortex are deaf. In lower species, such lesions may be without significant effect on the behavioral pure tone audiogram (Kryter and Ades, 1943; Ravizza and Masterton, 1972). In primates, generally (Heffner and Heffner, 1986a), and humans in particular (Jerger et al., 1969; Miceli, 1982) there may be some mild or moderate elevation of tone thresholds over some parts of the audible frequency range. In humans especially, thresholds for detecting very brief stimuli (<50 to 100 msec in duration) may be significantly elevated (Gersuni et al., 1971). There appear to be no deterioration of difference limens for frequency (Thompson, 1960) or intensity (Raab and Ades, 1946; Swisher, 1967), nor need there be deficits in the discrimination of the direction of frequency change in FM stimuli (Kelly and Whitfield, 1971). Animals with lesions invading the insular and ventral temporal fields, may have profound deficits in discrimination of long-term tone pulse patterns (Colavita et al., 1974; Kelly, 1973), although the extent to which these are acoustic and sensory deficits rather than deficits in acoustic memory (Chorazyna and Stepien, 1961) is unclear.

Primates with bilateral lesions of the primary auditory cortex show marked deficits in their ability to discriminate species-specific vocalizations (Hupfer et al., 1977), and recent evidence indicates that this behavioral capacity depends primarily on the functional integrity of the left auditory cortex (Heffner and Heffner, 1986b). In humans, a neurological syndrome termed *pure word deafness* has been attributed to bilateral primary auditory cortex damage (see Goldstein, 1974, for historical review). This diagnosis carries the implication that deficits in acoustic discriminations are restricted to linguistic stimulus material. Only recently, however, have the abilities of these patients to discriminate complex nonverbal sounds been systematically tested beyond confrontation with environmental sounds of (arguably) complex short-term spectrotemporal microstructure. With some possible exceptions (Metz-Lutz and Dahl, 1984), the more recent studies of this condition have suggested that the deficit is an acoustic sensory one in the processing of temporally complex signals per se, of which the failure to discriminate speech sounds is the most obvious and debilitating manifestation (Auerbach et al., 1982; Miceli, 1982; Saffran et al., 1976; von Stockert, 1982). In such cases, the deficit underlying the poor speech discrimination may therefore be at the prephonemic level (Auerbach et al., 1982; Saffran et al., 1976).

This notion, if confirmed by future studies, has a further ramification. It suggests that the primary auditory cortex contribution to speech perception in humans is a specifically acoustic one—in the coding of the complex spectra of speech sounds and their temporal variations. This view is compatible with an account of human speech perception mechanisms advocated most recently by Blumstein and Stevens (1979, 1980). Blumstein and Stevens drew on both physical acoustics and human psychophysics and revealed that it was possible to understand the acoustic perception of speech on the basis of the short-term spectra of speech sound components and their sequencing. This account requires that the auditory nervous system contain elements that, individually or collectively, are sensitive to spectral shape, spectral and amplitude changes, and the sequencing of such acoustic elements. The phonetic elaboration of the acoustic signals is presumably a later processing stage.

The studies of the coding of complex sounds by cortical auditory neurons have not even approached the degree of stimulus precision used by Blumstein and Stevens. Nevertheless, what is becoming clear is that the responses of cortical (and likely some subcortical) neurons to temporally varying signals may often be dominated by the short-term stimulus events, and since responses to these events are not independent, the ordering of stimulus elements is often a critical stimulus variable in determining responsivity to the complex sound as a whole. The neural machinery the animal studies have revealed to be available for encoding speech signals is thus of the form required by an acoustic model of speech perception.

ACKNOWLEDGMENTS

Special thanks are due to research collaborators who have influenced the research described in this chapter: J. F. Brugge, M. S. Cynader, S. E. Hall, J. B. Kelly, D. R. F. Irvine, and J. D. Pettigrew. Some of the research described here was supported by NIH International Research Fellowship F05 TW03102 and NSERC Grants U0442 and E2745 to the author. This chapter is dedicated to the memory of my father, who passed away in February 1987.

REFERENCES

Abeles, M., and Goldstein, M. H. Jr. (1972). Responses of single units in the primary auditory cortex of the cat to tones and to tone pairs. *Brain Res.* 42:337–352.

Aitkin, L. M., Irvine, D. R. F., and Webster, W. R. (1984). Central neural mechanisms of hearing. In: I. Darian-Smith (ed.), *Handbook of Physiology.* Section 1, Vol. 3, P. 2, pp. 675–737. American Physiological Society, G. Bethesda, MD.

Aitkin, L. M., Webster, W. R., Veale, J. L. and Crosby, D. C. (1975). Inferior colliculus. I. Comparison of response properties of neurons in central, pericentral, and external nuclei of adult cat. *J. Neurophysiol.* 38:1196–1207.

Andersen, R. A., Knight, P. L., and Merzenich, M. M. (1980). The thalamocortical connections of AI, AII and the anterior auditory field (AAF) in the cat: evidence for two largely segregated systems of connections. *J. Comp. Neurol.* 194:663–701.

Auerbach, S. H., Allard, T., Naeser, M., Alexander, M. P., and Albert, M. L. (1982). Pure word deafness. Analysis of a case with bilateral lesions and a defect at the prephonemic level. *Brain* 105:271–300.

Benson, D. A., and Teas, D. C. (1976). Single unit study of binaural interaction in the auditory cortex of the chinchilla. *Brain Res.* 103:313–338.

Benson, D. A., Heinz, R. D., and Goldstein, M. H., Jr. (1981). Single unit activity in the auditory cortex of monkeys actively localizing sound sources: spatial tuning and behavioral dependency. *Brain Res.* 219:249–267.

Blumstein, S. E., and Stevens, K. N. (1979). Acoustic invariance in speech production: evidence from measurements of the spectral characteristics of stop consonants. *J. Acoust. Soc. Am.* 66:1001–1017.

Blumstein, S. E., and Stevens, K. N. (1980). Perceptual invariance and onset spectra for stop consonants in different vowel environments. *J. Acoust. Soc. Am.* 67:648–662.

Brugge, J. F. (1975). Progress in neuroanatomy and neurophysiology of auditory cortex. In: D. B. Tower (ed.). *The Nervous System.* Vol. 3, pp. 97–111. New York: Raven Press.

Brugge, J. F., and Geisler, C. D. (1978). Auditory mechanisms of the lower brainstem. *Annu. Rev. Neurosci.* 1:363–394.

Brugge, J. F., and Merzenich, M. M. (1973). Responses of neurons in auditory cortex of the macaque monkey to monaural and binaural stimulation. *J. Neurophysiol.* 36:1138–1158.

Brugge, J. F., Anderson, D. J., and Aitkin, L. M. (1970). Responses of neurons in the dorsal nucleus of the lateral lemniscus of cat to binaural tonal stimulation. *J. Neurophysiol.* 33:441–458.

Brugge, J. F., Dubrovsky, N. A., Aitkin, L. M., and Anderson, D. J. (1969). Sensitivity of single neurons in auditory cortex of cat to binaural tonal stimulation: effects of varying interaural time and intensity. *J. Neurophysiol.* 32:1005–1024.

Calford, M. B., and Pettigrew, J. D. (1984). Frequency dependence of directional amplification of the cat's pinna. *Hearing Res.* 14:13–19.

Chorazyna, H., and Stepien, L. (1961). Impairment of auditory recent memory produced by cortical lesions in dogs. *Acta. Biol. Exp. Versovie.* 21:177–187.

Colavita, F. B., Szeligo, F. V., and Zimmer, S. D. (1974). Temporal pattern discrimination in cats with insular-temporal lesions. *Brain Res.* 79:153–156.

Costalupes, J. A., Young, E. D., and Gibson, D. J. (1984). Effects of continuous noise backgrounds on rate response of auditory nerve fibers in cat. *J. Neurophysiol.* 51:1326–1344.

Creutzfeldt, O., Hellweg, F.-C., and Schreiner, C. (1980). Thalamocortical transformation of responses to complex auditory stimuli. *Exp. Brain Res.* 39:87–104.

Evans, E. F. (1968). Cortical representation. In: A. V. S. de Reuck and J. Knight (eds.), *Hearing Mechanisms in Vertebrates*, pp. 272–295. Boston: Little, Brown.

Evans, E. F. (1975a). Normal and abnormal functioning of the cochlear nerve. *Symp. Zool. Soc. Lond.* 37:133–165.

Evans, E.F. (1975b). The cochlear nerve and cochlear nucleus. In: W. D. Keidel and W. D. Neff (eds.). *Handbook of Sensory Physiology.* vol. 5, pt. 2, pp. 1–108. Heidelberg: Springer-Verlag.

Evans, E. F., and Nelson, P. G. (1973). The responses of single neurons in the cochlear nucleus of the cat as a function of their location and anesthetic state. *Exp. Brain Res.* 17:402–427.

Fitzpatrick, K. A., and Imig, T. J. (1980). Auditory cortico-cortical connections in the owl monkey. *J. Comp. Neurol.* 192:589–610.

Funkenstein, H. H., and Winter, P. (1973). Responses to acoustic stimuli of units in the auditory cortex of awake squirrel monkeys. *Exp. Brain Res.* 18:464–488.

Fuzessery, Z. M., and Pollak, G. D. (1985). Determinants of sound location selectivity in bat inferior colliculus: a combined dichotic and free-field study. *J. Neurophysiol.* 54:757–781.

Gardner, E. P., and Costanzo, R. M. (1980). Neuronal mechanisms underlying directional sensitivity of somatosensory cortical neurons in awake monkeys. *J. Neurophysiol.* 43:1342–1354.

Geisler, C. D., and Sinex, D. G. (1980). Responses of primary auditory fibers to combined noise and tonal stimuli. *Hearing Res.*, 3:317–334.

Gersuni, G. V., Baru, A. V., Karaseva, T. A., and Tonkonogii, I. M. (1971). Effects of temporal lobe lesions on perception of sounds of short duration. In: G. V. Gersuni (ed.), *Sensory Processes at the Neuronal and Behavioral Levels*, pp. 287–300. New York: Acadmeic Press.

Gibson, D. J., Young, E. D., and Costalupes, J. A. (1985). Similarity of dynamic range adjustment in auditory nerve and cochlear nuclei. *J. Neurophysiol.* 53:940–958.

Glass, I., and Wollberg, Z. (1983). Responses of cells in the auditory cortex of awake squirrel monkeys to normal and reversed species-specific vocalizations. *Hearing Res.* 9:27–33.

Godfrey, D. A., Kiang, N. Y. S., and Norris, B. E. (1975). Single unit activity in the posteroventral cochlear nucleus of the cat. *J. Comp. Neurol.* 162:247–268.

Goldberg, J. M., and Brown, P. B. (1969). Response of binaural neurons of dog superior olivary complex to dichotic tonal stimulation: some physiological mechanisms of sound localization. *J. Neurophysiol.* 32:613–636.

Goldstein, M. H., Jr., and Kiang, N. Y. S. (1958). Synchrony of neural activity in electric responses evoked by transient acoustic stimuli. *J. Acoust. Soc. Am.* 30:107–114.

Goldstein, M. H., Jr., Hall, J. L., and Butterfield, B. O. (1968). Single unit activity in the primary auditory cortex of unanesthetized cats. *J. Acoust. Soc. Am.* 43:444–455.

Goldstein, M. N. (1974). Auditory agnosia for speech ("pure word-deafness"). *Brain Lang.* 1:195–204.

Goodwin, A. W., Henry, G. H., and Bishop, P. O. (1975). Direction selectivity of simple striate cells: properties and mechanism. *J. Neurophysiol.* 38:1500–1523.

Greenwood, D. D., and Maruyama, N. (1965). Excitatory and inhibitory response areas of auditory neurons in the cochlear nucleus. *J. Neurophysiol.* 28:863–892.

Hall, J. L., and Goldstein, M. H. Jr. (1968). Representation of binaural stimuli by single units in primary auditory cortex of unanesthetized cats. *J. Acoust. Soc. Am.* 43:456–461.

Harrison, R. V. (1981). Rate-vs-intensity functions and related AP responses in normal and pathological guinea pig and human cochleas. *J. Acoust. Soc. Am.* 70:1036–1044.

Harrison, R. V., and Evans, E. F. (1977). The effects of hair cell loss (restricted to outer hair cells) on the threshold and tuning properties of cochlear fibers in the guinea pig. *Les Collogues de l'Institute National de la Sante et de la Recherche Medicale* 68:105–124.

Heffner, H. E. (1978). Effect of auditory cortex ablation on localization and discrimination of brief sounds. *J. Neurophysiol.* 41:963–976.

Heffner, H. E., and Heffner, R. S. (1986a). Hearing loss in Japanese Macaques following bilateral auditory cortex lesions. *J. Neurophysiol.* 55:256–271.

Heffner, H. E., and Heffner, R. S. (1986b). Effect of unilateral and bilateral auditory cortex lesions on the discrimination of vocalizations by Japanese Macaques. *J. Neurophysiol.* 56:683–701.

Heffner, H. E., and Masterton, R. B. (1975). Contribution of auditory cortex to sound localization in the monkey (Macaca mulatta). *J. Neurophysiol.* 38:1340–1358.

Hupfer, K., Jurgens, U., and Ploog, D. (1977). The effect of superior temporal lesions on the recognition of species-specific calls in the squirrel monkey. *Exp. Brain Res.* 30:75–87.

Imig, T. J., and Adrian, H. O. (1977). Binaural columns in the primary field (AI) of cat auditory cortex. *Brain Res.* 138:241–257.

Imig T. J., and Brugge, J. F. (1978). Sources and terminations of callosal axons related to binaural and frequency maps in primary auditory cortex of the cat. *J. Comp. Neurol.* 182:637–660.

Imig, T. J., and Morel, A. (1983). Organization of the thalamocortical auditory system in the cat. *Annu. Rev. Neurosci.* 6:95–120.

Imig T. J., and Morel, A. (1984). Topographic and cytoarchitectonic organization of thalamic neurons related to their targets in low-, middle-, and high-frequency representations in cat auditory cortex. *J. Comp. Neurol.* 227:511–539.

Imig, T. J., and Morel, A. (1985a). Tonotopic organization in ventral nucleus of medial geniculate body in the cat. *J. Neurophysiol.* 53:309–340.

Imig, T. J., and Morel, A. (1985b). Tonotopic organization in lateral part of posterior group of thalamic nuclei in the cat. *J. Neurophysiol.* 53:836–851.

Imig, T. J., and Reale, R. A. (1980). Patterns of cortico-cortical connections related to tonotopic maps in cat auditory cortex. *J. Comp. Neurol.* 192:293–332.

Imig, T. J., and Reale, R. A. (1981). Ipsilateral corticocortical projections related to binaural columns in cat primary auditory cortex. *J. Comp. Neurol.* 203:1–14.

Imig, T. J., Ruggero, M. A., Kitzes, L. M., Javel, E., and Brugge, J. F. (1977). Organization of auditory cortex in the owl monkey (Aotus trivirgatus). *J. Comp. Neurol.* 171:111–128.

Jenkins, W. M., and Masterton, R. B. (1982). Sound localization: effects of unilateral lesions in central auditory pathways. *J. Neurophysiol.* 47:987–1016.

Jenkins, W. M., and Merzenich, M. M. (1984). Role of cat primary auditory cortex for sound localization behavior. *J. Neurophysiol.* 52:819–847.

Jerger, J., Weikers, N. J., Sharbrough, F. W., and Jerger, S. (1969). Bilateral lesions of the temporal lobe. A case study. *Acta Otolaryngol.* (Suppl.) 258:1–51.

Kavanagh, G. L., and Kelly, J. B. (1987). Contribution of auditory cortex to sound localization by the ferret (Mustela putorius). *J. Neurophysiol.* 57:1746–1766.

Kelly, J. B. (1973). The effects of insular and temporal lesions in cats on two types of auditory pattern discrimination. *Brain Res.* 62:71–87.

Kelly, J. B., and Whitfield, I. C. (1971). Effects of auditory cortical lesions on discriminations of rising and falling frequency-modulated tones. *J. Neurophysiol.* 34:802–816.

Kiang, N. Y. S., and Moxon, E. C. (1974). Tails of tuning curves of auditory nerve fibers. *J. Acoust. Soc. Am.* 55:620–630.

Kiang, N. Y. S., Watanabe, T., Thomas, E. C., and Clark, L. F. (1965). Discharge patterns of single fibers in the cat's auditory nerve. In: *MIT Research monograph no. 35.* Cambridge: The MIT Press.

Kitzes, L. M., Wrege, K. S., and Cassady, J. M. (1980). Patterns of responses of cortical cells to binaural stimulation. *J. Comp. Neurol.* 192:455–472.

Kryter, K. D., and Ades, H. W. (1943). Studies on the function of the higher acoustic nerve centers in the cat. *Am. J. Psychol.* 56:501–536.

Liberman, M. C. (1978). Auditory nerve responses from cats raised in a low-noise chamber. *J. Acoust. Soc. Am.* 63:442–455.

Masterton, R. B., and Imig, T. J. (1984). Neural mechanisms for sound localization. *Annu. Rev. Physiol.* 46:275–287.

McMullen, N. T., and Glaser, E. M. (1982). Tonotopic organization of rabbit auditory cortex. *Exp. Neurol.* 75:208–220.

Merzenich, M. M., and Brugge, J. F. (1973). Representation of the cochlear partition on the superior temporal plane of the macaque monkey. *Brain Res.* 50:275–296.

Merzenich, M. M., Kaas, J. H., and Roth, G. L. (1976). Auditory cortex in the grey squirrel: tonotopic organization and architectonic fields. *J. Comp. Neurol.* 166:387–402.

Merzenich, M. M., Knight, P. L., and Roth, G. L. (1975). Representation of cochlea within primary auditory cortex in the cat. *J. Neurophysiol.* 38:231–249.

Metz-Lutz, M. N., and Dahl, E. (1984). Analysis of word comprehension in a case of pure word deafness. *Brain Lang.* 23:13–25.

Miceli, G. (1982). The processing of speech sounds in a patient with cortical auditory disorder. *Neuropsychologia* 20:5–20.

Middlebrooks, J. C., and Knudsen, E. I. (1984). A neural code for auditory space in the cat's superior colliculus. *J. Neurosci.* 4:2621–2634.

Middlebrooks, J. C., and Pettigrew, J. D. (1981). Functional classes of neurons in primary auditory cortex of the cat distinguished by sensitivity to sound location. *J. Neurosci.* 1:107–120.

Middlebrooks, J. C., and Zook, J. M. (1983). Intrinsic organization of the cat's medial geniculate body identified by projections to binaural response-specific bands in the primary auditory cortex. *J. Neurophysiol.* 3:203–224.

Middlebrooks, J. C., Dykes, R. W., and Merzenich, M. M. (1980). Binaural response-specific bands in primary auditory cortex (AI) of the cat: topographical organization orthogonal to isofrequency contours. *Brain Res.* 181:31–48.

Møller, A. R. (1969). Unit responses in the rat cochlear nucleus to repetitive, transient sounds. *Acta Physiol. Scand.* 75:542–551.

Møller, A. R. (1976). Dynamic properties of primary auditory fibers compared with cells in the cochlear nucleus. *Acta Physiol. Scand.* 98:157–167.

Møller, A. R., and Rees, A. (1986). Dynamic properties of the responses of single neurons in the inferior colliculus of the rat. *Hearing Res.* 24:203–215.

Moore, D. R., Hutchings, M. E., Addison, P. D., Semple, M. N., and Aitkin, L. M. (1984). Properties of spatial receptive fields in the central nucleus of the cat inferior colliculus. II. Stimulus intensity effects. *Hearing Res.* 13:175–188.

Morest, D. K. (1965). The laminar structure of the medial geniculate body of the cat. *J. Anat.* 99:143–160.

Newman, J. D., and Symmes, D. (1974). Arousal affects on unit responsiveness to vocalizations in squirrel monkey auditory cortex. *Brain Res.* 78:125–138.

Newman, J. D., and Symmes, D. (1979). Feature detection by single units in squirrel monkey auditory cortex. *Exp. Brain Res.* (Suppl.) 2:140–145.

Newman, J. D., and Wollberg, D. (1973a). Multiple coding of species-specific vocalizations in the auditory cortex of squirrel monkeys. *Brain Res.* 54:287–304.

Newman, J. D., and Wollberg, Z. (1973b). Responses of single neurons in the auditory cortex of squirrel monkeys to variants of a single cell type. *Exp. Neurol.* 40:821–824.

Orman, S. S., and Phillips, D. P. (1984). Binaural interactions of single neurons in posterior field of cat auditory cortex. *J. Neurophysiol.* 51:1028–1039.

Palmer, A. R., and King, A. J. (1985). A monaural space map in the guinea pig superior colliculus. *Hearing Res.* 17:267–280.

Pfingst, B. E., and O'Connor, T. A. (1981). Characteristics of neurons in auditory cortex of monkeys performing a simple auditory task. *J. Neurophysiol.* 45:16–34.

Phillips, D. P. (1985). Temporal response features of cat auditory cortex neurons contributing to sensitivity to tones delivered in the presence of continuous noise. *Hearing Res.* 19:253–268.

Phillips, D. P. (1987). Stimulus intensity and loudness recruitment: neural correlates. *J. Acoust. Soc. Am.* 82:1–12.

Phillips, D. P. (1988a). Introduction to anatomy and physiology of the central auditory nervous system. In: A. F. Jahn and J. R. Santos-Sacchi (eds.). *Physiology of the Ear*, New York: Raven Press, in press.

Phillips, D. P. (1988b). Effect of tone-pulse rise time on rate-level functions of cat auditory cortex neurons: Excitatory and inhibitory processes shaping responses to tone onset. *J. Neurophysiol.* 59: in press.

Phillips, D. P., and Brugge, J. F. (1985). Progress in neurophysiology of sound localization. *Annu. Rev. Psychol.* 36:245–274.

Phillips, D. P., and Cynader, M. S. (1985). Some neural mechanisms in the cat's auditory cortex underlying sensitivity to combined tone and wide-spectrum noise stimuli. *Hearing Res.* 18:87–102.

Phillips, D. P., and Hall, S. E. (1986). Spike-rate-intensity functions of cat cortical neurons studied with combined tone-noise stimuli. *J. Acoust. Soc. Am.* 80:177–187.

Phillips, D. P., and Hall, S. E. (1987). Responses of single neurons in cat auditory cortex to time-varying stimuli: linear amplitude modulations. *Exp. Brain Res.* 67:479–492.

Phillips, D. P., and Irvine, D. R. F. (1981a). Responses of single neurons in physiologically defined primary auditory cortex (AI) of the cat: frequency tuning and responses to intensity. *J. Neurophysiol.* 45:48–58.

Phillips, D. P., and Irvine, D. R. F. (1981b). Responses of neurons in physiologically defined area AI of cat cerebral cortex: sensitivity to interaural intensity differences. *Hearing Res.* 4:299–307.

Phillips, D. P., Calford, M. B., Pettigrew, J. D., Aitkin, L. M., and Semple, M. N. (1982). Directionality of sound pressure transformation at the cat's pinna. *Hearing Res.* 8:13–28.

Phillips, D. P., and Orman, S. S. (1984). Responses of single neurons in posterior field of cat auditory cortex to tonal stimulation. *J. Neurophysiol.* 51:147–163.

Phillips, D. P., Mendelson, J. R., Cynader, M. S., and Douglas, R. M. (1985b). Responses of single neurones in cat auditory cortex to time-varying stimuli: frequency-modulated tones of narrow excursion. *Exp. Brain Res.* 58:443–454.

Phillips, D. P., Orman, S. S., Musicant, A. D., and Wilson, G. F. (1985a). Neurons in the cat's primary auditory cortex distinguished by their responses to tone and wide-spectrum noise. *Hearing Res.* 18:73–86.

Raab, D. H., and Ades, H. W. (1946). Cortical and midbrain mediation of a conditioned discrimination of acoustic intensities. *Am. J. Psychol.* 59:59–83.

Ravizza, R. J., and Masterton, R. B. (1972). Contribution of neocortex to sound localization in opossum (Didelphis virginiana). *J. Neurophysiol.* 35:344–356.

Reale, R. A., and Imig, T. J. (1980). Tonotopic organization in auditory cortex of the cat. *J. Comp. Neurol.* 192:265–291.

Reale, R. A., and Kettner, R. E. (1986). Topography of binaural organization in primary auditory cortex of the cat: effects of changing interaural intensity. *J. Neurophysiol.* 56:663–682.

Rees, A., and Møller, A. R. (1983). Responses of neurons in the inferior colliculus of the rat to AM and FM tones. *Hearing Res.* 10:301–330.

Rees, A., and Møller, A. R. (1987). Stimulus properties influencing the responses of inferior colliculus neurons to amplitude-modulated sounds. *Hearing Res.* 27:129–143.

Rhode, W. S., Geisler, C. D., and Kennedy, D. T. (1978). Auditory nerve fiber responses to wide-band noise and tone combinations. *J. Neurophysiol.* 41:692–704.

Ribaupierre, F. de, Goldstein M. H., Jr., and Yeni-Komshian, G. (1972). Cortical coding of repetitive acoustic pulses. *Brain Res.* 48:205–225.

Rose, J. E. (1949). The cellular structure of the auditory region of the cat. *J. Comp. Neurol.* 91:408–440.

Rose, J. E., and Woolsey, C. N. (1949). The relations of thalamic connections, cellular structure, and evocable electrical activity in the auditory region of the cat. *J. Comp. Neurol.* 91:441–466.

Rose, J. E., Greenwood, D. D., Goldberg, J. M., and Hind, J. E. (1963). Some discharge characteristics of single neurons in the inferior colliculus of the cat. I. Tonotopical organization, relation of spike counts to tone intensity, and firing patterns of single elements. *J. Neurophysiol.* 26:294–320.

Rose, J. E., Gross, N. B., Geisler, C. D., and Hind, J. E. (1966). Some neural mechanisms in the inferior colliculus of the cat which may be relevant to localization of a sound source. *J. Neurophysiol.* 29:288–314.

Rouiller, E., and Ribaupierre, F. de (1982). Neurons sensitive to narrow ranges of repetitive acoustic transients in the medial geniculate body of the cat. *Exp. Brain Res.* 48:323–326.

Ruggero, M. (1973). Responses to noise of auditory-nerve fibers in the squirrel monkey. *J. Neurophysiol.* 36:569–587.

Saffran, E. M., Marin, S. M., and Yeni-Komshian, G. H. (1976). An analysis of speech perception in word deafness. *Brain Lang.* 3:209–228.

Schreiner, C. E., and Cynader, M. S. (1984). Basic functional organization of the second auditory cortical field (AII) of the cat. *J. Neurophysiol.* 51:1284–1305.

Schreiner, C., and Urbas, J. V. (1986). Representation of amplitude modulation in the auditory cortex of the cat. I. The anterior auditory field (AAF). *Hearing Res.* 21:227–241.

Semple, M. N., and Aitkin, L. M. (1980). Physiology of pathway from dorsal cochlear nucleus to inferior colliculus revealed by electrical and auditory stimulation. *Exp. Brain Res.* 41:19–28.

Semple, M. N., Aitkin, L. M., Calford, M. B., Pettigrew, J. D., and Phillips, D. P. (1983). Spatial receptive fields in cat inferior colliculus. *Hearing Res.* 10:203–215.

Shamma, S. A., and Symmes, D. (1985). Patterns of inhibition in auditory cortical cells in awake squirrel monkeys. *Hearing Res.* 19:1–13.

Shofner, W. P., and Young, E. D. (1985). Excitatory/inhibitory response types in the cochlear nucleus: relationship to discharge patterns and responses to electrical stimulation of the auditory nerve. *J. Neurophysiol.* 54:917–940.

Smith, R. L., and Brachman, M. L. (1982). Adaptation in auditory nerve fibers: a revised model. *Biol. Cybern.* 44:107–120.

Sovijarvi, A. R. A. (1975). Detection of natural complex sounds by cells in the primary auditory cortex of the cat. *Acta Physiol. Scand.* 93:318–335.

Steinschneider, M., Arezzo, J., and Vaughan, H. G. (1982). Speech evoked activity in the auditory radiations and cortex of the awake monkey. *Brain Res.* 252:353–365.

Stevens, S. S., and Guirao, M. (1967). Loudness functions under inhibition. *Percept. Psychophys.* 2:459–466.

Strominger, N. L. (1969). Localization of sound after unilateral and bilateral ablation of auditory cortex. *Exp. Neurol.* 25:521–533.

Suga, N. (1965). Functional properties of auditory neurones in the cortex of echo-locating bats. *J. Physiol. (Lond.)* 181:671–700.

Suga, N. (1968). Analysis of frequency-modulated and complex sounds by single auditory neurones of bats. *J. Physiol. (Lond.)* 198:51–80.

Suga, N. (1982). Functional organization of the auditory cortex: representation beyond tonotopy in the bat. In: C. N. Woolsey (ed.). *Cortical Sensory Organization. Vol. 3. Multiple Auditory Areas*, pp. 157–218. Clifton, NJ: Humana.

Suga, N., Niwa, H., and Taniguchi, I. (1983). Representation of biosonar information in the auditory cortex of the mustached bat, with emphasis on target velocity information. In: J-P. Ewert and R. R. Capranica (eds.). *Advances in Vertebrate Neuroethology*, pp. 829–867. New York: Plenum Press.

Swisher, L. (1967). Auditory intensity discrimination in patients with temporal-lobe damage. *Cortex* 3:179–193.

Thompson, R. F. (1960). Function of auditory cortex in frequency discrimination. *J. Neurophysiol.* 23:321–334.

Tsuchitani, C. (1977). Functional organization of lateral cell groups of cat superior olivary complex. *J. Neurophysiol.* 40:296–318.

Viemeister, N. F. (1983). Auditory intensity discrimination at high frequencies in the presence of noise. *Science* 221:1206–1208.

Voigt, H. F., and Young, E. D. (1980). Evidence of inhibitory interactions between neurons in dorsal cochlear nucleus. *J. Neurophysiol.* 44:76–96.

von Stockert, T. R. (1982). On the structure of word deafness and mechanisms underlying the fluctuation of disturbances of higher cortical functions. *Brain Lang.* 16:133–146.

Whitfield, I. C., and Evans, E. F. (1965). Responses of auditory cortical neurones to stimuli of changing frequency. *J. Neurophysiol.* 28:655–672.

Winer, J. A. (1984). Anatomy of layer IV in cat primary auditory cortex (AI). *J. Comp. Neurol.* 224:535–567.

Winer, J. A. (1985). The medial geniculate body of the cat. *Adv. Anat. Embryol. Cell Biol.* 86:1–98.

Winter, P., and Funkenstein, H. H. (1973). The effect of species-specific vocalization on the discharge of auditory cortical cells in the awake squirrel monkey (Saimiri sciureus). *Exp. Brain Res.* 18:489–504.

Winter, P., Ploog, D., and Latta, J. (1966). Vocal repertoire of the squirrel monkey (Saimiri sciureus), its analysis and significance. *Exp. Brain Res.* 1:359–384.

Wise, L. Z., and Irvine, D. R. F. (1984). Neuronal sensitivity to interaural intensity difference based on facilitatory binaural interaction. *Hearing Res.* 16:181–187.

Wollberg, Z., and Newman, J. D. (1972). Auditory cortex of squirrel monkey: response patterns of single cells to species-specific vocalizations. *Science* 175:212–214.

Woolsey, C. N. (1960). Organization of the cortical auditory system: a review and synthesis. In: G. L. Rasmussen and W. F. Windle (eds.). *Neural Mechanisms of the Auditory and Vestibular Systems.* Springfield, IL: Charles C Thomas.

Yin, T. C. T., and Kuwada, S. (1984). Neuronal mechanisms of binaural interaction. In: G. M. Edelman, W. E. Gall, and W. M. Cowan (eds.). *Dynamic Aspects of Neocortical Function*, pp. 263–313. New York: John Wiley & Sons.

Young, E. D., and Brownell, W. E. (1976). Response to tones and noise of single cells in dorsal cochlear nucleus of unanesthetized cats. *J. Neurophysiol.* 39:282–300.

IV

VISION

The following five chapters deal with the visual system, perhaps the most widely studied of all the sensory pathways. The first three chapters cover aspects of the cortical analysis of vision; the next two examine studies of the relationships between vision and the vestibular and ocular motor systems. Although the visual receptor organ—the eye and its retina—are not discussed, it will be clear to readers that an understanding of the coding of the input developed within the retina is crucial to the experimenter's interpretation of central neural events. Vision involves a number of messages that travel over separate channels into the thalamus and other subcortical visual centers. These channels are the axons of different populations of retinal ganglion cells; even a single region, such as the lateral geniculate nucleus, receives input from several of these channels. The efforts of many investigators over the years have provided considerable information on their function and anatomical distribution. There is much evidence, for example, that each channel deals with different qualities of the visual world such as color, motion, or contrast. But we still have only a rudimentary grasp of the functional meaning of several well-defined fiber groups that relay information from retina to thalamus and then from thalamus to visual cortex.

Chapter 10 describes the anatomical mapping of these channels of input within the primary visual cortex of the macaque monkey, which is very similar to the same area in humans. The importance of understanding the physiological nature of each of the relays from the thalamus is clearly illustrated. Researchers are still struggling with this question, but considerable headway has been made in recent years, and the references for the first three chapters will lead the interested reader into this literature. The functional combination of the various anatomical channels of input within primary visual cortex and the degree to which each channel retains its identity when relayed within or between different cortical areas are important issues in studies of visual sensory functions. The same issues are relevant to all the sensory areas, but the most detailed anatomical studies have been done in the visual system. Thus, it is useful to examine the complexity of the anatomical organization found within the visual cortex and to ask whether the same features might also exist in other sensory cortical areas.

In Chapter 11 Poggio's discussion of binocular vision illustrates the sophistication that current studies of sensory systems have achieved. It is not always clearly understood that fusion of binocular images and depth perception are entirely neural events, achieved in mammals in the cerebral cortex; no description of the external conditions necessary for inducing the sensation of visual stereoscopy can clearly explain the event itself. The location of objects in three-dimensional space around the body through vision is an evolutionary tour-de-force, and it is scarcely surprising that, in animals with this capability, olfaction and even auditory input are often less important as locating senses. Stereoscopic vision, of course, has been achieved through evolution in a number of ways, and is found in such diverse groups as insects (see Collett, 1987) and cephalopods (e.g., the octopus), as well as vertebrates.

Chapter 12 brings us to striking new techniques of brain imaging. The imaging devices now used by clinicians are familiar. None of these currently reaches the resolution of the voltage-sensitive dye approach described by Blasdel, which permits the topography of function in the cerebral cortex to be visualized in the living organism without harm to the tissue. Studies of single neurons infused with these voltage-sensitive dyes have shown that the dyes change the brightness of their fluorescence as the neuron becomes active; in Blasdel and Salama's work on whole populations of cortical neurons infused with dye, the change in fluorescence or fractional change in light absorption can be interpreted as reflecting the average activity level in the population. The resulting images can then be compared to patterns of anatomical connectivity, physiological recording data, 2DG findings, and cytochrome oxidase levels, in an effort to chart the anatomical basis for functional events in the sensory cortical neuropil.

The role of the vestibular system that defines the position of the head in space, together with the rest of the somatosensory and motor control of body posture, is important in the use of vision or sound for locating objects. In Chapter 13 David Tomko points out how dependent the brain is on systems other than vision for interpreting the position and motion of a visual image—factors that also influence the interpretation of auditory signals. It is clear that the brain can be deceived in its interpretation of events in the outside world, as all of us who have thought our train was moving out of the station, only to find that it was the train on the track beside us that was departing, can testify. The reader is referred to the work of von Holst and Mittelstaedt (1950) and Sperry (1950) for a historical perspective on research bearing on these interesting relationships.

Chapter 14 illustrates the difficulties in examining the link between sensory input and the motor control of elements essential to the functional use of the receptor. David Sparks describes experimental analysis of the eye movement system that guides the two eyes in concert. These eye movements enable us to keep track of moving stimuli, to trace over the outlines of stationary objects, and to keep the object of fixation centered on corresponding regions of the two retinas while the rest of the body and head move. The movements of the eyes can be guided by the visual input itself, by the vestibular input, and by input from other sensory modalities such as the auditory or somatosensory systems. Of course, there is also the aspect of volitional control, which we have not been bold enough to tackle in this volume. Chapter 14 examines with great care the neuropil of the superior colliculus and defines which aspects compose the functional map within the layers dealing with eye movements. It shows that the link between visual input and motor response can involve maps of extremely complex interactions that are not intuitively obvious. These functional maps present a considerable challenge to anatomists trying to determine the relationship of connectivity to the patterns of functional activity. When one realizes that some animals, such as cats, have independently movable ears as well as movable heads and bodies, it becomes apparent that auditory location of sounds can be an extremely complicated sensory-motor interaction.

REFERENCES

Collett, T. S. (1987). Binocular depth vision in arthropods. *Trends Neurosci.* 10:1–2.
Holst, E. von, and Mittelstaedt, H. (1950). Das Reafferenzprincip. *Naturwissenschaften* 37:464–476. [Reprinted and translated in P. C. Dodwell (ed.). *Perceptual Processing: Stimulus Equivalence and Pattern Recognition.* New York: Appleton, 1971.]
Sperry, R. W. (1950). Neural basis of the spontaneous optokinetic response produced by visual inversion. *J. Comp. Physiol. Psychol.* 43:482–489.

10

Mapping Strategies of Monkey Primary Visual Cortex

JENNIFER S. LUND

The mammalian cerebral cortex has a forbiddingly complex anatomy, physiology, and biochemistry. Yet it is of such strategic importance to the processing of sensory information that sensory system investigators continue to pursue their cortical studies despite the difficulties of the task. Our clearest understanding of the relationship between cortical structure and function comes from investigations of the cat and monkey visual cortex. The success of these studies of visual function in the cortex has depended on the concurrent exploration of visual function at other levels: anatomy and physiology of retina, midbrain, and thalamus; visual behavior and psychophysics; and developmental studies of visual system function and structure. A review of this body of work on the visual system by Schiller (1986) illustrates the point that the questions asked in sensory cortical studies usually derive from an understanding of the nature of the afferent information and an appreciation of what types of analysis of sensory information are left for the cortical neuropil. The questions cortical sensory investigators are asking have a certain commonality, and the basic features in the cortical organization of the visual system will probably apply as well to cortical regions concerned with the other sensory modalities.

CONCEPT OF CORTICAL MAPS

An important feature of sensory cortical organization is its parcellation into clusters of separate areas that all serve a single sensory modality; each area is organized anatomically and functionally in a maplike fashion. The nature and interrelations of these maps are all-important to current cortical investigations. This chapter addresses map structure in primary visual cortex of the monkey; however, map structure is also important in the many extrastriate visual areas as well as in somatosensory, motor, auditory, and association cortex. The basic structure of the cerebral cortex—a sheet of neuropil with limited depth (perhaps constrained by the reach of dendrites of single neurons—suggests that the areal extent of cortical neuropil devoted to a particular map is important in its function; certainly the more complex and important a sensory function is behaviorally, the more cortical area seems to be devoted both to single maps and to different maps representing the sensory modality in the cortex.

Konishi (1986) makes an important general point about cortical sensory maps: these maps may topographically represent not only the spatial relationships of the peripheral sensory epithelium (e.g., retinotopic, body surface, or tonotopic for the auditory system), but also (or instead) the functions that are synthesized within the cortex and that have no representation at the periphery—for instance, line-orientation maps in visual cortex or sound-source locations in auditory cortex. Key questions that currently beset sensory cortex investigators include the following: What function or functions are represented in a particular map, and has it produced some new information? How many basic kinds of sensory information have been combined to form the map? Have any of these sources of information come from another map, and therefore be products of cortical synthesis themselves? Can the map's function be determined through its individual components or individual efferent neurons, or is its function derived from a more global interaction of its neurons? How much of the map's total area is needed for efficient functioning—is every part critical and unique, or is its function distributed so that even a small part can provide an adequate substrate? Are the maps used in serial or parallel fashion? How do maps interact? What is the nature of feedback from one map to another? What controls the flow of information through these maps? Can one map inhibit another? Can they be used simultaneously? How many different efferent channels leave a particular map, and are their functions different? How is the information from one map redistributed spatially to construct a new map? To illustrate and clarify these points, we now discuss the anatomical and physiological organization of the primate primary visual cortex in terms of such map concepts.

VISUAL CORTEX

Thalamic Inputs

The primary visual cortex of the primate (also called striate cortex, area 17 or V1), is best regarded as a set of maps, rather than a single map, which are stacked in cortical depth between pia and white matter. Although each map has a common "landscape"—in this case the retinal surface and its view of the external world—each is a specialized survey of that landscape, representing only a subset of the features to be seen there. From the investigator's point of view, the crucial maps in the stack are those whose properties are determined primarily by direct relays from the eyes through the visual thalamus—the lateral geniculate nucleus (LGN); an understanding of the primary input maps provides a firm footing for tracing the evolution of new mapping strategies in other maps in the stack. The relays from the LGN are not all the same type; when examined closely, they clearly differ from one another in functional properties. These relays reflect the characteristics of different populations of neurons in the retina that sample the incoming visually illicited receptor responses in different ways—extracting information concerning color, relative luminance levels, spatial frequency of contrast, or motion. Anatomically, these different thalamic inputs are channeled to terminate in strata at different depths in primary visual cortex; these strata often coincide with particular cell or fiber groupings called layers, or laminae (see Figure 10-1). Microelectrode recordings from the visual cortex often detect these input strata through the response properties of the neurons in them, which in some cases differ little from the properties of neurons in the LGN or even the retinal ganglion cell layer. Usually the topography of the retinal surface and the external world it surveys is represented in an orderly point-by-point fashion in these input strata (Hubel and Wiesel, 1968, 1978; Blasdel and Fitzpatrick, 1984; Hawken and Parker, 1984).

However, this picture rapidly becomes more anatomically and physiologically complex, even in these primary input maps. Figure 10-2 is a summary diagram of thalamic input pathways to primary visual cortex in the monkey; so far, at least six different input channels have been recognized, most of which are segregated in their termination to different laminae (Hubel and Wiesel, 1972; Hendrickson et al., 1978; Livingstone and Hubel, 1982; Blasdel and Lund, 1983). In humans and old world monkeys, relays of like property from the two eyes enter the cortex and end in the same lamina. However, rather than terminating in the same territory

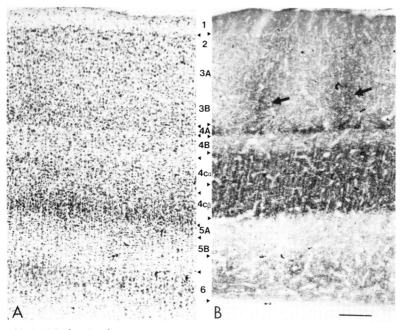

Figure 10-1 Nissl-stained section (A), and a section reacted for the mitochondrial enzyme cytochrome oxidase (B), through striate visual cortex in the macaque. The nissl stain reveals the changes in packing density of neurons between the pial surface and the white matter; the cell sparse regions of layers 5, upper 4cα and 4B have heavy bands of horizontally oriented fibers. The dark bands of heavy cytochrome oxidase (in B) seen in layers 4C, 4A, 6 and blobs in layer 3B (arrows) mark regions of thalamic fiber inputs; these input zones have a higher activity level than the rest of the tissue. Scale bar = 200 μm. [Reproduced with permission from Fitzpatrick et al., 1985.]

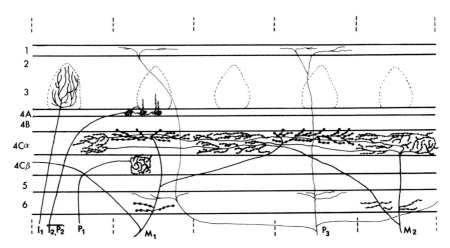

Figure 10-2 Summary diagram of the pattern of thalamic input pathways, coming from the lateral geniculate nucleus and terminating in the macaque monkey striate visual cortex. I_1, I_2, inputs from intercalated layers; P_1, P_2, and P_3, inputs from parvocellular layers; M_1, M_2 inputs from magnocellular layers. The borders of ocular dominance bands are indicated as dashed lines 400 to 500 μm apart. Note that all of these inputs would occur in every ocular dominance band, but here they have been separated laterally for ease of representation. [Data for the diagram derived from Hubel and Wiesel (1972); Hendrickson et al. (1978); Blasdel and Lund (1983); Fitzpatrick et al. (1983). Diagram modified from Fitzpatrick et al. (1985). Reproduced with permission.]

as might be expected (since the eyes survey largely the same areas of visual space) these relays remain spatially segregated from one another within their common lamina of termination. The spatial distribution pattern of relays from the two eyes differs depending on the lamina in which the particular relay terminates; looking at Figure 10-2 the inputs from the two eyes to laminae 6, 4C, and 4A separate out during development as alternating stripes of left eye and right eye inputs (Hubel and Wiesel, 1969; LeVay et al., 1980), like the black and white stripes of a zebra. In laminae 2 and 3, however, the right and left eye inputs form alternating rows of patches or blobs of inputs (see Figure 10-3; see also Livingstone and Hubel, 1982; Horton, 1984). This means that the initial representation of topography for each retina is broken into discrete bands in the laminar maps of 6, 4C, and 4A, and into patches in laminae 2 and 3. The total map of the retina and visual space is also distorted in its relay to cortex, with more area of primary visual cortex devoted to the retina's central region (which is used to fixate objects in the outside world) than to its periphery; this may reflect, in part, the much greater density of retinal receptors and ganglion cells that serve central vision (Van Essen et al., 1984). But distortion of retinal topography and fragmentation of the retinotopy of each eye are not the only mapping features shown in these laminae receiving direct input; in one lamina (4A), shown in Figure 10-3, the thalamic fibers are distributed in a discontinuous, honeycomblike array as a fine pattern within each eye's ocular dominance stripes. It has taken time to uncover these various anatomical patterns of thalamic fiber distribution and their physiological correlates, but they now form an immensely useful and critical framework in which

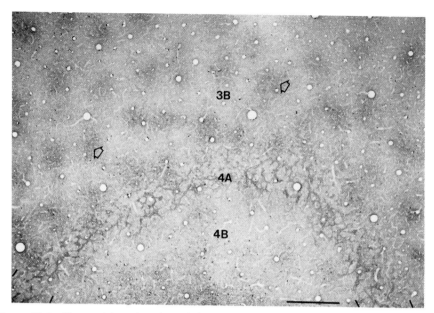

Figure 10-3 Tangential section through layers 3B, 4A, and 4B of macaque monkey striate visual cortex stained for cytochrome oxidase. The darker patches or blobs in layer 3B (*arrows*) are the I_1 thalamic input regions, shown in diagrammatic form in Figure 10-2; the reticular nature of the I_2–P_2 thalamic inputs to layer 4A is clearly evident as a darker staining network. Scale bar = 500 μm. [Reproduced with permission from Fitzpatrick et al. (1985).]

to analyze the workings of the region. The laminar segregation of different thalamic relays aids the physiologist in the initial search to recognize the nature of each of the relays; the lateral spatial segregation patterns within each lamina require the physiologist to explore the topography within each lamina—not only of retinotopy but of segregation of functional properties—a factor that we will see becoming more and more important in other maps in the stack and which is still worthy of further exploration in the primary input laminae.

The primary input maps, in some cases, appear to generate new response properties. Their constituent neurons appear to sample the basic properties of the thalamic relays selectively, to generate functions that, as Konishi indicated, are entirely new, and have no representation at the periphery. These new properties may be represented in a particular topography within the map, adding to, displacing, or disrupting the original retinal topography that the thalamic axons provide by their carefully ordered pattern of termination. This is seen in lamina 4Cα, which receives thalamic input from the magnocellular division of the LGN. Within its substance the population of neurons shows gradation between single neurons with the properties of thalamic neurons (with circular receptive fields and responsiveness to all line orientations and directions of motion) and neurons tuned to respond to only a narrow range of line orientation and a single direction of motion (Parker and Hawken, 1984). The circuitry involved in this transformation of properties is not understood but is a topic of extraordinary interest to the anatomists and physiologists working on the region. It has also interested the theoretical modeling community, and their approach is a growing field (von der Malsburg, 1973; Linsker, 1986a, 1986b, 1986b; and Chapters 12, 16 and 17). In other cases, the neurons of a thalamic input lamina may show little change from the order and functional properties of the thalamic input. Lamina 4Cβ, which receives input from the parvocellular LGN laminae, is an input map that shows little change from thalamic response properties in its constituent neurons.

Efferent Neurons

Another set of maps occurs within the depth of primary visual cortex. These arrays are those of different populations of efferent neurons (Lund et al., 1975; Van Essen, 1984). Again, when carefully examined (e.g., Movshon et al., 1985), each population of neurons projecting to an external destination appears to have its own particular functional properties; these efferent neuron populations are segregated to particular laminae (see Figure 10-4) and usually constitute only a limited subset of the neurons present in the lamina. The efferent neurons projecting to particular destinations are not always represented throughout the whole of the retinal topography in a lamina; for instance, they may occur only in the region representing central vision (to area V4; Yukie and Iwai, 1985), only in that representing retinal periphery but not central vision (efferents to PO; Colby et al., 1983), or lower but not upper visual field representation (efferents to V3, Van Essen et al., 1986). Single laminae may contain several different populations of efferent neurons (see Figure 10-4) with different functional properties as well as anatomical features such as size and frequency. One can presume that these efferent neurons carry from the primary visual cortex a particular synthesis of visual information, such as direction of motion, combined with a whole or partial topographic representation of the retina. The properties of these efferent maps then become crucial to the function of entirely new areas of cortex (or perhaps subcortical regions) where they become the inputs to primary map laminae.

It has become clear in recent years that not only can several different efferent neuron populations exist in single laminae, but functionally different populations can segregate themselves spatially from one another within a single lamina. In laminae 2 and 3, neurons projecting to neighboring visual area V2 are laterally separated into two populations. One population occurs within the regularly arranged blob territories of thalamic inputs and the other in patches sandwiched between the first population in regions without thalamic input (Livingstone and Hubel, 1984a). The two populations differ functionally in that the blob population seems to be color-coded but lacks line-orientation specificity, whereas the second efferent population has orientation but not color specificity (Livingstone and Hubel, 1984a; DeYoe and Van Essen, 1985; Shipp and Zeki, 1985). Thus, at least two efferent neuron maps are interdigitated within the substance of laminae 2 and 3. Topographic representation of the retinal surface is thus no longer a point-by-point representation in laminae 2 and 3, because each point may be repeated several times within a local area; each of these efferent maps probably contains a

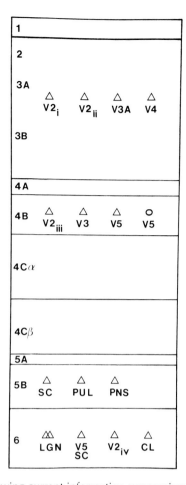

Figure 10-4 Diagram showing current information concerning the distribution of efferent neuron populations in different layers of striate visual cortex in the macaque monkey. Abbreviations: $V2_i$, pyramidal neuron population projecting to visual area 2 (area 18); the cell bodies occur within the cytochrome-rich blob regions of the superficial layers in striate cortex and project to narrow, cytochrome-rich stripes in V2, physiologically the $V2_i$ cells carry color-coded information. Efferent neuron population $V2_{ii}$ occurs in cytochrome-poor regions of the superficial layers and projects to cytochrome-poor stripes in area V2; physiologically the $V2_{ii}$ neurons recognize different orientations of line. Populations V3A and V_4 project to cortical area V3A and V_4, but their physiological characteristics have not been investigated. In layer 4B a third population of neurons projecting to cortical area V2 ($V2_{iii}$) is found; this population projects to broad, cytochrome-rich stripes in V2 and is physiologically characterized by direction specific responses to motion. Neurons in layer 4B also project to cortical areas V3 and V5 (also known as MT or STS) the projection to V5 arises from at least two different cell types—stellate neurons and pyramidal neurons. Layer 5B has neurons projecting to the superior colliculus (SC), the pulvinar nucleus (PUl), and the pons (PNS); some of these neurons may serve more than one of these destinations. In layer 6 a variety of pyramidal neurons project to the lateral geniculate nucleus (LGN). A different population, including giant pyramidal neurons, projects to both cortical area V5 and the superior colliculus; another population ($V2_{iv}$) projects to cortical area V2—its physiological nature and terminal distribution within V2 are not known; and another set of neurons (CL) projects to the claustrum. [Information compiled from numerous studies; see reviews by Van Essen (1984); Van Essen et al. (1986); Lund (1988).]

complete retinotopic map, but is highly specialized in the quality of the stimulus to which its neurons react. The same laminae have neurons projecting to destinations other than V2 (areas V3A, V4, and PO), and it remains to be seen if each of these is served by separate neuron populations.

Other laminae of primary visual cortex have also been found to contribute separate efferent maps to V2. Lamina 4B has an efferent neuron population projecting to V2 (Rockland and Pandya, 1979; Lund et al., 1981; Van Essen et al., 1986) that carries a map characterized by motion detection and direction selectivity; lamina 6 also has an efferent neuron population projecting to V2 whose properties have not yet been defined. It is interesting to note that, for at least the two projections from laminae 2 and 3 and that from laminae 4B to V2, the efferent neuron axon relays from V1 are segregated from one another in their terminal zones in V2 as laterally separated stripelike arrays—a pattern that none of the efferent neuron populations shows in primary visual cortex (i.e., each of these representations is spatially reordered). Moreover, it seems that the localized patches of efferent neurons in primary visual cortex may each send axons to several stripes within V2, indicating an areal divergence as well as a restructuring, of overall pattern of the maps (Livingstone and Hubel, 1984a; DeYoe and Van Essen, 1985; Shipp and Zeki, 1985). The functional importance of these mapping patterns and spatial restructuring between areas is not known and is an extremely interesting question for experimental study.

Other points about the efferent maps within primary visual cortex can be made. It is known that single efferent maps can project to more than one area. For instance, lamina 6 has a population of giant pyramidal neurons, rather limited in number with cell bodies spaced about 150 μm apart. These neurons are known to project both to another visual cortical area (area MT, Spatz, 1975; Lund et al., 1975) and subcortically to the superior colliculus (Fries and Distel, 1983; Fries, 1984). Thus, a single efferent map need not serve just a single endpoint. Another interesting anatomical feature about these large layer 6 neurons is their extensive dendrites; the basal dentrites of single neurons can spread horizontally as much as 1.3 mm within layer 6. They therefore contrast, in the way they sample the local environment or map in layer 6, with the discrete groups of efferent neurons in layers 2 and 3, whose dendrites spread far less (probably no more than 0.3 mm), with no dendritic field overlap between the separate clusters composing a single map in the complex array of interdigitating maps. Looking at Figure 10-4, one can see that other efferent neuron populations, distinct from these giant cells, serve the same destinations of the superior colliculus (from layer 5) and cortical area MT (from layer 4B). This emphasizes the point, made earlier in the case of projections to V2, that primary visual cortex can provide more than one efferent map to any external destination; one may also presume that these maps differ in function, although much work remains to be done in comparing properties of the efferent neuron populations. What is clear from work done so far is that many of the basic functional properties characteristic of each of the extrastriate visual areas have already been elaborated from thalamic inputs within primary visual cortex and then extracted by the various efferent neuron groups and sent on to the various extrastriate regions by single or multiple relays.

INTERLAMINAR RELAYS

Between the stacked laminar maps in the primary visual cortex pass interlaminar axon connections that travel vertically between pia and white matter to link the maps (Lund and Boothe, 1975; Fitzpatrick et al., 1985; Blasdel et al., 1985). These connections provide sources of information that enable new map constructions to be made, sometimes with considerable spatial anatomical reordering of topography between points in one map and points in the next, and this is accompanied by recognizable functional changes between the neuron properties in each map. Each lamina establishes elaborate sets of projections to other laminae, which serve only a particular set of the total stack of maps and omit projections to the others. These intermap (interlaminar) linkages are made by not only axon projections; the classic pyramidal neuron of the cerebral cortex is defined by its peculiar morphology, with one so-called apical dendrite stretching along the axis between pia and white matter reaching between maps while the rest of the neuron's dendrites (the basal field) remain within the lamina where the cell body is located. The apical dendrite can show elaborate patterns of side-branches off the vertical apical dendrite shaft only in certain laminae, presumably gathering information directly from some subset of the stack of maps that needs to be related to input to its basal field of dendrites (Lund and Boothe, 1975; Katz et al., 1984). No information is yet available on what computation single cells make with inputs onto these two parts of their dendritic arbor by any of the vast number of pyramidal neurons lying throughout the depth of the visual cortex; nor is it clear why an axon relay, traveling from the lamina sampled by the apical dendrite, to the basal dendritic field could not substitute for the apical-dendritic extension; however, the apical dendrites are so ubiquitous in the cerebral cortex neurons it seems clear that no substitution can be tolerated. It is interesting that the primary thalamic input zones of 4Cα and 4Cβ lack neurons with apical dendrites (Lund, 1973).

Interlaminar Relays of Parvocellular Recipient Layer—Lamina 4Cβ

We have discussed the fact that the neurons of lamina 4Cβ retain much of the physiological characteristics of the thalamic neurons projecting to the layer. The topographic map of the retina is exquisitely represented point-by-point in a very orderly, fine-grain fashion in the circular receptive fields of the 4Cβ neurons (Blasdel and Fitzpatrick, 1984). These neurons relay this information principally to two other laminae. One projection is to lamina 4A, which itself receives thalamic input from a separate set of LGN parvocellular division neurons from those projecting to 4Cβ. This 4Cβ-to-4A projection is point-to-point and vertical, with apparently little lateral spread occurring as the information passes from 4Cβ to 4A (Fitzpatrick et al., 1985). Indeed, recordings from lamina 4A show a fine-grain, spatially orderly map of the retina similar to that seen in 4Cβ (Blasdel and Fitzpatrick, 1984). Convergence of separate maps served directly by different sets of thalamic inputs is an interesting phenomenon that is seen several times in the depth of the cortex, but its implications have not been explored.

In contrast to its point-to-point connectivity with lamina 4A, the lamina 4Cβ projection to lamina 3B spreads laterally, with single cells in 4Cβ distributing their projection over a wide area, as well as making side-stepping projections with terminal fields offset from directly over the cell of origin in 4Cβ (Fitzpatrick et al., 1985; Blasdel et al., 1985). These spreading projections are accompanied by local disorder in the topography of the retinal map recorded in the neurons post-synaptic to the input in layer 3B. Visual receptive fields for a particular point on the retina are locally scattered somewhat haphazardly and repeat themselves.

Organization of the Superficial Layers—Laminae 2 and 3

The same 3B neurons that seem disorderly in their layout for retinal topography show a high degree of order when their responses are tested for recognition of different orientations of lines presented in the visual field (Hubel and Wiesel, 1968, 1978). Traveling horizontally (parallel to the pia) within lamina 3B, each successive neuron encountered responds specifically to a particular line orienta-tion, and each successive neuron shows a line preference just slightly shifted from its neighbor—with long sequences of neurons with continuous shift in orientation preference occurring from neuron to neuron, in clockwise or anticlockwise fash-ion. We now know (see Livingstone and Hubel, 1984a) that these sequences are interrupted by the so-called blob regions, which, as we noted earlier, receive di-rect thalamic input (from neurons lying between and beneath the main laminae of the LGN; Fitzpatrick et al., 1983). These blob regions contain neurons that lack orientation specificity, are largely monocularly driven, and have color-coded properties. The overall pattern of the orientation-specific map has recently been visualized with the use of voltage sensitive dye imaging techniques in the living monkey cortex by Blasdel and Salama (1986, and see Chapter 16); these authors compare the mapping strategy to a set of modules, within which orientation change occurs in smooth, gradual sequences, surrounded by boundaries where abrupt change in the orientation sequence occurs. It is at these boundaries that the tha-lamic input blobs occur; it should be noted, however, that thalamic input does not occupy all of the rims around these orientation modules.

The diverging arbors of the axons from the neurons in 4Cβ to the map of layer 3 presumably are spreading the information necessary to construct the orientation-specific responses in the neurons of the module postsynaptic to their input in layer 3. The exact manner in which the orientation-specific response is created is not understood and has been a matter of debate since Hubel and Wiesel (1962) first described orientation-specific responses (see also Hubel and Wiesel, 1968); the issue has more recently been taken up by theoretical modelers (e.g., von der Mals-burg, 1973; Linsker, 1986a, 1986b, 1986c). The disruption and repetition of local retinal topography in layer 3 can be viewed as a necessary consequence of con-struction of a complete range of orientation-specific neurons for each point in space (Blasdel et al., 1985).

Intrinsic, laterally spreading connections within maps is an important element in their construction. In 4Cβ laterally spreading local connections are present but apparently do not travel far. In contrast, in laminae 2 and 3 very widespreading lateral connections are found (Rockland and Lund, 1983). These lateral connections are remarkable for their high degree of patterning, and they seem to create a series of lattice connections, linking regions of common function. For instance, the blob zones of thalamic input appear cross-linked to one another, and zones between the blobs form a separate system, also linked together (Livingstone and Hubel, 1984b). Although it is not yet clear if these connections help construct the properties of the cell groups they link (see Chapter 16), or serve to coordinate already elaborated properties across the retinal map, it is clear that such links are a general feature of the superficial layers of the mammalian visual cortex, and probably serve to build up continuity between the spatially separated elements of each of the multiple interdigitated maps within layers 2 and 3 (Rockland et al., 1982; Rockland and Lund, 1983; Livingstone and Hubel, 1984a, 1984b). These lattice links are certainly made by the axon collaterals of pyramidal neurons contacting other pyramidal neurons (Gilbert and Wiesel, 1983; McGuire et al., 1984; Rockland, 1985), but the degree of participation of other varieties of neuron is uncertain. The specific relationship of these lattice arrays to efferent neuron populations is also unclear.

Interlaminar Relays from the Magnocellular Recipient Layer—Lamina 4Cα

The magnocellular LGN neurons send what is probably a dual projection to lamina 4Cα (see Figure 10-2), one input placed in the upper part of the lamina with partial overlap with an input that occupies the entire depth of the layer (Hubel and Wiesel, 1972; Blasdel and Lund, 1983). In contrast to the rather small terminal fields of the thalamic axons from parvocellular layers projecting to 4Cβ (which have terminal fields about 150 μm in diameter), the individual thalamic axons terminating in 4Cα can spread their arbors across up to three repeats of the ocular dominance stripes for a single eye (i.e., almost 2 mm; see Blasdel and Lund, 1983). Although the visual receptive fields of the postsynaptic neurons in 4Cα are much larger than those in 4Cβ, alignment of the retinotopic maps of nonoriented units in 4Cα and 4Cβ is perfect. However, as we discussed earlier, even within the depth of 4Cα, and certainly by the time its projections enter the immediately adjacent zone of 4B, all the properties of binocular fusion, orientation-specificity and direction-specificity have been fully developed, with a concomitent loss of precision of retinal topography—in a similar fashion to the loss in precision seen in layers 2 and 3 compared to 4Cβ (Hubel and Wiesel, 1978; Blasdel and Fitzpatrick, 1984; Livingstone and Hubel, 1984a; Hawken and Parker, 1984). This transformation is accompanied by wide-spreading lateral connections within upper 4Cα and 4B and the development of periodically organized lattice connections resembling those found in layers 2 and 3 (Rockland and Lund, 1983).

Relationship Between Lamina 4B and the Superficial Layers

The projections from the layer 4B map to the layer 2 and 3 map are made by both pyramidal and nonpyramidal neurons in a point-to-point fashion with almost no lateral divergence. (This suggests the maps of the two regions are in exact register with one another, even though they have apparently been constructed from different streams of thalamic information, each with its own scale of representation of retinal topography; see Fitzpatrick et al., 1985; Blasdel et al., 1985.) It also seems that at least some of the interdigitated maps connected by laterally spreading axon lattice arrays within layers 2 and 3 and within 4B may be individually interconnected between the two regions. It is of interest to note that the excitatory axon projections from layer 4B to layers 2 and 3 are almost entirely undirectional, for there is very little in the way of pyramidal neuron axon projections from layers 2 and 3 to layer 4B, although the apical dendrites of pyramidal neurons of 4B "listen in" to the superficial layers. Instead, layers 2 and 3 and lamina 4B project heavily on lamina 5B, with considerable lateral spreading of their axons once they have reached into the substance of lamina 5B, suggesting that the construction of new map functions is occurring within lamina 5B.

Inhibitory Interlaminar Relays

So far we have treated the intrinsic connections between the stacked maps of primary visual cortex as purely excitatory relays, made by the spine-bearing pyramidal or stellate neurons in the layers concerned. However, it is clear that neurons that are equated with inhibitory activity in the cerebral cortex (usually with smooth or sparsely spined dendrites and, in most cases, gamma-amino butyric acid—GABA—as a neurotransmitter; see Jones and Hendry, 1986) also project between the maps in highly ordered arrays (Lund, 1987). The destinations of these interlaminar inhibitory projections are highly specific for each neuron type, and each lamina contains a wealth of different neurons (see Figure 10-5, in which neurons of 4Cα are illustrated as an example). Some of these axon projections follow those of the excitatory relays, and others emphasize different links. For instance, these neurons clearly make axonal projections between the primary thalamic input layers 4Cβ and 4Cα (see Figure 10-5), and the question can be asked whether or not activity in one layer can inhibit the activity in another. Because certain types of visual information can be processed in parallel (e.g., line orientation and color) while others must be processed serially (e.g., color and direction of motion; see Treisman and Gelade, 1980; Nakayama and Silverman, 1986), it is likely that inhibitory activity between maps can indeed be a determinant of visual behavior.

In general the inhibitory projections between layers show a highly columnar form (i.e., point-to-point between maps, even when the excitatory relays may show considerably more divergence). This is an interesting phenomenon that deserves further attention; perhaps these links form a necessary framework around which new maps are constructed. Even within laminae the local axon fields of most varieties of these neurons remain quite narrowly focused. However, there are some important exceptions when inhibitory neurons in certain laminae have very wide-spreading axon projections. In the monkey these laterally spreading projections occur in upper layer 4Cα, layer 4B, and at the border of laminae 5 and 6 (Lund, 1987, and ongoing observations). These regions are associated with the development of directionally selective responses in which neurons respond to only one direction of motion in the outside world (Livingstone and Hubel, 1984a).

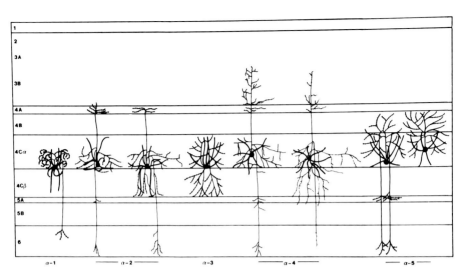

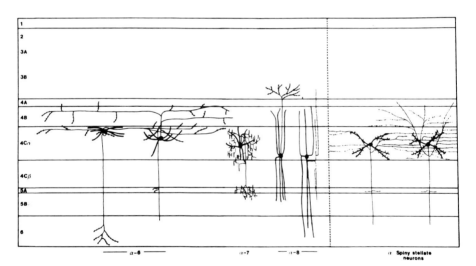

Figure 10-5 Diagrams summarizing the varieties of local circuit neurons (mainly GABAergic, and therefore presumed to make inhibitory connections) found within layer 4Cα. Axon projections of two varieties of spiny stellate neurons, believed to be excitatory, that form the dominant neuron populations in the layer are shown for comparison. [From Lund (1987). Reproduced with permission.]

These lateral inhibitory links crossing the surface of the laminar map may be part of the substrate for generating this important response property.

CONCLUSIONS

This chapter has briefly reviewed some of the major features of our current knowledge of the anatomical mapping strategies within the monkey primary visual cortex. No doubt this description is an oversimplification of the actual functional and structural complexities of the region (for further details, see Lund, 1988), and one must expect a wealth of new information to become available as further study of the region is made.

The importance of the early findings of Hubel and Wiesel cannot be over-emphasized, and their laboratories continue to make invaluable contributions to our understanding of the region. The early emphasis on topographic patterning, in both their physiological and anatomical studies, has provided invaluable clues for further experimentation. Future studies can now be firmly based in the structural framework for the region that these and other researchers have shown to exist, and this framework provides essential landmarks for further detailed physiological and anatomical studies. As a general lesson for all sensory cortex exploration, these studies on visual cortex have provided a series of cautions:

1. The investigator should not assume that thalamic input (or input from other cortical areas) is a single channel.
2. It is necessary to track down which layers are targeted by different inputs.
3. An even topographic distribution of afferent terminations within these targeted layers should not be assumed.
4. An even distribution of efferent neurons within a layer should not be assumed.
5. A single layer may contain several different efferent populations, perhaps spatially separated, with different properties.
6. It should not be assumed that the map of one lamina is transferred, without change in map function, to another lamina at a different depth in the same area or, when used as a primary input, to another area of cortex.
7. Links between laminae should be recognized as both inhibitory and excitatory.
8. Within single laminae, discontinuous mapping strategies may force a careful study of functional topography.

ACKNOWLEDGMENTS

My own studies are supported by Grant EY05282 from the National Eye Institute. I would like to thank Roberta Erickson for her preparation of the manuscript.

REFERENCES

Blasdel, G. G., and Fitzpatrick, D. (1984). Physiological organization of layer 4 in macaque striate cortex. *J. Neurosci.* 4:880–895.

Blasdel, G. G., and Lund, J. S. (1983). Termination of afferent axons in macaque striate cortex. *J. Neurosci.* 3:1389–1413.

Blasdel, G. G., Lund, J. S., and Fitzpatrick, D. (1985). Intrinsic connections of macaque striate cortex: axonal projections of cells outside lamina 4C. *J. Neurosci.* 5:3350–3369.

Blasdel, G. G., and Salama, G. (1986). Voltage-sensitive dyes reveal a modular organization in monkey striate cortex. *Nature* 321:579–585.

Colby, C. L,, Gattass, R., Olson, C. R., and Gross, C. G. (1983). Cortical afferents to visual area PO in the macaque. *Soc. Neurosci. Abstr.* 9:152.

DeYoe, E. A., and Van Essen, D. C. (1985). Segregation of efferent connections and receptive field properties in visual area V2 of the macaque. *Nature Lond.* 317:58–61.

Fitzpatrick, D., Itoh, K., and Diamond, I. T. (1983). The laminar organization of the lateral geniculate body and the striate cortex in the squirrel monkey (*Saimiri sciureus*). *J. Neurosci.* 3:673–702.

Fitzpatrick, D., Lund, J. S., and Blasdel, G. G. (1985). Intrinsic connections of macaque striate cortex. Afferent and efferent connections of lamina 4C. *J. Neurosci.* 5:3329–3349.

Fries, W. (1984). Cortical projections to the superior colliculus in the macaque monkey: a retrograde study using horseradish peroxidase. *J. Comp. Neurol.* 230:55–76.

Fries, W., and Distel, H. (1983). Large layer VI neurons of monkey striate cortex (Meynert cells) project to the superior colliculus. *Proc. R. Soc. Lond. B.* 219:53–59.

Gilbert, C. D., and Wiesel, T. N. (1983). Clustered intrinsic connections in cat visual cortex. *J. Neurosci.* 3:1116–1133.

Hawken, M. J., and Parker, A. J. (1984). Contrast sensitivity and orientation selectivity in lamina IV of the striate cortex of old-world monkeys. *Exp. Brain Res.* 54:367–373.

Hendrickson, A. E., Wilson, J. R., and Ogren, M. P. (1978). The neuroanatomical organization of pathways between the dorsal lateral geniculate nucleus and visual cortex in old world and new world primates. *J. Comp. Neurol.* 182:123–136.

Horton, J. C. (1984). Cytochrome oxidase patches: a new cytoarchitectonic feature of monkey visual cortex. *Philos. Trans. R. Soc. Lond. Biol.* 304:199–253.

Hubel, D. H., and Wiesel, T. N. (1962). Receptive fields, binocular interaction and functional architecture in the cat's visual cortex. *J. Physiol.* 160:106–154.

Hubel, D. H., and Wiesel, T. N. (1968). Receptive fields and functional architecture of monkey striate cortex. *J. Physiol. Lond.* 195:215–243.

Hubel, D. H., and Wiesel, T. N. (1969). Anatomical demonstration of columns in the monkey striate cortex. *Nature* 221:747–750.

Hubel, D. H., and Wiesel, T. N. (1972). Laminar and columnar distribution of geniculocortical fibers in the macaque monkey. *J. Comp. Neurol.* 146:421–450.

Hubel, D. H., and Wiesel, T. N. (1978). Functional architecture of macaque monkey visual cortex. Ferrier Lecture. *Proc. R. Soc. Lond. Biol.* 198:1–59.

Jones, E. G., and Hendry, S. H. C. (1986). Co-localization of GABA and neuropeptides in neocortical neurons. *Trends Neurosci.* 9:71–76.

Katz, L. C., Burkhalter, A., and Dreyer, W. J. (1984). Fluorescent latex microspheres as a retrograde neuronal marker for *in vivo* and *in vitro* studies of visual cortex. *Nature* 310:498–500.

Konishi, M. (1986). Centrally synthesized maps of sensory space. *Trends Neurosci.* 9:163–168.

LeVay, S., Wiesel, T. N., and Hubel, D. H. (1980). The development of ocular dominance columns in normal and visually deprived monkeys. *J. Comp. Neurol.* 191:1–51.

Linsker, R. (1986a). From basic network principles to neural architecture: emergence of spatial opponent cells. *Proc. Natl. Acad. Sci. USA* 83:7508–7512.

Linsker, R. (1986b). From basic network principles to neural architecture: emergence of orientation-selective cells. *Proc. Natl. Acad. Sci. USA* 83:8390–8394.

Linsker, R. (1986c). From basic network principles to neural architecture: emergence of orientation columns. *Proc. Natl. Acad. Sci. USA* 83:8779–8783.

Livingstone, M. S., and Hubel, D. H. (1982). Thalamic inputs to cytochrome oxidase-rich regions in monkey visual cortex. *Proc. Natl. Acad. Sci. USA* 79:6098–6101.

Livingstone, M. S., and Hubel, D. H. (1984a). Anatomy and physiology of a color system in the primate visual cortex. *J. Neurosci.* 4:309–356.

Livingstone, M. S., and Hubel, D. H. (1984b). Specificity of intrinsic connections in primate primary visual cortex. *J. Neurosci.* 4:2830–2835.

Lund, J. S. (1973. Organization of neurons in the visual cortex, area 17, of the monkey (*Macaca mulatta*). *J. Comp. Neurol.* 147:455–496.

Lund, J. S. (1987). Local circuit neurons of macaque monkey striate cortex: I. Neurons of laminae 4C and 5A. *J. Comp. Neurol.* 257:60–92.

Lund, J. S. (1988). Anatomical organization of macaque monkey striate visual cortex. *Ann. Rev. Neurosci.* 11:253–288.

Lund, J. S., and Boothe, R. (1975). Interlaminar connections and pyramidal neuron organization in the visual cortex, area 17, of the macaque monkey. *J. Comp. Neurol.* 159:305–334.

Lund, J. S., Hendrickson, A. E., Ogren, M. P., and Tobin, E. A. (1981). Anatomical organization of primate visual cortex area VII. *J. Comp. Neurol.* 202:19–45.

Lund, J. S., Lund, R. D., Hendrickson, A. E., Bunt, A. H., and Fuchs, A. F. (1975). The origin of efferent pathways from the primary visual cortex, area 17, of the macaque monkey. *J. Comp. Neurol.* 164:287–304.

McGuire, G. A., Hornung, P.-P., Gilbert, C. D., and Wiesel, T. N. (1984). Patterns of synaptic input to layer 4 of cat striate cortex. *J. Neurosci.* 4:3021–3033.

Movshon, J. A., Adelson, E. H., Gizzi, M. S., *Pattern Recognition Mechanisms*, and Newsome, W. T. (1985). The analysis of moving patterns. In: C. Chagas, R. Gattass, and C. Gross (eds.). *Exp. Brain Res.* (Suppl. 11) New York: Springer-Verlag.

Nakayama, K., and Silverman. (1986). Serial and parallel processing of visual feature conjunctions. *Nature* 320:264–265.

Rockland, K. S. (1985). Intrinsically projecting pyramidal neurons of monkey striate cortex: an EM-HRP study. *Soc. Neurosci. Abstr.* 11:17.

Rockland, K. S., and Lund, J. S. (1983). Intrinsic laminar lattice connections in primate visual cortex. *J. Comp. Neurol.* 216:303–318.

Rockland, K. S., Lund, J. S., and Humphrey, A. L. (1982). Anatomical banding of intrinsic connections in striate cortex of tree shrews (*Tupaia glis*). *J. Comp. Neurol.* 209:41–58.

Rockland, K. S., and Pandya, D. N. (1979). Laminar origins and terminations of cortical connections of the occipital lobe in the rhesus monkey. *Brain Res.* 179:3–20.

Schiller, P. H. (1986). The central visual system. *Vision Res.* 26:1351–1386.

Shipp, S., and Zeki, S. (1985). Segregation of pathways leading from area V2 to areas V4 and V5 of macaque monkey visual cortex. *Nature (Lond.)* 317:322–325.

Spatz, W. B. (1975). An efferent connection of the solitary cells of Meynert. A study with horseradish peroxidase in the marmoset *Callithrix*. *Brain Res.* 92:450–455.

Treisman, A. M., and Gelade, G. (1980). A feature integration theory of attention. *Cog. Psychol.* 12:97–136.

Van Essen, D. C. (1984). Functional organization of primate visual cortex. In: A. Peters and E. G. Jones (eds.). *Cerebral Cortex. Vol. 3. Visual Cortex*, New York: Plenum Press.

Van Essen, D. C., Newsome, W. T., and Maunsell, J. H. R. (1984). The visual field representation in the striate cortex of the macaque monkey: asymmetries, anisotropies and individual variability. *Vision Res.* 24:429–448.

Van Essen, D. C., Newsome, W. T., Maunsell, J. H. R., and Bixby, J. L. (1986). The projections from striate cortex (V1) to areas V2 and V3 in the Macaque monkey: asymmetries, areal boundaries and patchy connections. *J. Comp. Neurol.* 244:451– 480.

von der Malsburg, C. (1973). Self-organization of orientation sensitive cells in the striate cortex. *Kybernetik* 14:85–100.

Yukie, M., and Iwai, E. (1985). Laminar origin of direct projection from cortex area V1 to V4 in the rhesus monkey. *Brain Res.* 346:383–386.

11

Neural Responses Serving Stereopsis in the Visual Cortex of the Alert Macaque Monkey: Position-Disparity and Image-Correlation

GIAN F. POGGIO

In normal binocular vision, the fields of view of the left and right eyes are largely superimposed because of the convergence of the two visual axes (foveal projections) on the same point in space, the point of binocular fixation. Nearly identical images of the outside world are present simultaneously in the two eyes and are reflected in neural code in the visual brain. The brain uses the inputs from the two eyes to recover the relative depth of objects. Wheatstone (1838) first demonstrated that the difference in the relative horizontal position of an object's images in the two eyes can generate an impression of depth and solidity. Julesz (1960,1971) showed by means of random-dot stereograms that binocular disparity is the sufficient cue for stereoscopic depth perception.

A neuron in the visual cortex of the brain looks at the world through two windows, one in each eye—its receptive fields. These fields are composed of subfields of different sizes and properties (Hubel and Wiesel, 1962,1968; Schiller et al., 1976a; Movshon et al., 1978; Mullikin et al., 1984a,b; Camarda et al., 1985; Maske et al., 1986; Ohzawa and Freeman, 1986a,b; Emerson et al., 1987). As a rule, the associated fields of a single neuron in the two eyes have a very similar structure (Schiller et al., 1976b; Maske et al., 1984; Skottun and Freeman, 1984). The activity of these neurons reflects the interaction of inputs from the two eyes, and the outcome of this interaction is determined chiefly by two factors: (1) the internal configuration, or texture, of the image and (2) the internal configuration, or structure, of the neuron's receptive field. It follows that the activity of binocular neurons depends on the degree of mismatch between the retinal images "seen" by the two receptive fields.

Physiological evidence shows that substantial numbers of neurons in the visual cortex are sensitive to binocular disparity. These cells give differential responses to the relative horizontal position of the two images over the neuron's receptive fields in the two eyes (Barlow et al., 1967; Nikara et al., 1968; Poggio and Fischer,

1977; von der Heydt et al., 1978; Fischer and Krueger, 1979; Ferster, 1981; Maunsell and Van Essen, 1983; Poggio 1980,1984; Poggio and Talbot, 1981; Poggio et al., 1985b; Burkhalter and Van Essen, 1986, Felleman and Van Essen, 1987). We refer to these neurons collectively as "stereoscopic" neurons based on the conjecture that their response reflects the early cortical processing that leads to binocular depth perception (Poggio and Talbot, 1981; Poggio and Poggio, 1984).

The set of responses along the disparity domain that characterizes the disparity sensitivity of stereoscopic neurons includes (1) the response that obtains at the image disparity at which left and right images are in register with left and right receptive fields, and (2) the responses evoked at all other effective binocular disparities. The two conditions differ in a major way; in the former, texturally corresponding (identical) parts of the images occupy functionally corresponding parts of the neuron's receptive fields, a condition of *binocular correlation*, whereas in the latter there is no textural identity of the images over corresponding parts of the fields, a condition of *binocular uncorrelation*.

The initial steps toward unraveling the neural mechanisms of stereopsis require an understanding of how the binocular depth dimension is encoded in the impulse activity of nerve cells and where in the brain these stereoscopic neurons are located. The striate cortex, area V1, which is the first locale in the visual pathways where signals from the two eyes converge onto single binocular neurons, and the prestriate visual areas V2 and V3*, where all neurons are binocular, are the obvious candidates for exploration.

In alert rhesus macaques trained to maintain eye fixation, we have analyzed the response of single neurons in V1, V2, and V3 to binocularly correlated and uncorrelated dynamic random-dot stereopatterns and the relation between correlation sensitivity and the neuron's selectivity for horizontal positional disparity. We have found that neurons giving excitatory responses to stimuli at or about zero disparity are typically inhibited by stereo images with uncorrelated contrast texture, whereas all other types of disparity-sensitive neurons, as well as a proportion of disparity-insensitive ones, signal uncorrelation with excitation. These findings suggest that both the relative image position and image contrast in the two eyes, are interacting components for the early neural mechanisms of stereopsis in the visual cortex.

METHODS OF APPROACH

In our experiments male rhesus monkeys were trained to execute a fixation/detection task in an apparatus in which the animal's left and right eyes see separate display screens reflected by flat mirrors at a 45 degree angle (Figure 11-1). Stereoscopic and monocular patterns, unattended by the monkey, were used as stimuli, and the impulse activity of cortical neurons was recorded with Pt/Ir microclectrodes. Eye position was routinely monitored with an infrared corneal reflex system (Motter and Poggio, 1984). A PDP-11/34 minicomputer was used to reg-

---

*The term V3 is used in this chapter as a shorthand for the third visual complex of Zeki (1978), and includes area V3 and V3A in the cortex of the anterior bank of the lunate sulcus.

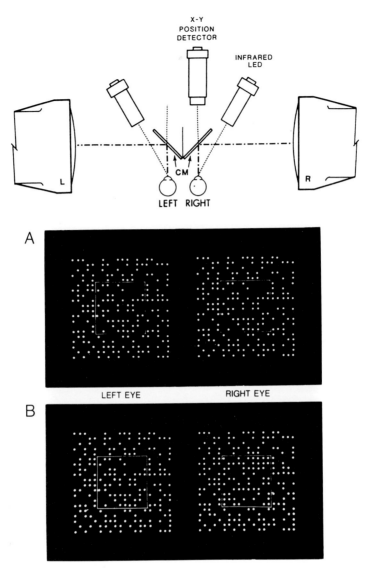

Figure 11-1 Experimental setup and random-dot correlograms. The monkey, with its head firmly held, views two screens (L,R) separately with each eye, each screen reflected by a heat-transmitting, flat mirror placed at 45 degree angles in front of each eye (CM). Eye position is monitored throughout the experiment by an infrared optoelectrical system measuring the corneal reflex of an infrared beam projected on each eye. Fields of dynamic random dots are generated on the two screens, and binocular correlograms are produced by making the dot patterns over a central area of each screen different, either by using two independent random-dot sequences or by switching the pattern on one screen to be the complement of the pattern on the other screen. The stereopatterns of Figure 1-11 illustrate this latter condition of negative correlation, or anticorrelation, with the test area outlined for emphasis by a white line: In *A*, the dot patterns seen by the left and right eyes are identical (correlation) throughout. In *B*, the patterns over a central square of the stereofields complement each other (uncorrelation); the surrounding background remains binocularly correlated. Neural sensitivity for binocular correlation is tested by alternating correlated and uncorrelated stereopatterns over the neuron's receptive fields for the duration of maintained eye fixation.

ulate and monitor the animal's behavioral task, to control visual stimulation, and to collect behavioral and electrophysiological data. The monkey was sacrificed with an overdose of pentobarbitone injected intravenously. The location and extent of the penetrations were reconstructed from thionin-stained serial sections (20 μm) of the cortex explored (see also Poggio et al., 1977; Poggio and Talbot, 1981).

Fields of dynamic random-dots were generated on two cathode-ray tube (CRT) displays (Hewlett-Packard, Mod. 1311), one for each eye, at 67 frames/sec (15-msec frame duration). The position of the dots on the two screens was controlled by an computer-driven digital display generator (Julesz et al., 1976). The display was constructed at a 100×100 dot raster subtending 10 degrees at the monkey's eye (0.1 degree inter-dot separation, 100 dots/deg^2). It was made up of "bright" (intensified) dots (0.02 degree at the eye) and "dark" pixels in different relative proportions. We identify these patterns by the percentage fraction of *bright* dots (dot density = 1 to 50%). The luminance of the display varied linearly with dot density between 0.6 cd/m^2 (dot density = 1%) and 14.4 cd/m^2 (dot density = 100%). The luminance of the blank screen was not measurable with our detector.

Bright contour figures (bars) of 100 percent dot density and selected size and orientation were used for each neuron to locate the minimum response field (Barlow et al., 1967) and define its two-dimensional properties. A few cells responding best to dark bars were not included in this sample. The neuron's disparity sensitivity was then determined with bar stereograms. The sensitivity to binocular correlation was assessed with dynamic random-dot correlograms (RDC) (Julesz et al., 1976; Tyler and Julesz, 1976, 1978) in which the dot pattern of a test area, centered over the neuron's left and right receptive fields, was made to alternate from identical in the two eyes (correlation) to different (uncorrelation). The test area was usually square (0.5 to 4.0 degrees side) and in the middle of the binocularly common background (10 degrees). Binocularly uncorrelated areas were generated in two ways: (1) with synchronized and independent pseudo-random-dot sequences, one for the left screen and one for the right: the average number of binocularly corresponding and noncorresponding dots is governed by the bright and dark dots' respective densities; (2) with complementary left and right dot sequences, each bright dot in one eye matched with a dark dot in the other eye (anticorrelation or negative correlation). This type of correlogram is illustrated in Figure 11-1.

STEREOSCOPIC PROPERTIES

The purpose of this study was to analyze the response properties of neurons in cortical visual areas V1, V2, and V3 of the macaque monkey to stereoscopic stimuli, to assess the dynamics of the cortical mechanisms leading to binocular depth perception. The areas studied are shown sketchily in Figure 11-2 (top), as they are encountered by a vertically penetrating microelectrode. Because the number of binocular neurons and the sensitivity to stereoscopic stimuli vary from one rhesus monkey to another, the three visual areas were explored in the same animal, with at least two areas sampled sequentially in the majority of microelectrode penetrations. Two male monkeys were studied in this way. The sample of units

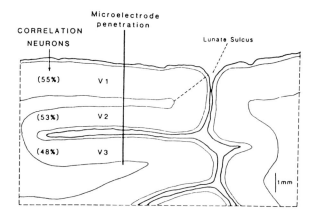

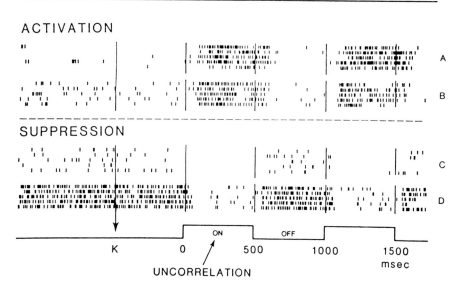

Figure 11-2 (*Above*) Outline of histological section of region of the brain explored to illustrate the positions of areas V1, V2, and V3 and their relation to a microelectrode penetration perpendicular to the surface of the cortex. The proportion of correlation-sensitive neurons found is given for each area. (*Below*) Replicas of impulse activity of four neurons (A through D) from monkey 41V during presentation of binocularly correlated and uncorrelated dynamic random-dot stereopatterns. In this and following similar illustrations, each line of record constitutes a single binocular fixation trial, and each upstroke marks the occurrence of an impulse. At *K* the monkey pulls the reward key to initiate the trial and at *t* = 0 the dot pattern in the central area of the display (see Figure 11-1) shifts from binocularly correlated to uncorrelated (ON) and alternates with a correlated one (OFF) while the monkey maintains eye fixation. Both activation (neurons A and B) and suppression (C and D) of activity in responses to uncorrelation are obvious and very similar for each type. Correlated stereopatterns may evoke responses opposite to those to uncorrelation (A and D) or may be virtually ineffective (B and C). The examples shown are representative of the three cortical areas explored. Correlograms of different size and dot-density were used for each neuron: A = area V2, stimulus size 2 × 2 degrees, density 20%, B = V3, 8 × 1 degrees, 20%; C = V2, 2 × 2 degrees, 5%; D = V1, 4 × 4 degrees, 50%.

analyzed quantitatively includes neurons in V1 (101), V2(90), and V3(68) for a total of 259 neurons tested for disparity sensitivity. A subset of this sample (141 neurons) was also tested for correlation sensitivity with dynamic random-dot correlograms. Additional supportive data are available for 264 neurons in area V1 (in seven other animals) analyzed for both of these stereo properties.

Neural Sensitivity to Binocular Image-Texture Correlation

A substantial proportion of neurons in the visual cortex signal the state of binocular correlation by giving reciprocal responses, excitatory (*activation*) versus inhibitory (*suppression*), to the two conditions. The response to uncorrelation is usually more evident, and may be regarded as characteristic of the neuron's sensitivity to binocular correlation (Poggio et al., 1985a; Gonzalez et al., 1986).

Figure 11-2 (bottom) shows examples of responses to dynamic random-dot correlograms for four neurons from the same monkey but different cortical areas. The responses to uncorrelation (ON) are strong and evident: neuron A and B respond with activation, and neurons C and D with suppression. The responses to correlation, on the other hand, are less predictable: strong and clearly different in A (inhibitory) and in D (excitatory), but indistinguishable, possibly nonexistent, in B and C.

The proportion of neurons signaling the state of binocular correlation (*correlation neurons*) was similar in areas V1 (30/55), V2 (28/53), and V3 (16/33), with an average for the three areas of 52 percent (74/141). Of the sensitive neurons, 69 percent responded to uncorrelation with activation and 31 percent with suppression, again in similar proportion in the three areas. No differential distribution of correlation neurons appears to exist in different cytoarchitectonic layers of V1, and probably not in those of V2 and V3.

The random-dot sequences used to generate the correlograms contained different ratios of bright and dark dots, and the fraction of uncorrelated dots and the average luminance of the display varied accordingly. Correlation sensitivity was tested with negatively correlated stereopatterns (anticorrelation; dot density 50%; luminance 7.0 cd/m^2), as well as with randomly uncorrelated patterns whose bright dot density varied from 1 (0.6 cd/m^2) to 50 percent (7.0 cd/m^2). Correlograms of different sizes were used, from the size estimate of the neuron's receptive field, as defined by narrow contrast bars, to arbitrarly large sizes. The results are independent of stimulus size and luminance over a wide range of dot densities: responses of similar magnitude are frequently obtained over a range of uncorrelated elements and associated screen luminance; this is especially evident for the inhibitory effects of uncorrelation, and often two lonely uncorrelated dots/deg^2/frame, one in each eye, repeating at 67 frames/sec are sufficient to suppress completely the neuron's activity. The response field associated with correlation responses is usually larger than the receptive field for narrow contrast bars. Indeed, an uncorrelated pattern the size of the bar has little or no effect, whereas fields several times that size evoke strong responses. Beyond a minimum image dimension, the response appears to be independent of size. An upper size limit exists for some cells, however, and too large an area of uncorrelation is ineffective.

A classification as "complex" and "simple" was obtained for 75 neurons on the basis of the spatial distribution of the binocular responses evoked by narrow bars of opposite contrast moving across the neuron's receptive field (Bishop et al., 1971; Schiller et al., 1976a; Kulikowski et al., 1981). Of these 75 neurons, 34 responded to uncorrelation and complex cells were nearly six times more frequent among them than simple cells (29 and 5). Complex (23) and simple (18) cells occurred in the proportion expected for an unbiased sample, about 1.5:1 (see also Schiller et al., 1976a) among the 41 correlation-insensitive cells. A similar preponderance of complex cells among those signaling binocular correlation was found for a larger sample of 223 neurons (177 in V1) in seven other monkeys.

Positional Disparity and Texture Correlation

In earlier studies with bar stereograms (Poggio and Fischer, 1977; Poggio and Talbot, 1981) and cyclopean random-dot stereograms (Poggio, 1980,1984; Poggio et al., 1985b), we found that a substantial proportion of neurons in cortical visual areas V1 and V2 of the alert macaque respond differentially to horizontal disparity. The present study adds neurons in V3 to the disparity-sensitive population. In all areas, two major types of stereoscopic sensitivity were observed: tuned and reciprocal. The response profile of *tuned* neurons is characterized by binocular facilitation (tuned excitatory, TE) or binocular suppression (tuned inhibitory, TI) over a narrow range of disparities, often with opposite responses at either side of maximum. TE neurons are not uniform in their sensitivity to binocular correlation (see following). Two groups may be recognized: neurons that respond maximally to bar stereograms with zero or near-zero disparities (± 0.05 degree), termed *tuned zero* (T0), and neurons responding to larger disparities, crossed or uncrossed, termed *tuned near* (TN) and *tuned far* (TF), respectively. The response profile of *reciprocal* neurons, is typically symmetrical about the zero disparity. Some cells give excitatory responses over a more or less extensive range of crossed disparities and inhibitory responses over a similar range of uncrossed disparities (near neurons, NE); other cells respond in the reverse manner, with facilitation to uncrossed and suppression to crossed disparities (far neurons, FA). A small number of disparity-sensitive cells has response profiles that cannot be assigned with certainty to one of these types (non-classified, nc). The term *flat* (FL) describes those neurons that give the same response at all disparities. Representative disparity sensitivity profiles are shown in Figure 11-3.

The proportion of disparity-sensitive neurons for the two monkeys of this study was 53 percent of all neurons classified in V1, 66 percent in V2, and 78 percent in V3. Although one has the impression that binocular interactions controls and specifies the activity of the neural population in prestriate areas more than it does in V1, the preceding proportions must be interpreted cautiously because of segregation of neurons with different functional properties in different cortical areas (Tootell et al., 1983; Livingston and Hubel, 1984; de Yoe and Van Essen, 1985; Shipp and Zeki, 1985; Van Essen et al., 1986). We found stereoscopic neurons in V2 to be distributed in zones extending across all cortical layers, and those in V3 to be grouped in clusters, whereas in V1 stereoscopic neurons were scattered irregularly.

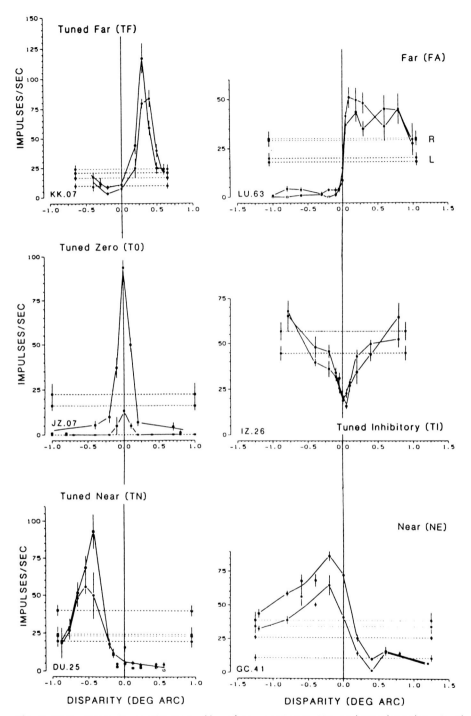

Figure 11-3 Disparity-sensitivity profiles of neurons in V1, V2, and V3 of monkey visual cortex. All the profiles were constructed from responses to stereograms of a bright, narrow bar adjusted for optimal spatiotemporal configuration for the neuron, and moving bidirectionally across the neuron's receptive field. For each panel, the two curves describe the response profile obtained with the diametrically opposite directions of stimulus motion. Horizontal dashed lines mark the left (L) and right (R) monocular responses to the two stimulus directions.

Table 11-1 Disparity Sensitivity and Correlation Sensitivity for Neurons in Visual Areas
V1, V2, and V3

Disparity Sensitivity			Responses to Uncorrelation		Total Neurons Responding/ Tested
			Activation	Suppression	
Tuned excitatory					
peak disparity	<0.05°	(T0)	—	16	16/19
	0.01°–0.2°		2	5	7/12
		(TN,TF)			
	>0.2°		14	—	14/21
			16	21	37/52 = 71%
Other stereo neurons					
(tuned inhibitory, near, far)			15	1	16/31 = 52%
Not disparity-sensitive					
(flat)			20	1	21/51 = 41%
	Total		51	23	74/134 = 55%

Sensitivity to binocular correlation, especially the signaling of uncorrelation,
was observed in more than half of the neurons, with no differences in the proportion of correlation-sensitive neurons in the three visual areas explored. Correlation sensitivity cuts across the sensitivity for horizontal disparity, and may be
a functional property of both depth neurons and flat neurons. It is, however, the
TE neuron that is most frequently sensitive to binocular correlation (71%); among
neurons with other types of disparity-sensitivity profiles (TI near and far), as well
as among the disparity-insensitive, flat neurons, fewer cells are sensitive to correlation and its absence (Table 11-1).

The results of this study reveal that disparity sensitivity and correlation sensitivity are not independent properties; although they may occur separately, both
are often part of the functional repertoire of binocular cortical neurons. Moreover,
a remarkably consistent association was observed between peak disparity sensi-

Figure 11-4 Data obtained from a tuned zero neuron (T0) in area V1, supragranular
layers (4b?), complex receptive field with center at 1.7 degrees from vertical meridian in
lower left quadrant. Responses to three conditions of stimulation are illustrated: (1) stimulation with bright bar (2.4 × 0.5 degrees, orientation 15 degrees from vertical) presented
against a dark background, and moving bidirectionally at 4 degrees/sec; (2) stimulation
of left or right eye alone (MONOCULAR) has little effect on the neuron's activity, whereas
binocular stimulation (STEREOGRAMS) evokes strong excitatory responses over a very
narrow disparity range with a peak at 0 degree disparity. Stimuli with larger crossed (−)
or uncrossed (+) disparities have inhibitory effects; (3) stimulation with dynamic random-
dot stereopatterns (CORRELOGRAMS), 2 × 2 degrees in size, centered on the neuron's
receptive field. The dot patterns are alternating between binocular uncorrelation (ON)
and correlation (OFF). Three sets of responses are shown with different times of uncorrelation–correlation: 250 to 750 msec; 500 to 500 msec; and 750 to 250 msec. The
neuron's activity is completely suppressed by uncorrelation and rebounds strongly to a
sustained discharge on presentation of correlated patterns. The latencies of both the inhibitory and the excitatory responses are very similar (60 to 70 msec) and are not affected
by the duration of the preceding opposite activity.

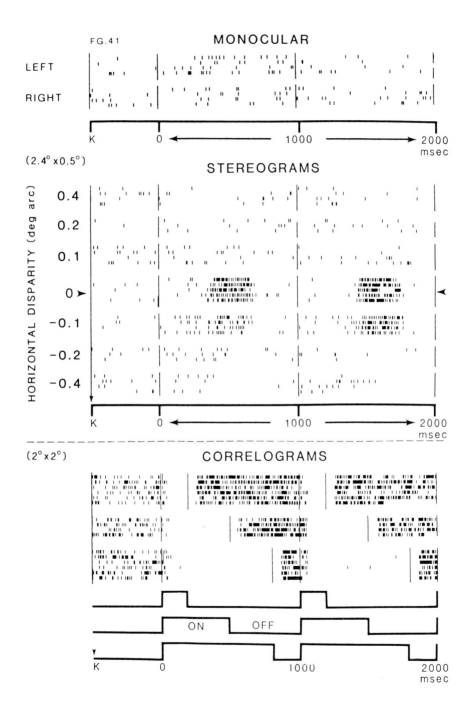

FG.41

MONOCULAR

LEFT

RIGHT

K   0 ←————————→ 1000 ————————→ 2000
                                        msec

(2.4° x0.5°)

STEREOGRAMS

HORIZONTAL DISPARITY (deg arc)

0.4

0.2

0.1

0 ►

-0.1

-0.2

-0.4

K   0 ←————————→ 1000 ————————→ 2000
                                        msec

(2° x2°)

CORRELOGRAMS

ON   OFF

K   0   1000   2000
                   msec

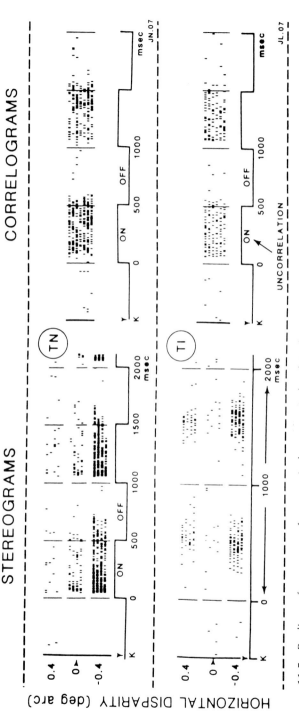

Figure 11-5 Replicas of responses of two cortical neurons isolated in the same microelectrode penetration to binocular disparity (stereograms) and binocular correlation (correlograms). (*Left*) Three set of responses are shown for each neuron to stereograms of different disparities: 0.4 crossed (−), 0 (zero), and 0.4 degree uncrossed. Stereograms were constructed with bright narrow bars (luminance = 14.4 cd/m²) of selected spatiotemporal configuration, appearing against a background of dynamic random-dot noise (density 20%; luminance 2.7 cd/m²). TN = tuned inhibitory neuron in area V2; peak disparity sensitivity at −0.4 degree; flashing bar 2.0 × 0.5 degree, orientation 40 degrees from vertical. TI = tuned inhibitory neuron in area V1; peak disparity sensitivity at 0 degree; moving bar 1.7 × 0.3 degree, speed 4 degrees/sec; orientation 55 degrees from vertical. (*Right*) Binocular uncorrelation evokes strong excitatory responses (activation) from either neuron, whereas correlation has no evident effects. Correlograms were constructed with a square area (2 × 2 degrees) centered on the neuron's receptive fields), within which the pattern of dots (density 50%; luminance 7.0 cd/m²) alternate between binocular correlation (prestimulation period, between K and t = 0, and OFF periods) and uncorrelation (ON), and are surrounded by a 10 degree background of correlated dots of the same density.

tivity and correlation sensitivity among the TE neurons. The extent of this association is shown in Table 11-1. Of the TE neurons with peak responses at or very close to zero disparity (T0 neurons), 16/19 responded to uncorrelation, and *all with suppression*. Conversely, of the TE neurons with peak responses at disparities greater than ±0.2 degree (TN and TF neurons), 14 out of 21 responded to uncorrelation, and *all with activation*. For TE neurons with peak responses around ±0.1, ±0.2 degree (also classified as TN or TF) either activation or suppression was observed.

Figure 11-4 illustrates the typical responses of a T0 neuron. Narrow bright bars of optimal two-dimensional configuration evoke strong excitatory responses only when presented binocularly with zero disparity (stereograms); the range of binocular facilitation is narrow and bar stimuli with larger disparities, either crossed or uncrossed, have inhibitory effects. Stimulation of either eye alone (monocular) evokes no response, as is frequently observed for this type of neuron (Hubel and Wiesel, 1970; Poggio and Fischer, 1977; Poggio, 1984). Stimulation with dynamic random-dot stereopatterns (correlograms) has a strong effect on the activity of this neuron. Presentation of a correlated pattern elicits a sustained activity that is completely suppressed when a binocularly uncorrelated pattern is made to appear over the neuron's receptive field (ON). The activity rebounds strongly to a sustained discharge as the uncorrelated pattern is reversed to a binocularly correlated one (OFF). Suppression and rebound activation have similar latencies, and the amplitude of both responses is independent of the duration of the stimulus evoking them.

All other types of correlation-sensitive stereoscopic neurons—including a majority of TN and TF cells, and the tuned inhibitory, the near, the far, and a number of disparity-insensitive neurons—respond to uncorrelation with activation. Figure 11-5 shows examples of this response behavior for a tuned near neuron (TN) and for a tuned inhibitory neuron (TI). The responses to bar stereograms are on the left and those to dynamic random-dot correlograms on the right. For all of these neurons, the excitatory response to uncorrelation was nearly always smaller, and never larger, than the excitatory response to binocular bar stimuli with the preferred disparity.

CONCLUSIONS

The results of these experiments confirm that the impulse activity in about 50 percent of the neurons in the visual cortex (V1, V2, and V3) of the alert macaque, signals the direction and magnitude of horizontal disparity of binocular images. In addition, cortical visual neurons are often sensitive to the textural identity of the images in the two eyes, and give opposite responses, excitatory and inhibitory, to binocularly correlated and uncorrelated patterns. Disparity sensitivity and correlation sensitivity may occur independently, or may both be present and interact to determine the functional profile of a single neuron.

It has long been known that during binocular vision the identical images of a single object in space may or may not fall on topographically corresponding retinal positions in the two eyes. A corollary is that functionally associated retinal positions in the two eyes, whether corresponding or disparate, receive either identical or different images. It is reasonable to assume that the activity of binocular neurons depends on the two-dimensional spatiotemporal configuration of the visual image and its relation to the structural organization of the receptive fields, as well as on the degree of mismatch between the images "seen" by the pair of receptive fields in the two eyes.

Our findings indicate that a large majority of TE neurons with peak binocular facilitation at or close to zero disparity (± 3 minarc) are strongly suppressed by binocularly uncorrelated patterns. These neurons are activated by objects at about the horopter distance—possibly within Panum's fusional area—and are silenced, in the absence of such objects, by the uncorrelated images of other objects at other depths. Their activity may provide a reference for the stereo system, of the distance of convergent fixation, and may represent the essential neural substrate for binocular single vision, carrying information for both sensory and motor fusional mechanisms.

A second group of TE neurons with peak binocular facilitation at disparities larger than 12 minarc—possibly outside Panum's area—respond to uncorrelation with activation and never with suppression. Similar effects are also evoked by uncorrelation in other kinds of stereoscopic neurons (near and far). This excitatory response, however, is consistently smaller than that obtained with correlated patterns of optimal disparity, which suggests that these neurons are specifically sensitive to positional disparity. The activity of these neurons may provide signals for coarse stereopsis (Bishop and Henry, 1971) as well as for the system controlling binocular vergence. Moreover, near and far neurons may also contribute to fine stereopsis (Bishop and Henry, 1971) by signaling small disparity changes through their finely graded response transition from suppression to facilitation about the fixation distance (Poggio, 1984). A group of disparity-insensitive neurons (flat), about 40 percent in the sample of this study, also signal uncorrelation with an increase in impulse activity; their role in stereopsis, if any, is unclear.

These findings indicate that the response selectivity of the cortical neuron for stimuli in depth depends on the interaction evoked by two binocular components of the visual stimulus, the relative horizontal position and the relative contrast of the images over the neuron's receptive fields in the two eyes.

ACKNOWLEDGMENTS

The research described in this paper was supported by Grant 5 R01 EY02966 from the National Institute of Health, U.S. Public Health Service, and by a grant from the A. P. Sloan Foundation.

REFERENCES

Barlow, H., Blakemore, C., and Pettigrew, J. D. (1967). The neural mechanism of binocular depth discrimination. *J. Physiol. (Lond.)* 193:327–342.

Bishop, P. O., and Henry, G. H. (1971). Spatial vision. *Annu. Rev. Psychol.* 22:119–161.

Bishop, P. O., Henry, G. H., and Smith, C. J. (1971). Binocular interaction fields of single units in the cat striate cortex. *J. Physiol. (Lond.)* 216:39-68.

Burkhalter, A., and Van Essen, D. C. (1986). Processing of color, form and disparity information in visual areas VP and V2 of ventral extrastriate cortex in the macaque monkey. *J. Neurosci.* 6:2327-2351.

Camarda, R. M., Peterhans, E., and Bishop, P. O. (1985). Spatial organizations of subregions in receptive fields of simple cells in cat striate cortex as revealed by stationary flashing bars and moving edges. *Exp. Brain Res.* 60:136–150.

De Yoe, E. A., and Van Essen, D. C. (1985). Segregation of afferent connections and receptive field properties of visual area V2 of the macaque. *Nature* 317:58–61.

Emerson, R. C., Citron, M. C., Vaughn, W. J., and Klein, S. A. (1987). Nonlinear directionally selective subunits in complex cells of cat striate cortex. *J. Neurophysiol.* 58:33–65.

Felleman, D. J., and Van Essen, D. C. (1987). Receptive field properties of neurons in area V3 of macaque monkey extrastriate cortex. *J. Neurophysiol.* 57:889–920.

Ferster, D. (1981). A comparison of binocular depth mechanisms in area 17 and 18 of the cat visual cortex. *J. Physiol. (Lond.)* 311:623–655.

Fischer, B., and Krueger, J. (1979). Disparity tuning and binocularity of single neurons in the cat visual cortex. *Exp. Brain Res.* 35:1–8.

Gonzalez, F., Krause, F., and Poggio, G. F. (1986). Sensitivity of cortical visual neurons to binocular correlation of dynamic random-dot stereo patterns. *Invest. Ophthalmol. Vis. Sci.* 27(Suppl.):244.

Hubel, D. H., and Wiesel, T. N. (1962). Receptive fields, binocular interaction and functional architecture in the cat's visual cortex. *J. Physiol. (Lond.)* 160:106–154.

Hubel, D. H., and Wiesel, T. N. (1968). Receptive fields and functional architecture of monkey striate cortex. *J. Physiol. (Lond.)* 195:215–243.

Hubel, D. H., and Wiesel, T. N. (1970). Cells sensitive to binocular depth in area 18 of the macaque monkey cortex. *Nature* 225:41–42.

Julesz, B. (1960). Binocular depth perception of computer-generated patterns. *Bell Syst. Tech. J.* 39:1125–1162.

Julesz, B. (1971). *Foundations of Cyclopean Perception.* Chicago: University of Chicago Press.

Julesz, B., Breitmeyer, B., and Kropfl, W. (1976). Binocular-disparity-dependent upper-lower hemifield isotropy as revealed by dynamic random-dot stereograms. *Perception* 5:129–141.

Kulikowski, J. J., Bishop, P. O., and Kato, H. (1981). Spatial arrangements of responses by cells in the cat visual cortex to light and dark bar and edges. *Exp. Brain Res.* 44:371–385.

Livingstone, M. S., and Hubel, D. H. (1984). Anatomy and physiology of a color system in the primate visual cortex. *J. Neurosci.* 4:309–356.

Maske, R., Yamane, S., and Bishop. P. O. (1984). Binocular simple cells for local stereopsis: comparison of the receptive field organization for the two eyes. *Vision Res.* 24:1921–1929.

Maske, R., Yamane, S., and Bishop, P. O. (1986). Stereoscopic mechanisms: binocular responses of the striate cells of cats to moving light and dark bars. *Proc. R. Soc. Lond. B* 229:227–256.

Maunsell, J. H. R., and Van Essen, D. C. (1983). Functional properties of neurons in middle temporal visual area of the macaque monkey. II. Binocular interaction and sensitivity to binocular disparity. *J. Neurophysiol.* 49:1148–1167.

Motter, B. C., and Poggio, G. F. (1984). Binocular fixation in the rhesus monkey: spatial and temporal characteristics. *Exp. Brain Res.* 54:304–314.

Movshon, J. A., Thompson, I. D., and Tolhurst, D. J. (1978). Receptive field organization of complex cells in the cat's striate cortex. *J. Physiol. (Lond.)* 283:79–99.

Mullikin, W. H., Jones, J. J., and Palmer, L. A. (1984a). Receptive-field properties and laminar distribution of X-like and Y-like simple cells in cat area 17. *J. Neurophysiol.* 52:350–371.

Mullikin, W.H., Jones, J. J., and Palmer, L. A. (1984b). Periodic simple cells in cat area 17. *J. Neurophysiol.* 52:372–387.

Nikara, T., Bishop, P. O., and Pettigrew, J. D. (1968). Analysis of retinal correspondence by studying receptive fields of binocular single units in cat striate cortex. *Exp. Brain Res.* 6:353–372.

Ohzawa, I., and Freeman, R. D. (1986a). The binocular organization of simple cells in the cat's visual cortex. *J. Neurophysiol.* 56:221–242.

Ohzawa, I., and Freeman, R. D. (1986b). The binocular organization of complex cells in the cat's visual cortex. *J. Neurophysiol.* 56:243–259.

Poggio, G. F. (1980). Neurons sensitive to dynamic random-dot stereograms in areas 17 and 18 of the rhesus monkey cortex. *Soc. Neurosci. Abstr.* 6:672.

Poggio, G. F. (1984). Processing of stereoscopic information in monkey visual cortex. In: Edelman G. M., Gall, W. E., Cowan, W. M. (eds.). *Dynamic Aspects of Neocortical Function*, pp. 613–635. New York: John Wiley & Sons.

Poggio, G. F., and Fischer, B. (1977). Binocular interaction and depth sensitivity in striate and prestriate cortex of behaving rhesus monkeys. *J. Neurophysiol.* 40:1392–1407.

Poggio, G. F., and Poggio, T. (1984). The analysis of stereopsis. *Annu. Rev. Neurosci.* 7:379–412.

Poggio, G. F., and Talbot, W. H. (1981). Neural mechanisms of static and dynamic stereopsis in foveal striate cortex of rhesus monkeys. *J. Physiol. (Lond.)* 315:469–492.

Poggio, G. F., Doty, R. W., and Talbot, W. H. (1977). Foveal striate cortex of behaving monkey: single-neuron responses to square-wave gratings during fixation of gaze. *J. Neurophysiol.* 40:1369–1391.

Poggio, G. F., Gonzalez, F., and Krause, F. (1985a). Binocular correlation system in monkey visual cortex. *Soc. Neurosci. Abstr.* 11:17.

Poggio, G. F., Motter, B. C., Squatrito, S., and Trotter, Y. (1985b). Responses of neurons in visual cortex (V1 and V2) of the alert macaque to dynamic random-dot stereograms. *Vision Res.* 25:397–406.

Schiller, P. H., Finlay, B. L., and Volman, S. F. (1976a). Quantitative studies of single-cell properties in monkey striate cortex. I. Spatiotemporal organization of receptive fields. *J. Neurophysiol.* 39:1288–1319.

Schiller, P. H., Finlay, B. L., and Volman, S. F. (1976b). Quantitative studies of single-cell properties in monkey striate cortex. II. Orientation specificity and ocular dominance. *J. Neurophysiol.* 39:1320–1333.

Shipp, S., and Zeki, S. (1985). Segregation of pathways leading from area V2 to areas V4 nd V5 of macaque monkey visual cortex. *Nature* 315:322–325.

Skottun, B. C., and Freeman, R. D. (1984). Stimulus specificity of binocular cells in cat's visual cortex: ocular dominance and the matching of left and right eyes. *Exp. Brain Res.* 56:209–216.

Tootell, R. B. H., Silverman, M. S., DeValois, R. L., and Jacobs, G. H. (1983). Functional organization of the second cortical visual area in primates. *Science* 220:737–739.

Tyler, C. W., and Julesz, B. (1976). The neural transfer characteristics (neurontropy) for binocular stochastic stimulation. *Biol. Cybern.* 23:33–37.

Tyler, C. W., and Julesz, B. (1978). Binocular cross-correlation in time and space. *Vision Res.* 18:101–105.

Van Essen, D. C., Newsome, W. T., Maunsell, J. H. R., and Bixby, J. L. (1986). The projections from striate cortex (V1) to visual areas V2 and V3 in the macaque monkey: asymmetries, areal boundaries, and patchy connections. *J. Comp. Neurol.* 244:451–480.

Von der Heydt, R., Adorjani, Cs., Haenny, P., and Baumgartner, G. (1978). Disparity sensitivity and receptive field incongruity of units in cat striate cortex. *Exp. Brain Res.* 31:523–545.

Wheatstone, C. (1838). Contributions to the physiology of vision. Part the first. On some remarkable, and hitherto unobserved, phenomena of binocular vision. *Phil. Trans. Roy. Soc.* 2:371–393.

Zeki, S. M. (1978). The third visual complex of rhesus monkey prestriate cortex. *J. Physiol. (Lond.)* 277:245–272.

12

Topography of Visual Function as Shown with Voltage-Sensitive Dyes

GARY G. BLASDEL

We now know more about primary visual cortex (area 17, V1, striate cortex) than about any other part of the neocortex, particularly with regard to its functional layout and underlying anatomy. Hubel and Wiesel, who did much of the classic work in this area, are renowned for their careful correlation of structure with function. At the heart of their work lie two response properties—ocular dominance and orientation selectivity—that emerge for the first time in striate cortex and help set it apart from all preceding stages of visual processing.

OCULAR DOMINANCE

Ocular dominance refers to one aspect of the binocular responses that characterize most cortical cells (Hubel and Wiesel, 1962, 1965, 1968, 1972). Before this level of the cortex, in the retina and lateral geniculate nucleus (LGN) for example, pathways corresponding to the two eyes are separate. Among old world primates, including macaque monkeys, chimpanzees, and humans, they remain separate in lamina 4C of striate cortex (the main input lamina) as well, where it is known, from the results of many complementary studies, that geniculate afferents segregate into an interdigitating system of right and left eye bands—each about 0.5 mm wide—that run perpendicularly into the edge of striate cortex at its boundary with area 18 (LeVay et al., 1975). In the adult, each band receives afferents exclusively from one eye, and neurons within it respond only to that eye. But after this the pathways converge, giving rise to large numbers of cells with binocular receptive fields.

The convergence is not uniform, so most of the cells outside lamina 4C still respond better to one eye than to the other, expressing a preference that could be regarded as a tendency for one eye to dominate. This tendency is most pronounced for cells in vertical register with the centers of layer 4C ocular dominance bands, and weakest for the cells in between. Accordingly, when projected through cortical depth, ocular dominance takes on the three-dimensional structure of slabs that are commonly referred to as 'columns' (Hubel and Wiesel, 1965).

ORIENTATION SELECTIVITY

Neurons in striate cortex, five or more synapses removed from the photoreceptor layer of the retina, are remarkable in their ability to select meaningful edges from the visual environment (Hubel and Wiesel, 1962, 1968). Since it is apparently difficult to do this equally well at all orientations, neighboring cells specialize in different angles, and they accommodate the range of possibilities by incorporating one or more complete sets of orientation into the cortical representation for each part of visual space. There is also a large amount of overlap in the angles covered by cells with different, but neighboring, preferences, because the absolute selectivity for any particular angle tends to be quite broad—on the order of 30 to 60 degrees.

Orientation selectivity has resisted numerous efforts directed at understanding its mechanisms, and there does not appear to be an easily recognizable anatomical substrate, as there is with ocular dominance, that can be exploited to visualize its organization. All currently available techniques for studying it therefore rely on some (direct or indirect) measure of physiological activity.

Our clearest insights still seem to come from the initial microelectrode studies of Hubel and Wiesel (1962), who noted that orientation preference, like eye preference, remains constant with depth. Accordingly, electrode penetrations that run vertically encounter cells with similar preferences (Figure 12-1A.) Penetrations that move laterally (Hubel and Wiesel, 1974), on the other hand, reveal a regular, almost mechanical, rotation in preferred orientation—one that occurs in many small increments and appears continuous over short intervals (0.5 to 1.0 mm). The continuity is broken over longer intervals, however, by abrupt changes as well as by sudden reversals in the direction of rotation.

Noting a variation in the rates at which orientation changed (over the short continuous segments), Hubel and Wiesel (1974) argued for a slablike organization, similar to that for ocular dominance. According to their model (illustrated in Figure 12-1B) each slab contains neurons with one orientational preference, and neighboring slabs contain cells with slightly shifted preferences. By stacking many such slabs together, one can generate a system in which the preferred orientation shifts continuously along one axis yet remains constant with depth and along the other axis, which, for reasons of symmetry, Hubel and Wiesel (1974) argued might intersect ocular dominance slabs at right angles.

Experiments with 2-deoxyglucose (2DG), a metabolic marker that labels active cells, at first appeared to support this model. But these results have been confounded more recently by the discovery of cytochrome oxidase blobs—small regions in the upper cortical layers containing high concentrations of the mitochondrial enzyme cytochrome oxidase (Wong-Riley, 1979)—that contain neurons responsive to all orientations (Horton and Hubel, 1980; Livingstone and Hubel, 1984 a,b). Thus, the model in Figure 12-1B must be updated to take this observation into account, one possibility being the scheme proposed by Livingstone and Hubel (Figure 12-1C).

But even if broad themes of Hubel and Wiesel's model are correct, and some regions of cortex are organized as stacks of parallel slabs, the large-scale structure

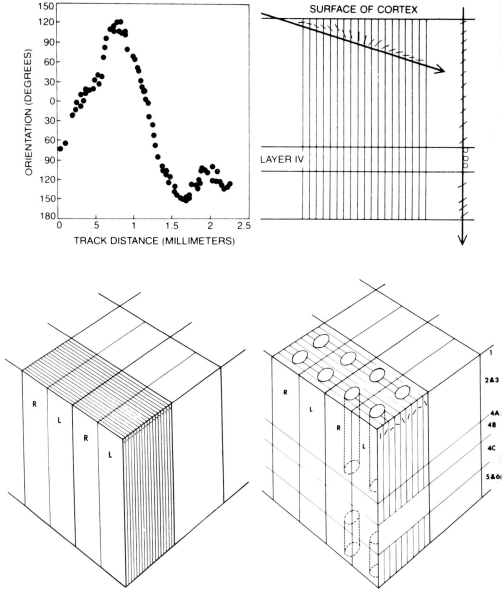

Figure 12-1 Organizational schemes for orientational preference deduced from micro-electrode penetrations. (*A*) Hubel and Wiesel's results for electrode penetrations conducted parallel and perpendicular to the cortical surface. For perpendicular penetrations, the preferred orientation remains constant with depth (except for laminae 4A and 4C, where orientational selectivity is weak or absent). For lateral penetrations, on the other hand, the angle of the preferred orientation rotates continuously. The continuity of this change is nevertheless disrupted at regular intervals by sudden reversals in the direction of rotation, as well as by abrupt absolute shifts. (*B*) Ice-cube model of striate cortex, proposed by Hubel and Wiesel (1974) on the basis of results from numerous electrode penetrations in which ocular dominance and orientation selectivity are organized as slabs that (in this model) intersect one another at right angles. (*C*) Ice-cube model (as in *B*) modified by Livingstone and Hubel (1984a) to take into account the unique physiology of cytochrome oxidase blobs where orientation selectivity is either weak or absent.

obviously contains more complexity than anyone had imagined; and it is unlikely that this complexity can ever be resolved with microelectrodes or 2DG. Electrodes are too limited in scope to reveal much more than they already have about large-scale features; and 2DG still can be used only once, making it impossible, for example, to see how the same piece of tissue responds to different stimuli, or even to do a control experiment.

OPTICAL TECHNIQUES

The first suggestion of an approach that might be used repeatedly to study activity patterns in the same piece of tissue came with reports that some correlates of neuronal activity can be monitored optically (Cohen et al., 1968, 1972). The bi-refringence in a squid giant axon, for example, varies with the passage of an action potential, and accordingly can be exploited to monitor activity. The signal is weak, however, and many averages are required. Cohen et al. (1974) showed that larger optical signals can be obtained if the nervous tissue is first stained with certain impermeable, membrane-bound dyes that function as optical transucers of membrane potential. Even though the potential across a cellular membrane is small (50 to 70 mV), because the membrane is thin (on the order of 50 nm) changes in the electric field strength that result from an action potential are enormous. Accordingly, they can easily alter the optical properties of an appropriately struc-tured and situated molecule. Although most of the dyes Cohen used initially proved toxic, many newer varieties have been developed over the last 12 years, and some of these have proved quite useful (Grinvald et al., 1983).

In 1986 Guy Salama and I reported the first successful exploitation of an op-tical technique based on one such dye (merocyanin-oxazalone) to reveal well-established patterns of functional organization in monkey striate cortex (Blasdel and Salama, 1986). The features we examined (bands corresponding to ocular dominance and to different orientations) are known, but they had never before been visualized in living animals, nor had they ever been demonstrated simulta-neously in the same piece of tissue. They provided new insights into many of the correlations made previously between structure and function. The major findings of this work, along with a detailed description of the technique, were reported previously (Blasdel and Salama, 1986), but for convenience are described here.

EXPERIMENTAL METHODS

Each animal is prepared for physiological recording according to established pro-cedures (Blasdel and Fitzpatrick, 1984). After induction of anesthesia with ke-tamine and xylazine, the animal is intubated and provided with an intravenous catheter for the infusion of electrolytes and drugs. Following this the animal breaths a 2:1 mixture of nitrous oxide and oxygen (for analgesia) and receives pentothal intravenously to maintain anesthesia as the ketamine wears off.

When these procedures are completed, the animal is placed on a heated wa-terbed with its head positioned in a stereotaxic head holder. The scalp is reflected

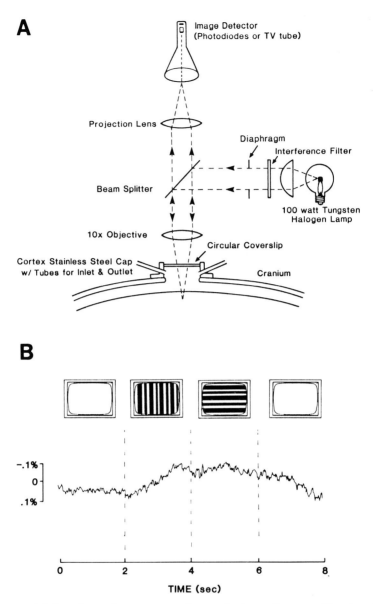

Figure 12-2 (*A*) Experimental arrangement for monitoring voltage-sensitive dyes in monkey striate cortex. A hole (25 mm in diameter) was cut in the cranium, a stainless steel ring cemented to the skull, and the dura removed. The ring had stainless steel inlet and outlet tubes to bathe the cortical surface with dye solution (0.1% in balanced salt solution). A second ring containing a round glass coverslip was screwed into the first ring to form a watertight compartment. The position of the glass could be adjusted to dampen cortical pulsations and provide an optically smooth window for viewing the cortex. A condenser lens collimated light from a 100-W tungsten halogen lamp and directed it through an interference filter (720 +/− nm, Omega Optical, Brattleboro, Vermont), an aperture, and then to the reflective surface of a beam splitter. Reflected light passed through an objective lens and glass coverslip onto the cortex. Reflected and scattered light from the cortex passed through the coverslip, to the objective lens, through a projection lens, to the image detector (photodiodes or the TV camera). (*B*) Averaged response to 10 suc-

and a 1-inch-diameter trephine is used to bore a hole just behind the lunate sulcus, as close as possible to the saggital suture. A stainless steel chamber is then inserted and cemented to the cranium (Figure 12-2A). This chamber contains an 18-mm-diameter window, through which the cortical surface can be viewed at high resolution, as well as inlet and outlet ports that allow superfusion of the cortex. When these ports are sealed, the chamber forms a rigid hydraulic barrier that resists pulsation.

At the time of physiological recording the animals' eyes are fitted with hard contact lenses and refracted so that they focus on the screen of a 19-inch color TV monitor 2 m away. The animal is then partially paralyzed with pancuronium bromide while pentothal (0.5 to 2 mg/kg/hour) is administered continuously to maintain anesthesia. Because paralysis is incomplete, the adequacy of anesthesia can be assessed most easily from the presence of reflexes. But the electroencephalogram (EEG), electrocardiogram (ECG), and end-tidal carbon dioxide (CO_2) are monitored in any case.

The optical apparatus is illustrated in Figure 12-2A. Light from a tungsten halogen lamp passes through a condensor, interference filter, and diaphragm, and is then deflected, by either a beam splitter or a partially offset mirror, through an objective lens onto the cortex. Light that is not scattered, or absorbed, is reflected back up through the glass window and objective lens, past the beam splitter, and through a lens to form an image of the cortical surface on a 2×2 array of photodiodes or on to the Newvicon element of a TV camera (Cohu Model 5300).

The signal from one photodiode, monitoring the absorbence by a small patch of cortex (approximately 100 μm across) of light at 720 nm, appears in Figure 12-2B. Dashed vertical lines indicate 2-sec intervals. As one can see from this figure, a baseline is established during the prestimulus period while the animal views a blank, gray screen. It then undergoes a sharp inflection 150 msec after the appearance of a drifting vertical grating, and rises steadily for the next second or more. A small inflection occurs when the vertical grating disappears, to be replaced by a horizontal one. And the signal decays back to baseline when the horizontal grating disappears.

The large size of the signal, as well as its long duration, suggest that these changes can also be monitored with a television camera that, if sufficiently sensitive, should provide better spatial resolution. The chief obstacle to the implementation of a TV system appeared to stem from the poor dynamic range that characterizes most commercially available cameras (typically on the order of 200). Digitization steps, introduced by analogue to digital (A/D) conversion, seemed to present a second problem, since most commercially available video A/D con-

---

cessive presentations of vertical and horizontal gratings, as seen by one photodiode that surveyed a 100-μm patch of cortex. At the onset of a drifting vertical grating, there is a sudden increase in absorption for this part of cortex. A small glitch appears where the grating changes from vertical to horizontal, and the trace returns to baseline when the horizontal grating fades back again into a blank screen. Sequential presentations were synchronized to both the electrocardiogram and the respiration, so that movement artifacts produced by these events could be subtracted, as previously described by Orbach et al. (1985). [From Blasdel and Salama (1986).]

vertors use only 8 bits (and therefore have only 256 gray levels), which at first glance, appears inadequate to register the 0.1 percent changes in radiance that one might expect, under the best circumstances, to indicate differing levels of neuronal activity.

Fortunately, both problems have solutions. Some of the digitization problem can be overcome through a simple DC subtraction (followed by AC amplification) of the video signal before it reaches the A/D convertor. And although the remaining signal is still too small to register within 256 gray levels, the electrical noise that limits dynamic range is not; this noise causes the least significant bits of the A/D convertor to change randomly, with the subthreshold components of the analog signal retained as the probability that these bits are set at any particular time. The noise is then removed through averaging, leaving the signal—a process that electrical engineers refer to as *dithering*. When 1800 frames are averaged, the video and image-processing system used to collect the signals reported in this paper resolve radiance changes down to 0.01 percent (1 part in 10,000).

The strategy for monitoring optical changes associated with neuronal activity entailed the accumulation of two sets of images—one each for two different stimulus conditions. Frame collection began 0.5 sec after stimulus onset and usually lasted for 1.5 sec. Periods longer than this usually resulted in unacceptably large amounts of noise from the cortical vasculature, the perfusion and optical density of which were found to change constantly (though slower than the optical signals that interested us). Some of these changes undoubtedly derive from the metabolic requirements of stimulus-induced activity, but some appear to result from more general, complicated homeostatic adjustments that are difficult to predict. Since the resulting artifacts are too complicated to remove through subsequent processing, they are minimized at the outset by adjusting the duration of frame accumulation (usually 2 sec or less), as well as by restricting comparisons to cortex undergoing comparable levels of activity. Accordingly, those images acquired during visual stimulation of one eye or one orientation are best compared with images acquired at roughly the same time with visual stimulation of the other eye, or the orthogonal orientation. In this manner, intensity differences deriving from the nonspecific activation of cortex cancel out, and patterns generated by changes in ocularity and orientation are emphasized.

We recorded from a number of cortical cells before and after each optical experiment, to ensure that their responses were healthy and had not been affected by the dyes. These recordings also provided us with conventionally determined values of ocular dominance and orientation for comparison with the optical results.

DATA COLLECTION AND RESULTS

Figure 12-3 shows the outline of a stainless steel chamber affixed to a monkey's skull. The dashed lines indicate the two regions of cortex studied. In Figure 12-4A one can see patch *a* as it appeared through the video system. Blood vessels appear out of focus because of the relatively large numerical aperture of the objective lens, which was focused beneath the cortical suface. In Figure 12-4C one

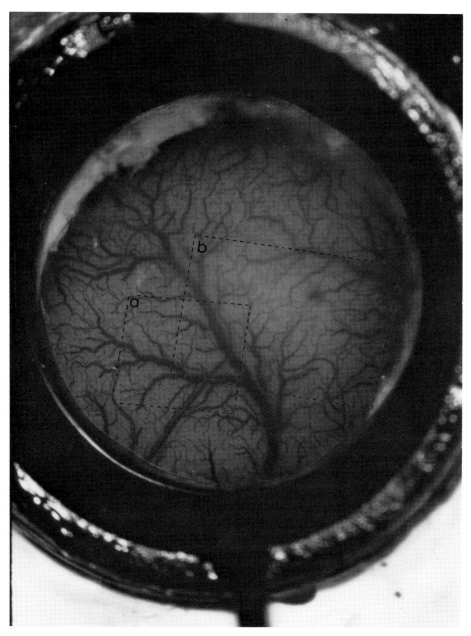

Figure 12-3 View of cortex surounded by a stainless steel chamber that has been cemented to monkey's right cranium. The saggital suture runs vertically just to the right of the chamber's right border, and the lunate sulcus runs horizontally under the lower rim. The two regions (labeled *a* and *b*) outlined in dotted lines correspond to the two cortical regions featured in subsequent images. Both frames have their lower boundaries aligned with the lunate sulcus, which means they are also aligned with the 17/18 border. But *a* is taken at higher magnification than *b*, and the width of *a* is 5.5 mm whereas the horizontal width of *b* is 8 mm.

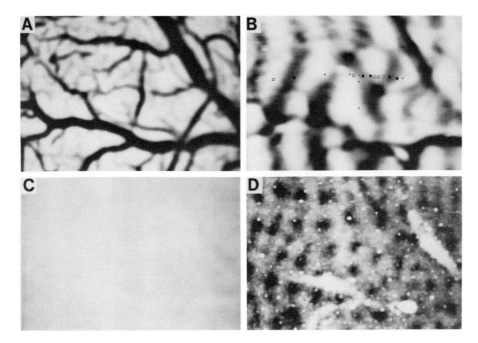

Figure 12-4 (A) Cortical blood vessels viewed under white light. They appear out of focus because the optical system was focused 300 μm below the pial surface. This patch of cortex, which corresponds to *a* in Figure 12-3 was located just behind the lunate sulcus in the right hemisphere. Posterior, anterior, medial, and lateral correspond to the top, bottom, right, and left of the image frame, respectively. The border between areas 17 and 18 lies parallel to and just below the lower border of this frame. The height and width of this image corresponds to a 4 × 5-mm patch of cortex. (B) The familiar pattern of ocular dominance bands is produced by subtracting images accumulated during visual stimulation of the contralateral eye from images accumulated during stimulation of the ipsilateral eye (all orientations). This produces an intensity difference of 0.2 percent (amplified electronically) between the dark and light bands. The bands have an average width of 0.5 mm, run vertically into the area 17/18 border, and correspond to the rows of cytochrome oxidase blobs, which appear in (D). The spots in *B* illustrate the location of 16 electrode penetrations conducted in this region of cortex before and after measurements are taken with voltage-sensitive dyes. The small black dots indicate sites that yielded binocular responses, open squares indicate sites dominated by the contralateral eye, and closed squares indicate sites dominated by the ipsilateral eye. In agreement with the photodiode results, the correspondence between the dark bands and closed squares indicates that stained cortical tissue darkens with increased electrical activity (see text). (C) The result of subtracting two time-averaged images while the monkey viewed a blank screen. The same algorithm was used as in *B*. (D) At the end of the experiment, the same piece of cortex was sectioned tangentially and stained for cytochrome oxidase. The section shown here was aligned and expanded (necessary because of tissue shrinkage) according to the positions of blood vessels so that they laid in register with the images in *A* to *C*. Note that during examination of the histological preparations, no evidence was found of tissue damage attributable to repeated applications of the dye over more than 8 weeks. [From Blasdel and Salama, 1986.]

can see the result of subtracting one image, derived from 1800 video frames accumulated while the animal viewed a blank gray screen, from 1800 video frames also accumulated while the animal viewed a blank screen—a control image that should appear blank except for noise from such inescapable sources as electrical, vascular, and neuronal components.

Figure 12-4B shows the result of collecting two sets of frames while the right and left eyes were stimulated alternately with high contrast, multifrequency gratings moving at all orientations. The alternating dark and light bands, seen to run vertically through this image, appear strikingly similar to the ocular dominance bands visualized with other techniques. They have the correct width (0.5 mm)—a fact that is appreciated best by noting the presence of five pairs within a horizontal distance of 5.5 mm—and run perpendicularly into the boundary of striate cortex that lies parallel to, and just below, the lower border of the image. The optically determined ocular dominance values also correlate with the results of 16 electrode penetrations (indicated by open and filled squares and circles). And the light and dark bands also coincide with the cytochrome oxidase blobs that were revealed through subsequent histochemical processing (Figure 12-4D).

The promise of this approach lies less in the revelation of ocular dominance bands, the general organization of which we know, than in providing information about orientation selectivity. Figure 12-5 reveals activity patterns produced by four different orientations in the same part of cortex. Each image represents the difference in activities elicited by each orientation and its orthogonal and has a contrast comparable to that of the ocular dominance bands in Figure 12-4B. The patterns are clearly different though, and tend to resemble the isoorientation bands seen earlier with 2DG (Hubel et al., 1978).

Because at least two parameters (response strength and angle) are required to describe the orientation-dependent response of any particular cortical neuron, the response, and hence the activity pattern, any single orientation generates contains ambiguities that are difficult to interpret and that can be resolved only by comparing it with responses or activity patterns generated at other orientations.

As one can see from Figure 12-5, however, the activity patterns generated at several different orientations do not appear to reveal much more, even though collectively they should contain all the necessary information. This is because it is difficult to compare them when they are displayed separately. Since the patterns were acquired at roughly the same time, however, the response intensity at every location can be extracted from each image to yield activity as a function of orientation—information that is easily converted into values for orientational preference and selectivity.

In practice this is accomplished most easily with a computer by taking every scaler pixel value and turning it into a vector (r, θ) where r represents intensity and θ corresponds to twice the angle of the bidirectionally moving contours used to generate a response. Since the presented visual stimuli moved bidirectionally (to avoid confusion between orientation and motion selectivities), one complete orientation cycle consisted of only 180 degrees, an angle that must be doubled to make it correspond with the 360 degrees that describe a complete cycle in polar coordinates. Since the inversion of r in polar coordinates is formally equivalent to a rotation through 180 degrees for θ (through 90 degrees for stimulus orien-

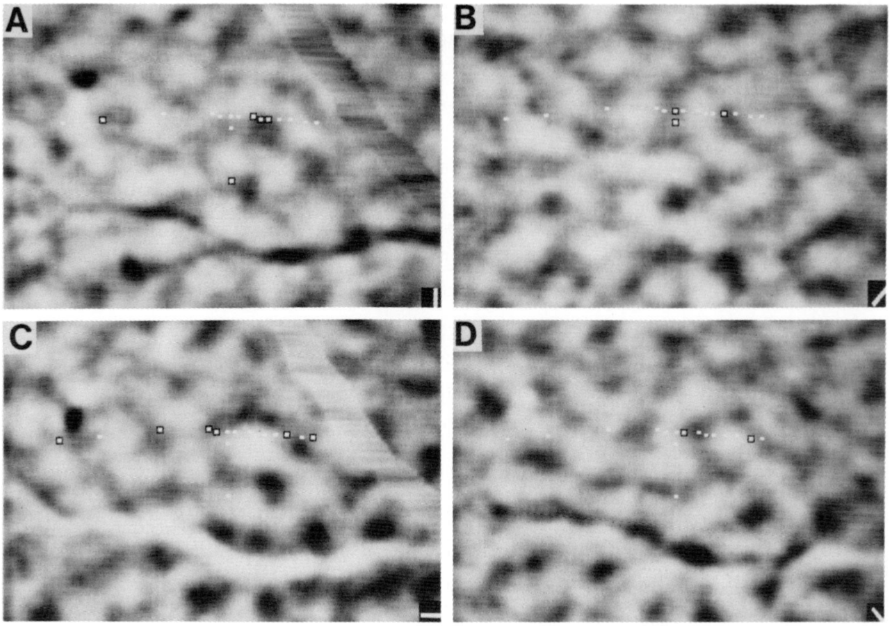

Figure 12-5 Isoorientation activity patterns produced by four separate orientations—vertical, right oblique, horizontal, and left oblique. Each image was compiled separately and related to a reference image generated by the orthogonal orientation. This subtraction was necessary to minimize the effects of activity-dependent hyperemia. For example, images of cortex taken during horizontal presentations were subtracted from images taken during vertical presentations to produce the pattern in *A*. The observation that vertical and horizontal patterns shown in *A* and *C* are complementary merely underscores the consistency of the result. The white dots indicate the positions of electrode recording sites. Those outlined in black denote locations where sampled cells preferred orientations within 22.5 degrees of that presented. These correlate with the darker parts of each image and never overlap with the lighter ones. Major blood vessels, which sometimes produce shadows caused by uncontrollable dilation during image acquisition, have been blocked out to prevent unwanted influences on adjacent regions of the same image during subsequent normalization and image processing. [From Blasdel and Salama, 1986.]

tation), this conversion has the additional desirable effect that similar responses to orthogonal stimuli (i.e., ones offset by 90 degrees) produce vectors (offset by 180 degrees) that cancel. One might note that the input data (Figure 12-5), in which each pixel reflects the difference between responses generated by one orientation and its orthogonal, already exist in this format.

Once the intensity arrays of Figure 12-5 have been turned into vector arrays, they are added to yield a new array of vectors in which magnitude r indicates orientational selectivity, and angle θ corresponds to twice the angle of preferred orientation. Responses at any particular orientation have now been normalized with respect to those generated at all orientations; and most of the artifacts that may have been introduced by the initial collection of response differences are thereby removed. Most important, however, this operation extracts precise information about the distribution of orientational preferences and selectivities from the results of many different scans (Figure 12-6).

ORIENTATIONAL PREFERENCE (THETA/2)

The orientation component appears in gray scale in Figure 12-6A, and in false color in Figure 12-7B. The gray-scale rendition gives a better indication of continuity—one can easily find Hubel and Wiesel's result of orderly, clocklike shifts in preferred orientation for any arbitrarily placed, imaginary electrode penetration—but it suffers from a serious inability to depict more than a single orientation cycle continuously. Consequently, the brightest edges in this figure correspond to places where orientation shifts smoothly between 180 and 0 degrees. The color-coded rendition in Figure 12-7B overcomes this difficulty but has other problems (deriving mostly from limitations in our color perception) that frustrate attempts to illustrate continuity.

Figure 12-7B depicts the distribution of orientational preferences across the surface of a small cortical patch; from this one can extract orientation values along single scan lines, and plot them directly as a function of distance. This is done twice in Figure 12-8 for scan lines taken from Figure 12-7B: once for a line that runs through 3 vertically aligned recording sites, and once for a line that intersects 14 horizontally aligned sites. As one can see from the illustrations in Figure 12-8, the orientational preferences that are computed from the results of optical recordings correspond closely to the results obtained with convention

al microelectrode recordings. Moreover the plots in this figure resemble those reported initially by Hubel and Wiesel (1974). Short sequences in which the preferred orientation shifts continuously are readily apparent (see Figure 12-1A, as are points where this continuity is broken by abrupt shifts and sudden reversals in rotation direction.

From the two-dimensional maps of orientational preference in Figures 12-6 and 12-7, however, one can see that this property distributes in a fashion that is much more elaborate than one might have expected from stacks of isoorientation slabs (see Figure 12-1B). For Hubel and Wiesel's model to apply generally, the images in Figure 12-7, should appear streaked, like stacks of rainbows. But this only happens in small circumscribed areas, patches less than 0.5 to 1.0 mm across; for larger areas the streaked appearance is fragmented.

The fragmentation becomes more apparent if one ignores absolute values of orientation and focuses instead on the continuity with which they change, a property that is visualized most easily by computing the gradient of orientation at every point in the image and extracting its magnitude (a scaler) which indicates the rate of orientation change with respect to distance. From the result of this operation (Figure 12-6E) one can see that continuous regions aggregate in two dimensions to form patches, and noncontinuous regions (evident as white lines) aggregate in one dimension to form boundaries between them. Viewed along any arbitrary axis, that is, in a single dimension, it is easy to find short line segments for which the rate of orientation change is constant. When the reference frame switches back from one to two dimensions, though, the continuous regions expand laterally to

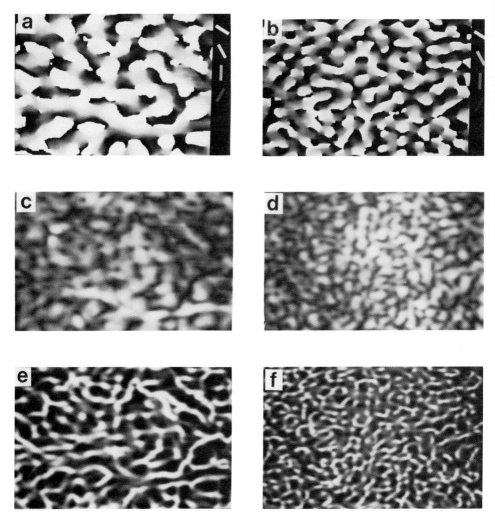

Figure 12-6 (*A* and *B*) Gray-scale values of orientational preference (θ/2) for two different patches of cortex examined at different magnifications (corresponding to *a* and *b* in Figure 12-3, respectively), as determined from multiple scans at different orientations. These renditions illustrate the continuity in changing orientation preference that characterizes most of the cortical surface—from either one it is easy to imagine short (less than 1.0 mm) trajectories at which the orientation preference changes continuously—but they suffer from an inability to depict more than a single orientation cycle. Accordingly, the sharpest edges in this image (those separating the lightest whites from the darkest blacks) actually correspond to regions where orientation preference shifts smoothly through 180 degrees and back to zero. The same images appear (with orientation encoded in color) in Figure 12-7. (*C* and *D*) The magnitude component (*r*) of the vectors used to deduce orientation preference. The lighter the image, the greater the apparent orientation dependency. Darker regions, on the other hand, indicate regions with little or no selectivity. (*E* and *F*) Gray-scale renditions of the magnitude component of the orientation gradient. The brighter the image, the more rapidly orientation is changing with respect to distance in that part of cortical space. Note that the brightest regions aggregate into lines (indicating border zones where orientation changes abruptly) whereas the darker ones, where orientation shifts more slowly and continuously, aggregate to form patches.

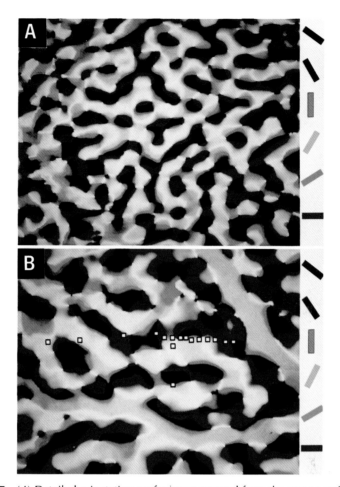

Figure 12-7 (A) Detailed orientation preferences mapped from the same patch of cortex using successive presentations of 12 separate orientations. Like the images in Figures 12-6A and B, these patterns were generated by collecting two sets of frames with orthogonal orientations, as the animal viewed bidirectionally moving gratings, and subtracting one from the other. The process was then repeated with different orthogonal pairs. The computer (LSI 11/73) used the 12 resulting images to analyze orientation selectivity at every point in the image. Every pixel in each image was transformed into a vector; the magnitude of each vector was the pixel value (response intensity), and its angle was twice the orientation of the stimulus grating. This representation ensured that, for each pixel, vectors representing equivalent responses at orthogonal orientations would cancel out. The 12 vectorial images were then summed to produce an output image consisting of two parts, one corresponding to vector magnitudes, reflecting the degree of orientation selectivity for each part of cortex, and one corresponding to half the vector angle, reflecting orientation preferences. This figure shows color-coded representations of optimal orientations. Actual orientation values appear graphically in Figure 12-8. It is important not to overinterpret areas devoted to any particular orientation, because most of the apparent differences derive from nonlinearities in color processing (particularly with respect to blue). (B) Color-coded distribution of orientation, similar to that shown in Figure 12-7A. If orientation selectivity were really organized into parallel slabs, this image would look like a rainbow, or stacks of rainbows. Such color distributions are present only on a small scale, however. [From Blasdel and Salama, 1986.]

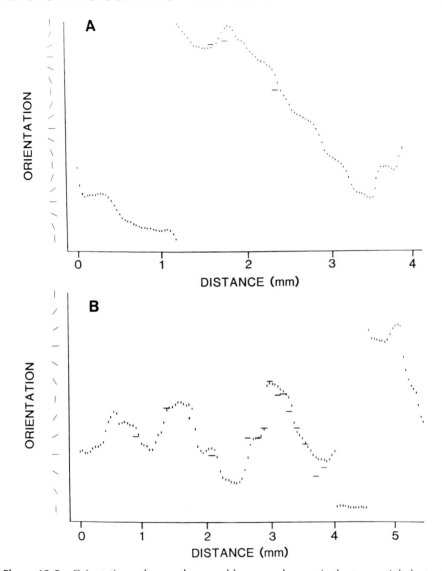

Figure 12-8 Orientation values as they would appear along a single, tangential electrode penetration. This is done twice. (A) Orientation values along a vertical scan line running through the three vertically aligned sites in the center of Figure 12-7B. (B) Orientation values along a single horizontal scan line lying along the 14 horizontally aligned sites in Figure 12-7B. Short horizontal lines indicate the actual orientation values determined (by averaging orientation preferences determined at three different depths) for each of the sites. [From Blasdel and Salama, 1986.]

form two-dimensional patches; and the discontinuous points propagate along a single axis to form lines that segregate the patches and sometimes circumscribe them entirely. The regularity of this pattern is particularly apparent in Figure 12-6F, in which, because of lower magnification, there are more repetitions.

Hubel and Wiesel's decision to emphasize continuity, as opposed to discon-

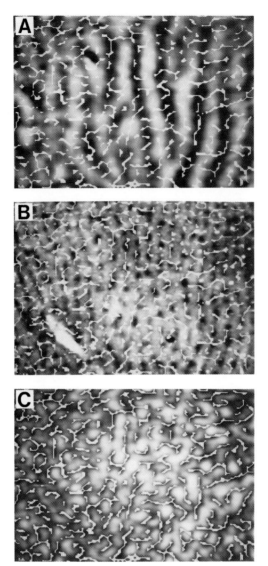

Figure 12-9 Three images depicting the most abrupt transitions in preferred orientation (where orientation preference jumps through an angle greater than 45 degrees within a distance of 40 μm) superimposed on (A) the pattern of ocular dominance depicted in Figure 12-10B, (B) the pattern of blobs revealed by subsequent histochemical staining for the mitochondrial enzyme cytochrome oxidase, and (C) the selectivity for orientation as determined by imaging vector magnitude (as done previously in Figure 12-6E). (A) Here one can see that the most abrupt changes in orientation express a tendency, in this bit of cortex, to propagate down the centers of ocular dominance columns. When they intersect the boundaries between adjacent ocular bands, they tend to do so at right angles. (B) Because they run down ocular dominance centers, the fractures also tend to overlap with cytochrome oxidase blobs, as one can see from this image. Apparent mismatches are most likely the result of differential tissue shrinkage, encountered during histochemical processing for cytochrome oxidase, which makes it difficult to superimpose exactly images obtained before and afterward. (C) Barring contributions from any of the artifacts

tinuous breaks, in orientation mapping probably resulted from the short one-dimensional tracks through cortex they were forced to examine; the regularity of the breaks first becomes apparent when orientation is examined in two dimensions, at low magnification.

ORIENTATIONAL SELECTIVITY (r)

Lighter and darker portions in Figure 12-6C and D correspond, respectively, to regions of cortex with more and less selectivity for orientation. As one can see from this image, selectivity remains constant over larger distances (more then 2 mm) but varies locally. Deferring for the moment any discussion of artifacts (see next section), the simplest explanation for this variation appears to be that these variations reflect changes in orientational selectivity—a possibility that fits well with Livingstone and Hubel's (1984a) observation of reduced orientation selectivity near cytochrome oxidase blobs, especially since the dark regions in Figures 12-6C and D correlate well with the centers of ocular dominance bands as well as with the positions of histochemically determined cytochrome oxidase blobs (see Figure 12-9). But if one accepts that Figures 12-6C and D depict only the selectivity for orientation, these results go a bit further by suggesting that some inter-blob regions, particularly those populating ocular dominance centers, lack orientation selectivity as well.

It is conceivable, however, that the dark regions in Figure 12-6C and D derive from something more complicated than a simple loss in orientation selectivity. Since they correlate strongly with discontinuities in the orientation gradient (Figure 12-6E, F), they may derive from some optical artefact arising from the juxtaposition of neighboring cortical regions with incompatible preferences in orientation.

It is also possible that the dark regions in Figure 6C and D reflect some other parameter—one that affects cortical responsiveness to visual stimuli, for example, which in these experiments consisted of a bidirectionally drifting black and white, multifrequency square wave grating. If some bits of cortex respond less well than others to this stimulus—for example, if they contain disproportionately large num-

---

that are difficult to rule out at this time (see the text for further discussion), this image suggests that the regions mapped most continuously for orientation also express the greatest selectivity. Given the observation in B, that fractures overlap extensively with cytochrome oxidase blobs, the apparent loss of orientation selectivity in the vicinity of fractures coincides with Livingstone and Hubel's (1984a) finding that orientation selectivity is weak or absent near blobs. The results in this figure go a step farther, however, since they also suggest that orientation selectivity is weak or absent for many interblob regions as well—particularly for those lying in the middle of ocular dominance bands. There are other explanations for these results, however, because a drop in absolute responsiveness would also be interpreted (by this sort of analysis) as an apparent lack of orientation selectivity.

bers of end-stopped or color-selective cells—then orientationally dependent changes in their responses might be less pronounced. This would darken the corresponding regions in Figure 6C and D.

VASCULAR ARTIFACTS

At this point it is useful to consider possible sources of artifact in the images we have discussed. There is a concern, expressed most recently by Grinvald et al. (1986), that some of the patterns reported previously (Blasdel and Salama, 1986) derive from blurred portions of the vasculature, if not the activity patterns themselves, then some of those that are derived using simple algorithms—the orientation gradient, for example (as suggested by Van Essen and Orbach, 1986).

Any possible relationship between patterns of optically determined activity and the vasculature is dealt with most easily by comparing images of the two. From Figures 12-10A and B one can see that, except for regions near the largest blood vessels (which would be excluded in any case), images of the ocular dominance bands in Figure 12-10B (or of any other activity patterns from this part of cortex, for that matter) could never be obtained (by blurring or any other convolution) from the vasculature in Figure 12-10A. Even if there was a relationship, this would indicate only that the vasculature correlates with ocular dominance (which it does not), since the bands in Figure 12-10B are known (on the basis of numerous correlative studies) to represent ocular dominance.

The possibility that orientation fractures (see Figure 12-6E) derive from subtle vascular events seems more plausible, particularly since the two patterns look alike. But a simple understanding of the gradient operator, and how it works (like a first derivative), rules out this possibility because any perturbation of the gradient by a blood vessel has to occur twice, once on either side. Thus, perturbations in the gradient that derive from blood vessels should appear in pairs, one on either side of each vessel, as indeed they do (Figure 12-10C). In the hypothetical case in which an offending blood vessel is so small that the two disruptions converge into a single pixel location, they disappear entirely because the gradient values encoding them are opposite in polarity and cancel each other. In short, the fracture lines in Figures 12-6B or 12-7A do not correlate with blood vessels, and wherever a correspondence appears evident, one can verify directly (from the orientation map in Figure 12-6D in this case) that the implicated vessel lies over a boundary separating radically different orientation schemes that extend from the blood vessel for some distance.

The possibility that blood vessels might occlude, and thereby reduce the apparent orientation selectivity of, various underlying regions of cortex must be taken seriously because any attenuation of the optical signals could artificially reduce the apparent difference in signals elicited by optimally and nonoptimally oriented stimuli. From Figure 12-10D, however, one can see that, except in the vicinity of very large vessels, patterns of depressed orientation selectivity bear little or no resemblance to the vasculature.

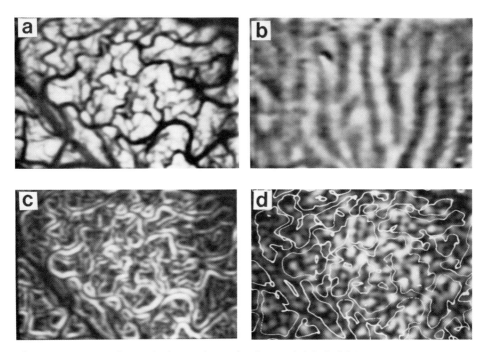

Figure 12-10   (A) The cortical vasculature for the patch labeled b in Figure 12-3. To see that there is no way for the patterns reported previously (Blasdel and Salama, 1986) to derive from a simple blurring of this pattern (as suggested by Grinvald et al., 1986) one has only to compare it with the ocular dominance pattern (B) determined for the same patch of cortex by alternately shuttering the two eyes. Addressing the concerns of Van Essen and Orbach (1986) that the fracture lines in the orientation gradient (Figures 12-6E,F result from artifacts associated with the vasculature, the gradient of vascular pattern itself is presented in C, where it is easy to verify that there is little in common between this image and that appearing in Figure 12-6F. The gradient operator is such that any disturbance deriving from a blood vessel should appear twice, forming two parallel lines that correspond to each side of the offending vessel, as indeed it does. From (D), where an outline of the vasculature appears superimposed on the image presented in Figure 12-6D, where intensity may correspond to orientation selectivity, one sees that there is little correspondence between the two.

RELATIONSHIP BETWEEN OCULAR DOMINANCE AND ORIENTATION SELECTIVITY

Striate cortex is organized over a large scale with respect to visual space. But on a smaller scale this representation is broken apart and distorted to allow the insertion of locally ordered representations of ocular dominance and orientation— a fragmentation that, in effect, adds dimensions to the initially two-dimensional topography. At the smaller scale, ocular dominance and orientation selectivity both repeat their representations within the relatively short distance of 1 to 2 mm, which corresponds approximately to the distance separating retinotopically distinct parts of visual space (Hubel and Wiesel, 1974). Given the similarity in this repeat distance, as well as the expectation of slablike organizations for orientation and ocular dominance, it would seem most reasonable, as Hubel and Wiesel (1974) originally suggested, for the slabs encoding these two modalities to intersect at right angles. The representations would then be orthogonal and less likely to interfere. When Hubel et al. (1978) tested this prediction, however, they found the relationship between ocular dominance and vertical isoorientation bands to be random. In light of the ambiguities introduced by the discovery of cytochrome oxidase blobs (Horton and Hubel, 1980), these experiments probably need to be reinterpreted. But blobs clearly are not a factor in the vertical isoorientation bands of Figure 12-5A, and a simple relationship between these bands and those of ocular dominance still fails to materialize.

If one compares the most abrupt changes in orientation (highest values in the orientation gradient, indicated by the brightest regions in Figure 12-6F) with ocular dominance, however, a correlation does become apparent (Figure 12-9A): the most abrupt shifts in orientation either run parallel to and down the centers of ocular dominance bands, so that they overlie cytochrome oxidase blobs as well (Figure 12-9B), or else they intersect the ocular dominance bands at right angles. From this analysis it is clear that the regions of cortex that extend between adjacent ocular dominance centers are very unlikely to be marred by sudden jumps in preferred orientation. Accordingly, the patches where orientation changes continuously lie exactly where they must to receive balanced inputs from both eyes for one part of visual space. And, since ocular dominance also shifts continuously between the centers of adjacent ocular domiancne bands (for all noninput laminae) this arrangement brings patches of orientation and ocular-dominance continuity into correspondence with one another.

To explore further the relationship between ocular dominance bands, cytochrome oxidase blobs, and other activity patterns, the positions of ocular dominance centers were derived from Figure 12-9B and 12-10B and used to construct a grid, as shown in Figure 12-11A, and superimposed upon images that appear in Figures 12-6D and F.

Figure 12-11B shows the relationship between this grid and the orientation gradient. As in Figure 12-9A, continuity fractures show a clear tendency to propagate down the centers of ocular dominance bands. Periodically, though, they give rise to singularities that correlate closely (91% of the time) with the positions of cytochrome oxidase blobs. By comparing the position of any particular singularity with the orientation map in Figure 12-7A, one can see that it corresponds to a point where a complete cycle of orientations converges.

In Figure 12-11C, the grid appears superimposed with the image of vector magnitude r (see Figure 12-6D), which may represent orientation selectivity. As one can see from this comparison, the two correlate well—orientation selectivity is most pronounced in those regions where shifts in preferred orientation are most continuous. This correspondence agrees well with the observation of Livingstone and Hubel (1984b) that orientation selectivity slackens and frequently disappears entirely as an electrode approaches, and then passes through, a cytochrome oxidase blob. However, as noted earlier, it also suggests that orientation selectivity may be weak or absent for cells that lie between blobs. This result is not an artifact

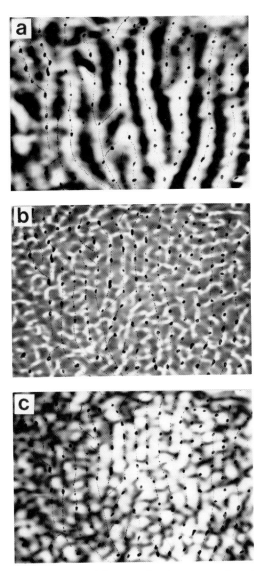

Figure 12-11 Assuming some special function for the centers of ocular dominance bands, a grid corresponding to ocular dominance has been constructed (with a felt-tipped pen to mark the surface of a video monitor) from the ocular dominance bands appearing in Figure 12-10B. (A) Dashed lines have been drawn down the centers of both the light and the dark bands; the black dots correspond to the positions of cytochrome oxidase blobs, deduced from the histologically processed cortex (see Figure 12-9B). In B this grid is superimposed on the orientation gradient. As in Figure 12-9A, there is a clear correspondence between the fractures and the centers of ocular dominance bands. In C the grid is superimposed on an image of vector magnitude (r). As one can see, dips in the apparent orientation selectivity correlate strongly with the centers of ocular dominance bands—a result similar to that shown in Figure 12-9C.

of the image manipulations used to achieve Figures 12-6D and 12-11C, because a similar phenomenon occurs with single-activity patterns. In Figure 12-12A, for example, it is easy to find places where the vertical isoorientation bands weaken, and these places correspond to the lines indicating ocular dominance column centers. A similar phenomenon is apparent for oblique isoorientation bands (Figure 12-12B).

From Figure 12-12 it is apparent that the weakness in the orientation selectivity correlates with a general lack of stimulus-related change in the images used to generate it. The absence of a change could derive from a number of factors, the most obvious of which is a reduction in orientation selectivity (since all of the isoorientation images were obtained as the difference in activity produced by each orientation and its orthogonal). However, it is also possible that the phenomenon derives from regional variations in responsiveness to the visual stimulus. If cells occupying ocular dominance centers do not respond well to black and white contours, for example, or to extended contours, one might expect submaximal responses. Given their location (along the centers of ocular dominance bands) it is also possible (even likely) that these cells are slightly inhibited, or less facilitated, by binocular stimulation.

CONCLUSIONS

There is little question that optical techniques allow one to visualize activity patterns repeatedly in the exposed cortex of a living animal. They have made it possible, for the first time, to visualize bands of ocular dominance and isoorientation. Moreover, they have made it possible to visualize patterns generated in the same piece of cortex by many different stimuli. Because of its enormous efficiency and potential, this approach promises to unleash a flood of detailed new information about striate cortex and its organization.

On the basis of one simple analysis, it appears that orientation preference changes continuously within localized patches of cortex that are 0.5 to 1.0 mm across and separated from one another by abrupt borders. The patches where orientation shifts continuously appear well modeled by a system of parallel slabs (Hubel and Wiesel, 1974), and tend to lie between adjacent ocular dominance centers.

Functional Anatomy

The real strength of this work derives from the numerous correlations that can be made with known anatomical features of striate cortex. Hence, the ocular dominance bands are convincing (even without microelectrode recordings) because they look like patterns seen previously with other techniques. They have appropriate widths (0.5 mm) and run perpendicularly into the 17/18 border. By contrast, the lack of any obvious correlation between these bands and blood vessels makes it unlikely that they derive from an artifact (Grinvald et al., 1986).

The revelation of orientation fractures, abrupt changes in orientation selectivity

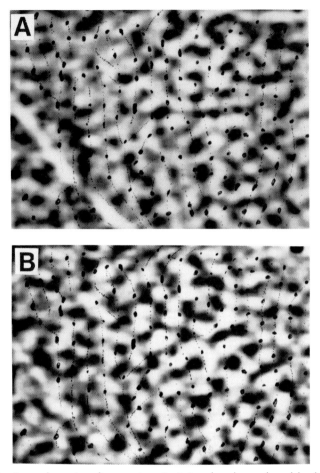

Figure 12-12 From these two figures, isoorientation bands produced by horizontal (*A*) and left oblique (*B*) orientations (and normalized with respect to vertical and right oblique, respectively), one can see that the apparent dips in orientation selectivity in Figure 12-11D also characterize the individual scans that were used to compute it. Following isoorientation bands across ocular dominance boundaries, one sees that response intensity weakens considerably in the centers of the ocular dominance bands, where one eye's dominance is most pronounced. Although this effect could derive from reduced orientation selectivity, it could also result from a reduced overall responsiveness of cortical neurons to the visual stimuli that were used.

that separate modules of orientation continuity, draws anatomical support for similar reasons. If they occurred randomly or overlapped extensively with the vasculature, one might wonder about their authenticity; but their close association with the centers of ocular dominance bands and cytochrome oxidase blobs makes it likely that they are real.

Given the close correlation between modules of orientation continuity and other established structures in striate cortex, one naturally wonders about the anatomical patterns of lateral connectivity that might link everything together. Preliminary studies with microinjections of HRP (Blasdel and Lund, unpublished observations) suggest that neurons lying within fractures project to neurons in other fractures, whereas neurons centered within modules (and away from fractures) appear not to project far at all. Since fractures extensively overlap cytochrome oxidase blobs, these observations are partially consistent with those of Livingstone and Hubel (1984b), who found that blobs project to other blobs. But they are even more consistent with the observations of Rockland and Lund (1983), who found

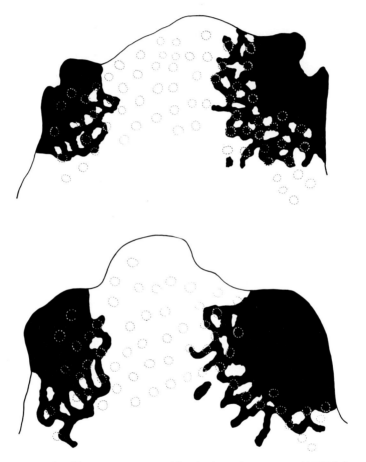

Figure 12-13 Lattice like patterns created by the lateral transport of HRP following large (greater than 1 μl) microinjections into macaque striate cortex. From these patterns it seems likely that neurons located within lattice walls interconnect over larger distances than neurons located within holes. Interestingly, the histochemical processing of adjacent sections for the mitochondrial enzyme cytochrome oxidase reveals that the blobs (indicated in the diagrams by broken circles) lie within the walls of the lattice—the regions with the longest ranging interconnections. This correspondence, along with the qualitative similarity in appearance between the lattice and the fracture patterns in Figures 12-6E and F, raises the possibility that the two correspond to one another. [From Rockland and Lund (1983).]

that HRP transported to the fringes of large injection sites produced latticelike patterns similar to that illustrated in Figure 12-13, which suggests that neurons located within the walls of the lattice project farther than cells located within the lattice holes. The lattice walls in these patterns resemble the orientation fracture lines in Figure 12-6E and F and this resemblance is strengthened by the observation that the walls of the lattice, like the fracture lines in the orientation gradient, overlap extensively with cytochrome oxidase blobs (Rockland and Lund, 1983).

Speculations

Given that neurons in fractures and blobs project farther than neurons in modules, what might this signify? There are many possible explanations, but one seems particularly intriguing and therefore deserving of attention. This is the possibility that striate cortex is subdivided laterally, in yet another fashion that integrates known structures such as ocular dominance, orientation selectivity, cytochrome oxidase blobs, and retinotopic position. In its simplest form this lateral subdivision might be thought to consist of two components—one specializing in surface properties, the other in the location and certainty of object boundaries. The fundamental task of vision, after all, is to detect real objects in the real world. And in this context, the distinction between edges that derive from surface properties (like texture) and those that define object boundaries is crucial. Accordingly, it should not be surprising to find large areas of primary visual cortex devoted to this task.

One way to simplify the distinction between surfaces and boundaries entails looking at the distribution of edges in neighboring parts of the visual field. Those that derive from surface properties, as a rule, are accompanied by similar ones nearby, whereas those deriving from boundaries tend to be singular. Accordingly, regions that specialize in surface property detection (in this case, texture detection) might interconnect profitably over wide areas of cortex, in a way that allows them to support and resonate with neighbors responding to similar features. Regions that specialize in boundary assertion, on the other hand, need know only that an edge exists that does not correlate with a textured surface (indicated in the preceding example by a lack of resonant activity over long-range interconnections).

One can combine this reasoning with some of the preceding findings to advance a fairly simple model of cortical organization. According to this model (illustrated in Figure 12-14), modular patches consisting of stacked parallel isoorientation slabs, and extending between adjacent ocular dominance centers, specialize in boundary detection. Less extensive regions of cortex that lie in between (and include blobs and fractures) specialize in surface property detection. Both systems interdigitate with and inhibit one another in such a way that a single edge, exciting modular cells, would act indirectly to inhibit the activity of network cells attempting to detect surface properties. Multiple edges, on the other hand (a textured surface or grating, e.g.) would excite these cells at many points in the network, and thereby enable them to overpower local inhibition and assert the presence of a textured surface, an event that would then discourage boundary detecting cells in the modules from "seeing" boundaries within the texture.

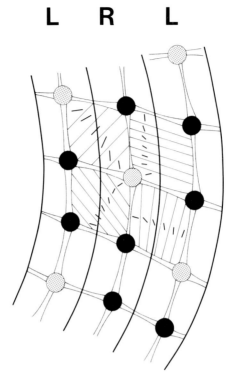

Figure 12-14   Idealized and speculative scheme (that is almost certainly too simplistic) for the lateral organization of cortical regions specializing in boundary and surface detection. Cytochrome oxidase blobs, along with the narrow linear regions interconnecting them, are indicated by black and stippled zones, and the modules, where orientation shifts continuously, are indicated by stacks of parallel isoorientation slabs. According to the model advanced in the text, the black and stippled zones form a network that specializes in the analysis of surface properties—color and texture, for example—whereas the modules specialize in the detection of object boundaries. The surface property network profits from long-distance interconnecttions because the presence of similar adjacent properties increases the probability that the property under consideration is a surface. Accordingly, HRP that is transported over great distances goes preferentially to neurons located within lattice walls. Since an edge is much more likely to indicate a boundary if it occurs in isolation, the modules do not require such long-ranging connections. Consistent with the results presented in this and previous publications (Blasdel and Salama, 1986), the modules lie between the centers of adjacent ocular dominance columns in positions where orientation and ocular dominance shift most continuously.

Although this model ties together a number of disparate observations from anatomical, physiological, psychophysical, and artificial intelligence perspectives in a manner that has great practical significance, it is almost certainly too simple. It nevertheless results in a number of clear predictions that can be tested in future experiments.

ACKNOWLEDGMENTS

This work was supported by grants from the National Eye Institute (EY05403, EY06586-01). My thanks to Jenny Lund for helpful comments on the manuscript as well as to Suzanne Holbach and Jacqueline Mack, who provided excellent technical assistance during various phases of this work, and Caroline Collins who typed the manuscript.

REFERENCES

Blasdel, G. G., and Fitzpatrick, D. (1984). Physiological organization of Layer 4 in macaque striate cortex. *J. Neuroscience* 4:880–895.

Blasdel, G. G., and Salama, G. (1986) Voltage-sensitive dyes reveal a modular organization in the monkey striate cortex. *Nature* 321:579–585.

Cohen, L. B., Keynes, R. D., and Hille, B. (1968). Light scattering and birefringence changes during nerve activity. *Nature* 218:438–441.

Cohen, L. B., Keynes, R. D., and Landowne, D. (1972). Changes in axon light scattering that accompany the action potential: current dependent components. *J. Physiol.* 224:727–752.

Cohen, L. B., Salzberg, B. M., Davila, H. V., Ross, W. N., Landowne, D., et al. (1974). Changes in axon fluorescence during activity: molecular probes of membrane potential. *J. Membr. Biol.* 19:1–36.

Grinvald, A., Anglister, L., Hildesheim, R., and Freeman, J. A. (1983). Optical monitoring of naturally evoked dynamic patterns of neural activity from the intact frog optic tectum using a photodiode array. *Soc. Neuroscience Abst.* 9(1):540.

Grinvald, A., Lieke, E., Frostig, R. P., Gilbert, C., and Wiesel, T. M. (1986). Functional architecture of cortex revealed by optical imaging of intrinsic signals. *Nature* 324:361–364.

Horton, J., and Hubel, D. H. (1980). Regular patchy distribution of cytochrome oxidase staining in primary visual cortex of macaque monkey. *Nature* 292:762–764.

Hubel, D. H., and Wiesel, T. N. (1962). Receptive fields, binocular interaction and functional architecture of monkey striate cortex. *J. Physiol.* (Lond). 160:106–154.

Hubel, D. H., and Wiesel, T. N. (1965). Binocular interaction in striate cortex of kittens reared with artificial squint. *J. Neurophysiol.* 28:1029–1040.

Hubel, D. H., and Wiesel, T. N. (1968). Receptive fields and functional architecture of monkey striate cortex. *J. Physiol. (Lond.)* 195:215–243.

Hubel, D. H., and Wiesel, T. N. (1972). Laminar and columnar distribution of geniculocortical fibers in the macaque monkey. *J. Comp. Neurol.* 146:421–450.

Hubel, D. H., and Wiesel, T. N. (1974). Uniformity of monkey striate cortex: a parallel relationship between field size, scatter and magnification factor. *J. Comp. Neurol.* 158:295–306.

Hubel, D. H., Wiesel, T. N., and Stryker, M. P. (1978). Anatomical demonstration of orientation columns in macaque monkey. *J. Comp. Neurol.* 177:361–380.

LeVay, S., Hubel, D. H., and Wiesel, T. N. (1975). The pattern of ocular dominance columns in macaque visual cortex revealed by a reduced silver stain. *J. Comp. Neurol.* 159:559–576.

Livingstone, M. S., and Hubel, D. H. (1984a). Anatomy and physiology of a color system in the primate visual cortex. *J. Neurosci.* 4:309–356.

Livingstone, M., and Hubel, D. H. (1984b). Specificity of intrinsic connections in primate primary visual cortex. *J. Neurosci.* 4:2830–2835.

Orbach, H. S., Cohen, L. B., and Grinvald, A. (1985). Optical mapping of electrical activity in rat somatosensory and visual cortex. *J. Neuroscience* 5:1886–1895.

Rockland, K. S., and Lund, J. S. (1983). Intrinsic laminar lattice connections in primate visual cortex. *J. Comp. Neurol.* 216:303–318.

Van Essen, D., and Orbach, H. S. (1986). Optical mapping of activity in primate visual cortex. *Nature* 321:564–565.

Wong-Riley, M. T. T. (1979). Changes in the visual system of monocularly sutured or enucleated cats demonstrable with cytochrome oxidase histochemistry. *Brain Res.* 171:11–28.

13

Coding the Movements and Position of the Head in Space

DAVID L. TOMKO

To function effectively, many perceptual and reflex mechanisms require accurate information about the position and movements of the head in space. The head movements that must be encoded by the nervous system may result from active (self-generated) motion, or from forces applied passively to the body, as in vehicular travel. In particular, the correct analysis and interpretation of data from the visual sense, auditory sense, and proprioceptors of axial body structures depend on the precise definition and neural analysis of the accelerations acting on the vestibular system end-organs. This chapter briefly describes the information provided by vestibular receptors to the nervous system, and considers two related phenomena, one reflexive and one perceptual, that illustrate how vestibular information might be used to help shape "the brain's view of the world."

One goal of neuroscience is to elucidate the basic mechanisms underlying perceptual phenomena. In the context of the perceptual consequences of vestibular system function, for instance, the oculogravic illusion and the related "elevator" illusion are of interest. In both, illusory displacements of visual targets result from the application of a linear acceleration to a human subject. Graybiel (1952) first described the oculogravic illusion that subjects experience when exposed to centripetal acceleration changes in a human centrifuge:

> . . . the subject is seated facing the center of rotation, fixating a collimated star directly before him in an otherwise dark room . . . and when . . . exposed to a change in direction of resultant force of 45 degrees in approximately three seconds, . . . the subject perceived an apparent change in body position, an apparent tilting of the supporting structures, and an apparent movement of the star. The apparent change in position consists of a curious feeling of being tilted smoothly, yet firmly, backwards. . . . the star, previously considered to be at or near the horizon, apparently rises; it is in the nature of a smooth glide, without any jerkiness. (Graybiel, 1952)

In the elevator illusion, similar visual phenomena result from a change in the magnitude of the applied acceleration vector without any change in its direction with respect to the subject (Cohen, 1973). The elevator illusion is mediated solely by the otoliths, since it is absent in labyrinthine-defective subjects; the oculogravic illusion exists in a modified form in labyrinthine-defective subjects (Graybiel and Clark, 1965).

During the oculogravic and elevator illusions, there is no actual change in the orientation of the subject's head with respect to the visual world. Even so, the subject perceives that visual stimuli are in different locations in the visual field than they are in reality. *Why should a linear acceleration that does not change the subject's position with respect to the visual field result in changes in the apparent position of stimuli in that visual field?*

In the oculogravic illusion, application of a G_x axis (chest to back or nose to occiput) acceleration elicits vestibular responses that normally occur during slow nose-up or nose-down head pitches. Such a stimulus activates vestibular gravity receptors in the otoliths, in particular. During head pitch over a broad bandwidth, signals from vestibular receptors produce vertical compensatory eye movements (e.g., Benson and Guedry, 1971; Darlot et al., 1981; Böhmer and Henn, 1983; Matsuo and Cohen, 1984; Correia et al., 1985; Tomko et al., 1988). Those vestibular signals are also available to neural perceptual circuits (e.g., Horn et al., 1972; Tomko et al., 1981a) to indicate that the head has tilted and to assist in maintaining a stable perception of the visual world during the movement. Because the response of otolith gravity sensors during G_x axis linear acceleration is similar to that which would *normally* result from tilting the head relative to gravity, it elicits reflexes and perceptions normally produced during head pitch. The visual illusions produced during application of a G_x vector in the oculogravic illusion therefore probably result from neural mechanisms interpreting the stimulus as resulting from a tilt of the head around the pitch axis, the closest physiological analog to the stimulus. I therefore hypothesize that phenomena such as the oculogravic illusion reflect an attempt by neural and perceptual mechanisms to interpret the presence of sensory signals produced in a behavioral context *different* from what the brain experienced in its evolution.

VESTIBULAR END-ORGAN FUNCTION: RAW DATA

The nonauditory portion of the mammalian labyrinth contains the receptors of the otolith organs, the sacculus and utriculus, and the three semicircular canal ampullae. The utricular maculae and the horizontal semicircular canals lie near the horizontal plane of the head; the saccular maculae and the anterior and posterior canals lie roughly perpendicular to that plane. It is generally agreed that the *otolith organ receptors* are responsible for the transduction of forces produced by linear acceleration and gravity, and the *ampullary receptors* of the semicircular ducts transduce forces produced during rotational acceleration (see Wilson and Melvill Jones, 1979, for a review).

Physiology of Statoreceptors

During the 1950s, electron microscopists began a detailed examination of the ciliated vestibular receptor cells of the semicircular canal cristae and the otolithic maculae. They showed that each receptor cell contains a bundle of stereocilia, arranged in an array of increasing length on its distal surface. On the side of the receptor surface, at the apex of the array of stereocilia, was a single, longer kinocilium. Microscopists termed this anatomic asymmetry *morphological polarization* and hypothesized that the physiologic response of the cell might have a corresponding functional asymmetry (e.g., Lowenstein and Wersäll, 1959). Each receptor cell makes synaptic contacts with processes of eighth-nerve neurons. The receptor cells themselves are inaccessible for recording in mammalian preparations because they are buried deep within the temporal bone; therefore, most of the physiological data on the receptors' function has been inferred from recordings of the electrical activity of eighth-nerve or vestibular nuclei neurons (e.g., Fernandez and Goldberg, 1971; Fernandez et al., 1972; Loe et al., 1973).

Early recording experiments (Fernandez and Goldberg, 1971; Fernandez et al., 1972; Loe et al., 1973) demonstrated that the anatomic polarization seems to have a functional corollary. Each statoreceptor cell recorded in the eighth nerve was excited by head tilt in one direction and inhibited by head tilt in the opposite direction. Statoreceptor responses were modulated sinusoidally as the head position was changed, leading to the conclusion that the primary afferents behaved as though they were simple linear accelerometers sensing the orientation of the skull to the acceleration of gravity.

Two important features of the statoreceptor afferent activity were defined in early studies. First, these cells were maximally excited when the gravity vector was in or near the plane of the macula (producing shear force), and they were not stimulated by forces orthogonal to the macula (either compression or extension). This was later proved more directly by Fernandez and Goldberg (1976a, 1976b), who used centrifugation to show that hair cell shear is the most effective macular stimulus and that compressional forces are not effective. Second, directional selectivity seemed to be a property of all these afferents, since they are excited by head tilt (linear acceleration) in one direction and inhibited by head tilt in the opposite direction.

Otolith afferents exhibit a resting discharge that may be anything from temporally highly regular (i.e., evenly spaced action potentials) to highly irregular. Regularly discharging afferents, for the most part, behaved tonically. When the head was moved from one position to another, a new discharge level was reached with little or no adaptation. Discharge frequency varied as a function of the sine of angular head position and was remarkably repeatable from one trial to another in these afferents. On the other hand, irregularly discharging afferents exhibited more phasic responses. Their activity on movement from one position to another showed marked adaptation, and discharge frequencies were frequently unrepeatable, with marked hysteresis and direction-dependent phenomena (Vidal et al., 1971; Loe et al., 1973; Fernandez and Goldberg, 1976a, 1976b, 1976c).

Fernandez et al. (1972) and Loe et al. (1973) compared physiological data to the outputs of a mathematical model developed by Werner et al. (1969). The model was based, first, on the anatomical data suggesting that hair cells might have a functional asymmetry and, second, on the assumption that tangential shearing was the adequate stimulus for hair cell excitation. With only minor changes in the model, the primary afferent data matched its predictions, and the comparison of data and model allowed the conclusion that the model's basic assumptions were correct. Otolith afferents basically behaved like linear accelerometers mounted in the skull to sense head orientation relative to gravity. Each afferent was stimulated maximally by positioning the head in one position and minimally by repositioning it by 180 degrees.

A spatial tuning curve was calculated for each regularly discharging afferent by measuring the discharge frequency after positioning the animal's head in many combinations of pitch and roll to yield three-dimensional relationships. Plotting the spatial tuning curve of each afferent permitted the subsequent estimation of a pair of polar coordinates in the animal's coordinate system, indicating its optimal stimulation direction. The calculated coordinates of the statoreceptor afferents lay roughly in the planes of the utriculus and sacculus (Fernandez et al., 1972; Loe et al., 1973; Fernandez and Goldberg, 1976b; Tomko et al., 1981b). An important functional consequence of this arrangement is that statoreceptor afferents that lie in the utricular plane are midway between their maximally excitatory and maximally inhibitory positions when the head is in its normal upright position. Thus, as a population, they are ideally suited to encode and signal slight deviations of the head from its normal position. Cells of this population with polarization vectors oriented toward the sides show a preferential response to roll. Those whose polarization vectors are oriented toward the front or back of the head show a large response to pitch and little or none to roll. It is this second group, statoreceptor afferents with polarization vectors pointing either toward the front or toward the back of the head, which will be maximally stimulated by the G_x linear accelerations that produce the oculogravic illusion.

Physiology of Ampullary Receptors

The activity of eighth-nerve afferents of semicircular canal origin has been recorded during rotational head accelerations in a variety of mammals with similar results (Goldberg and Fernandez, 1971a, 1971b; Fernandez and Goldberg, 1971; Blanks et al., 1975; Estes et al., 1975; Schneider and Anderson, 1976; Anderson et al., 1978; Tomko et al., 1981c). Cells have a resting discharge that is modulated maximally during rotational motions of the head in the plane of the canal. Each semicircular canal afferent is excited by rotational acceleration in one direction and inhibited by acceleration in the opposite direction. Individual cells have highly replicable response dynamics. Exact gain and phase characteristics vary widely from one cell to another and are correlated with the cell's regularity of discharge, as is the case for otolith afferent dynamics (Fernandez and Goldberg, 1976c). The considerable diversity of responses was shown by fitting transfer functions to the data of individual cells as well as to averaged gain and phase values for three subpopulations of cells, those with a high, intermediate, or low degree of resting discharge regularity.

In our work (Tomko et al., 1981c), transfer functions were fit to data from individual cells, allowing the estimation of a dominant time constant and up to three additional time constants for each cell. The gains and dominant time constants for the entire population of cells, as well as those of the three subpopulations, showed remarkable variation among afferent neurons in these basic measures. Although a wide variety of response types constitute the entire population of afferents (Tomko et al., 1981c) the most regularly discharging afferents generally provide a signal roughly proportional to head velocity over a significant portion of the bandwidth of normal head movements. The response properties of irregularly discharging afferents were more complex, in most instances, than those of the regularly discharging ones. Their low-frequency gain reductions and the development of relative high-frequency phase leads indicated their quicker adaptation during rotational acceleration. One of the major conclusions of this work (Tomko et al., 1981c), as with Fernandez and Goldberg's otolith dynamics study (1976c), was that a large variety of information encoded by parallel channels in the vestibular periphery is available to the central nervous system.

FUNCTION OF LABYRINTHINE OUTPUTS: CANAL/OTOLITH INTERACTIONS

The neural output of the vestibular end-organs serves as an input signal to oculomotor reflexes, postural mechanisms, and perceptual systems. All three of these input signals fulfill a single design purpose during movements in inertial space— to preserve *organismic self-stability* with respect to the outside world.

Most of the existing body of knowledge on the physiology of statoreceptors and semicircular canals is based on research that has deliberately treated the components of the labyrinth as discrete, isolated parts, almost as though the otoliths and canals were separate end-organs (see Wilson and Melvill Jones, 1979, for review). There is ample evidence that many central vestibular neurons are sensitive not only to both angular and linear acceleration (e.g., Duensing and Schaeffer, 1959), but also to somatosensory (e.g., Fredrickson et al., 1966) and visual (e.g., Henn et al., 1974) stimuli. The functional significance of this convergence is still speculative, and will be discussed later in this chapter.

One of the most interesting and significant contemporary developments in vestibular system physiology is an increasing emphasis on the study of *interactions* between the outputs of the semicircular canals and otoliths in the control of a variety of reflexive and perceptual functions (e.g., Collewijn et al., 1985; Viirre et al., 1986). In this section, I briefly describe studies that indicate potential interplay between gravity receptors and semicircular canals in the control of vertical eye movements and a study in progress on elicitation of vertical eye movements by linear acceleration.

Otolith–Semicircular Canal Interactions in Controlling Vertical Eye Movements

The vestibuloocular reflex (VOR) is widely used as rough measure of semicircular canal output (see Carpenter, 1977, for review). The horizontal VOR generated by yaw head movements is well understood. Generally, during head oscillations around the yaw axis (an axis perpendicular to the horizontal plane of the head), head movement to the right reflexively drives the eyes to the left and vice versa. If the movements of the eyes exactly compensate for the head movements, the reflex gain is unity. Without visual inputs, the horizontal VOR typically has a gain of about 0.9, and eye movements are also roughly in phase with head velocity over a bandwidth of below 0.1 Hz to about 2 or 3 Hz. The vertical VOR generated by head pitch (rotation about an interaural axis) works analogously.

The reflexive control of the eyes during vertical head movements is particularly interesting because it differs in fundamental ways from eye movement control during horizontal head movements. In the first place, during head yaw the *position* of the head relative to gravity is *not* changed, and the movement therefore activates only the semicircular canals. But in head pitch, which elicits the vertical VOR, the position of the head relative to gravity *does change,* activating the gravity-sensing otolith organs in addition to the canals. In the second place, yaw head movements are inherently symmetric; movements of the head to the left and to the right are mostly identical, cost the same amount of energy, and require the same motor control strategy. During pitch, however, upward head movementsrequire pulling the head's mass against gravity, while downward head movements require that the head's mass be braked against the fall induced by gravity. Head pitch is therefore inherently asymmetric, with upward and downward head movements requiring different motor strategies as well as different energy expenditures.

Unfortunately, the vertical VOR has been less extensively studied than the horizontal, although interest in it has recently increased (e.g., Anderson and Liston, 1983; Matsuo and Cohen, 1984; Tomko et al., 1984, 1988). In most experiments, pitch around the interaural axis is delivered to the subject while the subject lies on his side, stimulating only the vertical semicurcular canals (e.g., Benson and Guedry, 1971; Darlot et al., 1981; Matsuo and Cohen, 1984). Under such conditions, the gain of the vertical VOR is less than that of the horizontal VOR, and asymmetries in the upward and downward eye movements are produced.

A recent study (Tomko et al., 1984, 1988) directly compared the dynamic response characteristics of the vertical VOR during on-side pitch with those same response properties produced by *upright* pitch (with the subject sitting in a normal upright position). The subject lying on one side was subjected to rotational acceleration about the head's pitch axis, but the head did not change its orientation relative to gravity. In the upright pitch condition, the head experienced the *same* rotational pitch motions, but *its position changed relative to gravity* during stimulation. The gain of the vertical VOR was about 15 percent less when the subject was positioned on his side than when upright. The gain of the vertical VOR produced by upright pitch was near that of the horizontal VOR of the same subjects. In addition, the prominent asymmetries in the vertical VOR when elicited with the subject on-side were significantly reduced when the reflex was elicited with the subject in the upright position. These results suggested to us that the vertical VOR is normally controlled jointly by otoliths and canals, although further experiments are required to verify this. It seems logical to expect the involvement of two populations of otolith afferents. The first are otolith afferents with sensitivity orientations directed toward the nose or the back of the head. The second are saccular afferents, with upward- and downward-pointing vectors, and symmetric responses to pitch, since their activity is at a maximum or minimum when the head is upright.

Linear Acceleration Produces Vertical Eye Movements

In the experiment described in the preceding section, the stimulation of vertical semicircular canals and otoliths during head pitch was "uncoupled" by placing the animal on its side, producing a vertical VOR driven *only* by responses to rotational acceleration. To demonstrate that otolith stimulation can produce vertical eye movements, it is necessary to uncouple the rotational component from the linear component by removing the former. During head pitch, the orientation of the head and the otoliths changes with respect to the earth's gravity vector; during a 20-degree nose-up head pitch, the gravity vector swings 20 degrees further toward the back of the head than in the upright condition. The opposite is, of course, the case for a nose-down tilt. This swinging vector can be reproduced without the rotational component by applying a linear acceleration of the appropriate frequency and length along the nasooccipital axis of the upright animal. The roughly ± 0.35-g translational vector and earth gravity produce a swinging vector, which roughly duplicates the linear acceleration conditions seen by the otoliths during a ± 20-degree head pitch. This was accomplished by applying translational acceleration on the linear sled in NASA's Vestibular Research Facility while recording vertical eye movements (Paige and Tomko, 1987, and unpublished results). Eye movements were recorded using the scleral search coil technique (Robinson, 1963) in adult squirrel monkeys. Some preliminary data are shown in Figure 13-1.

The vertical eye movements produced by G_x accelerations (Figure 13-1, bottom eye position trace) are clearly present and have the correct phase relationship and amplitudes to account for the gain difference between upright and on-side pitch described in the preceding section, if the otolith and canal ocular reflexes sum to produce the normal vertical VOR. It seems probable that the vertical eye movements produced by G_x linear acceleration result from otolith stimulation, and we conclude that these dynamic otolith–ocular reflexes interact synergistically with vertical–canal mediated ones.

The data of Figure 13-1 showing a vertical otolith–ocular reflex during G_z axis stimulation (center eye position trace) may be of potential relevance to the elevator illusion, because it implies that G_z axis stimulation drives the eyes in a compensatory vertical movement. Since it has not been demonstrated that the eyes remain absolutely stationary during the production of the elevator illusion, one cannot rule out the possibility that the production of the illusion is entirely or partly due to reflexive repositioning of the eyes (and therefore the retina) because of the G_z axis acceleration.

SPACE AND THE VESTIBULAR SYSTEM

Illusory phenomena related to misinterpretation of otolith signals by nervous system reflexes and perceptual processes present a panoply of potential problems to persons who experience a change in *any* of the signals (visual, vestibular, or proprioceptive) that normally are delicately balanced to produce behaviorally normal

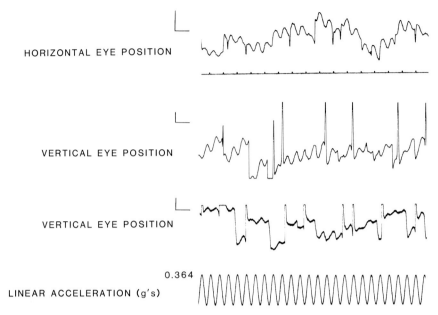

HORIZONTAL EYE POSITION

VERTICAL EYE POSITION

VERTICAL EYE POSITION

0.364

LINEAR ACCELERATION (g's)

Figure 13-1 Linear acceleration stimulus, as a function of time, (*bottom trace*). The eye position during linear acceleration is shown for three stimulus conditions. (*Top trace*) Horizontal eye position during interaural (G_y) linear acceleration. (*Center trace*) Vertical eye position during head-to-tail (G_Z) axis stimulation. (*Bottom trace*) Vertical eye position during chest-to-back (G_X) axis stimulation. The vertical calibration to the left of each eye position trace is 2.5 degrees of eye position; The horizontal calibration is 1 second.

reflexive and perceptual function. The altered reflex and perceptual phenomena elicited by otolith stimuli presented in an unusual context provide many potentially significant experimental questions for the neuroscientist: (1) What are the exact mechanisms whereby otolith and other proprioceptive signals are integrated to produce normal reflex and perceptual function? (2) Does the organism have adaptive capabilities that allow for "recalibration" of the affected mechanisms? If so, what is the exact time course and characteristics of the adaptation? Are there external factors that can be manipulated to alter the course of adaptation? (3) What neural mechanisms underlie the adaptation process?

It is easy enough on earth to demonstrate the delicate balance of visual, vestibular, and proprioceptive information that begins in the brainstem. There is a significant integration of sensory information from vestibular receptors and from neck proprioceptors in the descending vestibular nucleus (e.g., Fredrickson et al., 1966). Disruption of any of these signals, whether by experimental design or by natural causes, results in reflex and perceptual anomalies for the brain. For example, it has been demonstrated that lesions of the labyrinths (Carpenter et al., 1959), the neck proprioceptors (Cohen, 1961), or the descending vestibular nucleus (Carpenter et al., 1960; Uemura and Cohen, 1972) lead to postural deficits that are quite marked and quite similar including, at least for degeneration of neck proprioceptors, a conspicuous deficit in visually guided reaching out into extraocorporeal space (Cohen, 1961). Disease states in humans that damage either the musculature or the cervical dorsal roots of the neck lead to similar deficits in posture, to vertigo, and to nystagmus (Weeks and Travell, 1955; Cope and Ryan, 1959; Biemond and de Jong, 1969). Parts of the descending nucleus that give rise to few spinal projections (Pompeiano and Brodal, 1957; Carpenter et al., 1960) seem likely candidates as a relay to more rostral structures in sensing the position of the head in relation to the body's position. Cohen's (1961) observation that animals with the first three cervical dorsal roots either anesthetized or severed were unable "to locate accurately in space a point upon which they had fixed their gaze" (Cohen, 1961, p. 5) implies the integration with the visual sense, which has been studied extensively in the past 15 years or so (see Henn et al., 1974), especially with optokinetic stimuli. It seems likely that the output of such a neural system (integrating information from the vestibular system about head position, from the somatosensory system about neck position, and from the visual system) would be required as an input to cerebral cortical circuits defining position in space. It would be a necessary antecedent to the functional role cortical neurons of area 5 play in visually guided reaching behavior in the macaque (Lynch et al., 1973, 1977; Mountcastle et al., 1975; Mountcastle, 1975).

The delicate balance between vestibular, visual, and somatosensory mechanisms has implications for humans who are required to operate high-performance aircraft or perform in microgravity. Cohen et al. (1973) have reviewed the disorienting effects of aircraft catapult launchings, which produce large G_x linear accelerations. They conclude that "in the absence of external visual cues, the pilot is normally unable to differentiate the catapult launch forces into their inertial and gravitational components. Thus, the pilot may experience an illusory pitch-up attitude." This erroneous perception of body orientation can lead to incorrect reorientation of the plane and a "dark night take off accident" (Buley and Spelina, 1970).

More recently, Lentz and Guedry (1982) have reported a study based on reports from pilots of F-14 aircraft of "apparent visual bending or bowing of instrument horizons during and immediately following ascending roll maneuvers." They concluded that the VOR could be responsible for the misperception at high angular velocities.

Exposure to microgravity has demonstrable effects on both the vertical vestibuloocular and optokinetic reflexes (Clément et al., 1986; Viéville et al., 1986). In one subject, the gain of the vertical VOR during space flight was reduced for the first 4 days before it recovered its preflight value, and the optokinetic reflex underwent both a gain change and a reversal of its normal up–down asymmetry. Neither the exact mechanisms nor the long-term consequences of these findings are understood. One thing that is clear, however, is that eye movements and visual perception are profoundly influenced by and dependent on somatic and vestibular stimuli, particularly those from the otoliths.

CONCLUSIONS

Neural mechanisms are stressed by the machine age in which we live. We have a reasonable understanding of the normal function of the biological sensors of the vestibular system. We do not, however, have adequate data to know how vestibular information is combined with that of other sensory systems to produce effective motor control and perception. Biological sensors are exposed to a variety of vestibular and other stimuli that are within their range of operation, but occur in contexts that were rarely or never experienced during evolutionary history. The speed with which neural mechanisms adapt to most such short-acting changes to produce effective behavior is impressive. The adaptation mechanisms are not well understood. A better understanding of them should yield useful information about the interactions between neural reflexes and perception.

At this point, however, further research is required before we will be able fully to answer the question posed at the beginning of this chapter: Why should a linear acceleration that does not change the subject's position with respect to the visual field result in changes in the apparent position of stimuli in that visual field?

ACKNOWLEDGMENTS

I gratefully acknowledge the many discussions with Mal Cohen that have helped me sort out my own thoughts on the subject of this paper and for his thoughtful critique of an earlier version of this manuscript. The not yet published research on the linear VOR described here is a collaborative effort by Gary Paige and myself, and was supported by NASA Space Medicine Program Tasks 199-22-92-02 and 199-22-92-03.

REFERENCES

Anderson, J. H., and Liston, S. L. (1983). Asymmetry in the human vertical vestibulo-ocular reflex. *Abst. Assn. Res. Otolaryngol.* 6:82.
Anderson, J. H., Blanks, R. H. I., and Precht, W. (1978). Response characteristics of semicircular canal and otolith systems in cat. I. Dynamic responses of primary vestibular fibers. *Exp. Brain Res.* 32:491–507.
Benson, A. J., and Guedry, F. E. (1971). Comparison of tracking-task performance and nystagmus during sinusoidal oscillation in yaw and pitch. *Aerosp. Med.* 42:593–601.
Biemond, A., and de Jong, J. M. B. V. (1969). On cervical nystagmus and related disorders. *Brain* 92:437–458.
Blanks, R. H. I., Estes, M. S., and Markham, C. H. (1975). Physiologic characteristics of vestibular first-order canal neurons in the cat. II. Response to constant angular acceleration. *J. Neurophysiol.* 38:1250–1268.
Böhmer, A., and Henn, V. (1983). Horizontal and vertical vestibulo-ocular and cervico-ocular reflexes in the monkey during high frequency rotation. *Brain Res.* 277:241–248.
Buley, L. E., and Spelina, J. (1970). Physiological and psychological factors in "the dark night takeoff accident." *Aerosp. Med.* 41:553–556.
Carpenter, M. B., Fabrega, H., and Glinsman, W. (1959). Physiological deficits occurring with lesions of labyrinth and fastigial nuclei. *J. Neurophysiol.* 22:222–234.
Carpenter, M. B., Alling, F. A., and Bard, D. S. (1960). Lesions of the descending vestibular nucleus in the cat. *J. Comp. Neurol.* 14:39–50.

Carpenter, R. H. S. (1977). *Movements of the Eyes*. London: Pion.

Clément, G., Viéville, T., Lestienne, F., and Berthoz, A. (1986). Modifications of gain asymmetry and beating field of vertical optokinetic nystagmus in micro-gravity. *Neurosci. Lett.* 63:271–274.

Cohen, L. A. (1961). Role of eye and neck proprioceptive mechanisms in body orientation and motor coordination. *J. Neurophysiol.* 24:1–11.

Cohen, M. M. (1973). Elevator illusion: influences of otolith organ activity and neck proprioception. *Percept. Psychophysics* 14:401–406.

Cohen, M. M., Crosbie, R. J., and Blackburn, L. H. (1973). Disorienting effects of aircraft catapult launchings. *Aerosp. Med.* 44:37–39.

Collewijn, H., Van der Steen, J., Ferman, L., and Jansen, T. C. (1985). Human ocular counterroll: assessment of static and dynamic properties from electromagnetic scleral coil recordings. *Exp. Brain Res.* 59:185–196.

Cope, S., and Ryan, G. M. S. (1959). Cervical and otolith vertigo. *J. Laryngol. Otol.* 73:113–120.

Correia, M. J., Perachio, A. A., and Eden, A. R. (1985). The monkey vertical vestibulo-ocular response (VVOR): a frequency domain study. *J. Neurophysiol.* 54:532–548.

Darlot, C., Lopez-Barneo, J., and Tracey, D. (1981). Asymmetries of vertical vestibular nystagmus in the cat. *Exp. Brain Res.* 41:420–426.

Duensing, F., and Schaeffer, K. P. (1959). Über die Konvergenz verscheidener labyrinthärer Afferenzen auf enizelne Neurone des Vestibulariskerngebietes. *Arch. f. Psychiat. u. Zeitschr. f. d. ges. Neurologie.* 199:345–371.

Estes, M. S., Blanks, R. H. I., and Markham, C. H. (1975). Physiologic characteristics of vestibular first-order canal neurons in the cat. I. Response plane determination and resting discharge properties. *J. Neurophysiol.* 38:1232–1249.

Fernandez, C., and Goldberg, J. M. (1971). Physiology of peripheral neurons innervating semicircular canals of the squirrel monkey. II. Response to sinusoidal stimulation and dynamics of peripheral vestibular system. *J. Neurophysiol.* 34:661–675.

Fernandez, C., and Goldberg, J. M. (1976a). Physiology of peripheral neurons innervating otolith organs of the squirrel monkey. I. Response to static tilts and to long duration centrifugal force. *J. Neurophysiol.* 39:970–984.

Fernandez, C., and Goldberg, J. M. (1976b). Physiology of peripheral neurons innervating otolith organs of the squirrel monkey. II. Directional selectivity and force-response relations. *J. Neurophysiol.* 39:985–995.

Fernandez, C., and Goldberg, J. M. (1976c). Physiology of peripheral neurons innervating otolith organs of the squirrel monkey. III. Response dynamics. *J. Neurophysiol.* 39:996–1008.

Fernandez, C., Goldberg, J. M., and Abend, W. K. (1972). Response to static tilts of peripheral neurons innervating otolith organs of the squirrel monkey. *J. Neurophysiol.* 35:978–997.

Fredrickson, J. M., Schwarz, D., and Kornhuber, H. H. (1966). Convergence and interaction of vestibular and deep somatic afferents upon vestibular nuclei neurons in the cat. *Acta Otolaryngol.* 61:168–188.

Goldberg, J. M., and Fernandez, C. (1971a). Physiology of peripheral neurons innervating semicircular canals of the squirrel monkey. I. Resting discharge and response to constant angular accelerations. *J. Neurophysiol.* 34:635–660.

Goldberg, J. M., and Fernandez, C. (1971b). Physiology of peripheral neurons innervating semicircular canals of the squirrel monkey. III. Variations among units in their discharge properties. *J. Neurophysiol.* 34:676–684.

Graybiel, A. (1952). Oculogravic illusion. *AMA Arch. Ophthalmol.* 48:605–615.

Graybiel, A., and Clark, B. (1965). Validity of the oculogravic illusion as a specific indicator of otolith function. *Aerospace Med.* 36:1173–1181.

Henn, V., Young, L. R., and Finley, C. (1974). Vestibular nuclear units in alert monkeys are also influenced by moving visual fields. *Brain Res.* 71:144–149,

Horn, G., Stechler, G., and Hill, R. M. (1972). Receptive fields of units in the visual cortex of the cat in the presence and absence of bodily tilt. *Exp. Brain Res.* 15:113–132.

Lentz, J. M., and Guedry, F. E. (1982). Apparent instrument horizon deflection during and immediately following rolling maneuvers. *Aviat. Space Environ. Med* 53:549–553.

Loe, P. R., Tomko, D. L., and Werner, G. (1973). The neural signal of angular head position in primary afferent vestibular nerve axons. *J. Physiol. (Lond.)* 230:29–50.

Lowenstein, O., and Wersäll, J. (1959). A functional interpretation of the electron-microscopic structure of the sensory hairs in the crista of the elasmobranch *Raja clavata* in terms of directional sensitivity. *Nature (Lond.)* 184:1807–1810.

Lynch, J. C., Acuna, C., Sakata, H., Georgopoulos, A., and Mountcastle, V. B. (1973). The parietal association areas and immediate extrapersonal space. *Neurosci. Abstr.* 3:244.

Lynch, J. C., Mountcastle, V. B., Talbot, W. H., and Yin, T. C. T. (1977). Parietal lobe mechanisms for directed attention. *J. Neurophysiol.* 40:362–389.

Matsuo, V., and Cohen, B. (1984). Vertical optokinetic nystagmus and vestibular nystagmus in the monkey: up-down asymmetry and effects of gravity. *Exp. Brain Res.* 53:197–216.

Mountcastle, V. B. (1975). The world around us: neural command functions for selective attention, 1975 F. O. Schmitt Lecture in Neuroscience *NRP Bull.* 1–47.

Mountcastle, V. B., Lynch, J. C., Georgopoulos, A., Sakata, H., and Acuna, C. (1975). Posterior parietal association cortex of the monkey: command functions for operations within extracorporeal space. *J. Neurophysiol.* 38:871–908.

Paige, G. D., and Tomko, D. L. (1987). Canal-otolith interactions in the vestibulo-ocular reflex. *Invest. Ophthal. Vis. Sci. (ARVO Suppl.)* 28:332.

Pompeiano, O., and Brodal, A. (1957). The origin of vestibulospinal fibers in the cat. *Arch. Ital. Biol.* 95:166–195.

Robinson, D. A. (1963). A method of measuring eye movements using a scleral search coil in a magnetic field. *IEEE Trans. Biomed. Electron.* 10:137–145.

Schneider, L. W., and Anderson, D. J. (1976). Transfer characteristics of first and second order lateral canal vestibular neurons in gerbil. *Brain Res.* 112:61–76.

Tomko, D. L., Barbaro, N. M., and Ali, F. N. (1981a). Effect of body tilt on receptive field orientation of simple visual cortical neurons in unanesthetized cats. *Exp. Brain Res.* 43:309–314.

Tomko, D. L., Peterka, R. J., and Schor, R. H. (1981b). Responses to head tilt in cat eighth nerve afferents. *Exp. Brain Res.* 41:216–221.

Tomko, D. L., Peterka, R. J., Schor, R. H., and O'Leary, D. P. (1981c). Response dynamics of horizontal canal afferents in barbiturate anesthetized cats. *J. Neurophysiol.* 45:376–396.

Tomko, D. L., Wall, C. III, and Robinson, F. R. (1984). Vertical eye movements and vertical semicircular canal responses in cat during normal pitch and during pitch with the animal positioned on its side. *Neurosci. Abstr.* 10:539.

Tomko, D. L., Wall, C. III, Robinson, F. R., and Staab, J. P. (1988). Influence of gravity on cat vertical vestibulo-ocular reflex. *Exp. Brain Res.* 69:307–314.

Uemura, T., and Cohen, B. (1972). Vestibulo-ocular reflexes: effects of vestibular nuclear lesions. *Prog. Brain Res.* 37:515–518.

Vidal, J., Jeannerod, M., Lifschitz, W., Levitan, H., and Segundo, J. (1971). Static and dynamic properties of gravity-sensitive receptors in the cat vestibular system. *Kybernetik* 9:205–215.

Viéville, T., Clément, G., Viéville, T., Lestienne, F., and Berthoz, A. (1986). Adaptive modifications of the optokinetic and vestibulo-ocular reflexes in micro-gravity. In: E. Keller and D. Zee (eds.). *Adaptive Mechanisms in Visual and Oculomotor Systems*, pp. 97–106. New York: Pergamon Press.

Viirre, E., Tweed, D., Milner, K., and Vilis, T. (1986). A reexamination of the gain of the vestibuloocular reflex. *J. Neurophysiol.* 56:439–450.

Weeks, V. D., and Travell, J. (1985). Postural vertigo due to trigger areas in the sternocleidomastoid muscle. *J. Pediat.* 47:315–327.

Werner, G., Sacks, H., and Fierst, J. (1969). *Design and Evaluation of Experiments with Labyrinthine Statoreceptors*. Technical Report #2 to U.S. Air Force Office of Scientific Research, Arlington, VA.

Wilson, V. J., and Melvill Jones, G. (1979). *Mammalian Vestibular Physiology*. New York: Plenum Press.

14

The Coding of Information about Eye Movements in the Superior Colliculus of Rhesus Monkeys: Saccade Direction, Amplitude, and Velocity

DAVID L. SPARKS

This chapter summarizes the evidence implicating the superior colliculus (SC) in the control of saccadic eye movements. It discusses the problem of coding information about movement metrics with the use of neurons that have large movement fields, and it describes the rationale and preliminary results of experiments concerned with the extraction of motor signals from spatially (anatomically) encoded information.

THE SUPERIOR COLLICULUS AND SACCADIC EYE MOVEMENTS

There is now compelling evidence that, in rhesus monkeys, the deep layers of the superior colliculus (SC) are critical components of the neural circuitry that initiate and control the metrics (direction, amplitude, and velocity) of saccadic eye movements.

1. The deep layers receive inputs from brain areas involved in the analysis of visual, auditory, and somatosensory signals—sensory signals used to guide saccadic eye movements (see Huerta and Harting, 1984; Sparks, 1986, for references).
2. Neurons in the deep layers project to brainstem areas known to be important in generating saccadic eye movements and to nuclei containing neurons that project to the motoneuron pools that innervate the extraocular muscles (Harting, 1977; Weber and Harting, 1978; Huerta and Harting, 1984).
3. Many collicular neurons generate a high-frequency pulse of spike activity that precedes saccade onset by approximately 20 msec (Sparks, 1978). These neurons have movement fields (i.e., discharge before saccades having a particular range of directions and amplitudes) and, since they are organized

according to their movement fields, form a map of motor (saccadic) space (Wurtz and Goldberg, 1972; Sparks et al., 1976).

4. Microstimulation of a discrete point in the SC reliably produces a saccade with a particular direction and amplitude that is indistinguishable in velocity and trajectory from a visually guided saccade (Robinson, 1972; Schiller and Stryker, 1972). The latency of the stimulation-induced saccade is short (20 to 30 msec), and the current required to elicit a saccade is low (5 to 20 μamp for stimulus trains of 40 msec duration, 500 pulses/sec).

5. Reversible inactivation of collicular neurons impairs the ability of the animal to generate accurate saccades (Hikosaka and Wurtz, 1985, 1986); monkeys with combined lesions of the SC and frontal eye fields are unable to initiate saccades to visual targets (Schiller et al., 1980).

CODING OF INFORMATION ABOUT SACCADIC METRICS USING NEURONS WITH LARGE MOVEMENT FIELDS

The studies just summarized attest to the importance of the SC in the control of saccades, but they do not answer the question of how information about the direction, amplitude, and velocity of saccadic eye movements is coded. That question is addressed in this section.

Saccade Direction and Amplitude

As described in the preceding section, neurons with saccade-related activity are arranged topographically within the SC. Neurons discharging maximally before small saccades are located anteriorly and those firing maximally before large saccades posteriorly. Cells near the midline discharge before movements with up components and cells located laterally discharge maximally before movements with down components (Schiller and Stryker, 1972; Wurtz and Goldberg, 1972; Sparks et al., 1976). However, information concerning saccade direction and amplitude is not coded by the discharge of a single cell. Except for the maximal discharge that precedes saccades to the center of the movement field, the discharge of a collicular cell is ambiguous with respect to saccade direction or amplitude. Identical discharges precede many saccades with different directions and amplitudes (Sparks and Mays, 1980). Thus, collicular neurons do not generate specific rates of firing before saccades of different directions or amplitudes. Rather, information about the direction and amplitude of the movement is encoded anatomically—that is, by the location of the active population of neurons within the topographical map of movement fields.

Saccadic Velocity

Because of the small range in saccadic velocity observed when movements of the same direction and amplitude are made to continuously present visual targets, our earlier experiments (Sparks and Mays, 1980) revealed no relationship between the discharge of saccade-related burst neurons and saccadic velocity. Recently, we

noticed that saccades to remembered targets and those to auditory targets are reduced in velocity and, because of these variations in saccadic velocity, were able
to measure a relationship between the average firing rate of collicular cells and
saccadic velocity (see Figure 14-1; also, see Berthoz et al., 1986; Rohrer et al.,
1987).

In summary, based on currently available information, our working hypothesis
is that information about saccade direction and amplitude is encoded anatomically
by the location of the active neurons within the topographical map of movement
fields and that saccadic velocity is encoded by the level of activity within the
active population

EXTRACTION OF SIGNALS FROM ANATOMICAL CODES: HYPOTHESES AND PRELIMINARY EXPERIMENTS

The movement fields of most collicular neurons are large; a neuron that discharges
maximally before saccades 12 to 15 degrees in amplitude may fire for movements
to visual targets placed at almost any position in a quadrant of the visual field.
The observations that the movement fields of collicular neurons are large and the
activity of a single collicular neuron cannot specify the parameters of a saccade
led early investigators to conclude that the SC is not involved in coding the exact
location of a visual target or in specifying the exact amplitude and direction of
the saccade required to look to the target (Wurtz and Goldberg, 1972).

Subsequently, researchers have noted that, although the activity of a single
collicular neuron cannot accurately encode the direction and amplitude of a sac-

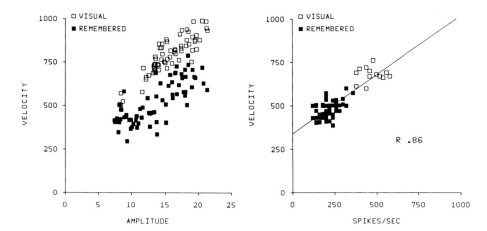

Figure 14-1 (*Left*) Amplitude/velocity relationship for saccades to continuously present
visual targets (*open symbols*) and to the remembered location of visual targets (*filled
symbols*). For clarity, data from only one of seven trials obtained for a single experimental
session are plotted. On average, the velocity of saccades to remembered targets is about
200 degrees/sec less than the velocity for saccades to visual targets. (*Right*) Relationship
between average spike frequency during saccade-related bursts and saccadic velocity for
a single neuron in the SC. All saccades had the same direction and amplitude.

cade, this information could be precisely defined by the location of the active population of neurons within the SC. Thus, the problem is one of developing biologically plausible and experimentally testable schemes by which information is extracted from an anatomical code. McIlwain (1975) and Sparks and colleagues (Sparks et al., 1976) have speculated how this might be accomplished. The proposal of Sparks and colleagues is described here.

Assume that the region of collicular neurons active before a given saccade occupies a symmetrical area, as shown by the small circle superimposed on Robinson's motor map in panel 1 of Figure 14-2 (left). Also, assume that neurons at

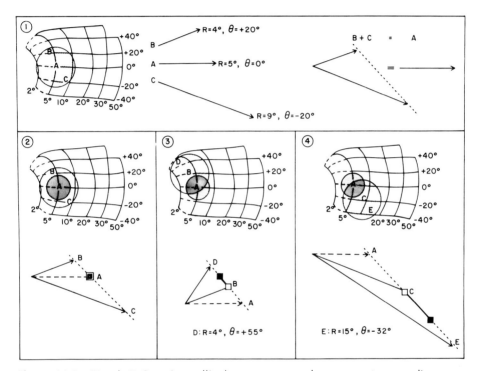

Figure 14-2 (Panel 1) Superior colliculus neurons produce accurate saccadic movements. (Left) A dorsal view of left superior colliculus showing the hypothetical region of neurons active (shaded area) before saccades to a target located 5 degrees to the right of a fixation stimulus. (Middle) Cells at locations A, B, and C fire most vigorously for the movements shown. (Right) Weighted averaging of activity at points B and C yields the same movement as activity at the center of the active population (A). See the text for further details. [From Sparks et al. (1976).] (Panels 2 to 4) The active population. The site of deactivation (darkly shaded circles) remains the same in each panel. However, the location of the active population (lightly shaded areas) is different in each panel because saccades to three different targets are required. Beneath each map are the saccade vectors associated with neural activity at each of the locations illustrated. The open square represents the vector of the intended, or programmed, saccade associated with activity in the lightly shaded area. The dashed line represents the vector of the movement tendency produced by neurons at the deactivated site. These neurons will not contribute to the metrics of the saccade, and a saccade to the approximate location of the filled square is predicted.

point A in the center of the activated region discharge maximally before straight right saccades 5 degrees in amplitude. According to the motor map, neurons located at point B discharge maximally before upward saccades (20-degree angle) 4 degrees in amplitude. Thus, spike activity occurring at point B, when a strictly horizontal eye movement has been programmed, could influence the movement to be made and result in an inaccurate saccade. Note, however, that neurons at point C, an equal distance from point A, are also active and, according to the motor map, will discharge maximally before downward saccades (-20-degree angle) 9 degrees in amplitude. The effect of neural activity at point B tending to produce a small upward movement is balanced by the effect of neural activity at point C tending to produce a larger, downward movement. Thus, for each subset of active neurons (B) producing a movement tendency with a direction and amplitude other than the programmed movement there is a second subset of active neurons (C) producing a movement tendency such that the combination of the two movements will have the programmed direction and amplitude. According to this hypothesis, each member of the active population contributes to the ensuing saccade; there is no need to "sharpen" the population response to generate an accurate saccade.

Recently, we completed experiments designed to test this hypothesis (Lee et al., 1987). The rationale for the experiment is shown in Figure 14-2 (panels 2 to 4). Referring to panel 2, suppose a small subset of collicular neurons discharging maximally before straight right saccades 5 degrees in amplitude was inactivated (shown as the darkly shaded area). If the animal were required to make a 5-degree rightward movement, saccade direction and amplitude should be unaffected. The sum of the movement tendencies produced by the unaffected neurons in the active population (shown as the larger lightly shaded circle) will result in a saccade with the correct direction and amplitude. Since the pool of inactivated neurons in the center of the active zone normally discharges earlier than those of the fringe of the zone, the latency of the saccade should increase slightly. However, referring to panel 3, with the same region of neurons inactivated, if the animal is now requested to make an upward (20-degree angle) 4-degree amplitude saccade, the location of the active population shifts, and the inactivated region falls on the periphery of the active zone. The inactivated cells discharge maximally before saccades with a larger amplitude and more of a downward component than the desired movement. Cells in the inactivated region will not balance the activity of cells discharging maximally before saccades that have a smaller amplitude and greater upward component than the programmed saccade. The animal should make a saccade that undershoots the target and has too much of an upward component. Similarly, referring to panel 4, if the animal attempts to make a downward (-20-degree direction) 9-degree amplitude saccade, the resulting movement would overshoot the target and have too much of a downward component. Thus, with the same region of colliculus inactivated, requiring the animal to make different saccades should result in different but predictable error patterns. Also, since according to our current working hypothesis, saccadic velocity is encoded by the level of activity in the population, it should be reduced after inactivation of a subset of the active population.

We used the method of Malpeli and Schiller (1979) to reversibly inactivate small, precisely localized regions of the SC with injections of the local anesthetic lidocaine hydrochloride. Results of these experiments confirm the findings of Hikosaka and Wurtz (1986) and support the major predictions of the model outlined previously. Dramatic reductions in saccadic velocity were observed, and inactivation of collicular neurons produced the predicted pattern of errors in the direction and amplitude of visually guided saccades. These findings support the hypothesis that saccadic accuracy results from averaging the movement tendencies produced by the entire active population rather than dicharging a small number of finely tuned cells. Moreover, since the contribution of each neuron to the direction and amplitude of the movement is relatively small, the effects of variability or noise in the discharge frequency of a particular neuron are minimized. Thus, the large movement fields (resulting in a large population of neurons active during a specific movement) may contribute to, rather than detract from, saccadic accuracy. Further research is required to determine how information about the direction, amplitude, and velocity of a saccade is extracted from the spatial and temporal pattern of activity of the large population of collicular neurons discharging before each saccadic eye movement.

ACKNOWLEDGMENTS

The research cited from the author's laboratory was supported by National Institutes of Health Grants R01-EY01189, R01-EY05486, and P30-EY03039.

REFERENCES

Berthoz, A., Grantyn, A., and Droulez, J. (1986). Some collicular efferent neurons code saccadic eye velocity. *Neurosc. Lett.* 72:289.

Harting, J. K. (1977). Descending pathways from the superior colliculus: an autoradiographic analysis in the rhesus monkey (*Macaca mulatta*) *J. Comp. Neurol.* 172:583.

Hikosaka, O., and Wurtz, R. H. (1985). Modification of saccadic eye movements by GABA–related substances. I. Effect of muscimol and bicuculline in monkey superior colliculus. *J. Neurophysiol.* 53:266.

Hikosaka, O., and Wurtz, R. H. (1986). Saccadic eye movements following injection of lidocaine into the superior colliculus. *Exp. Brain Res.* 61:531.

Huerta, M. F., and Harting, J. K. (1984). The mammalian superior colliculus: studies of its morphology and connections. In: H. Vanegas (ed.). *Comparative Neurology of the Optic Tectum*, pp. 687–773. New York: Plenum Press.

Lee, C., Rohrer, W. H., and Sparks, D. L. (1987). Effects of focal inactivation of superior colliculus on saccades: support for the hypothesis of vector averaging of neuronal population response. *Soc. Neurosc. Abstr.* 13:394.

Malpeli, J. G., and Schiller, P. H. (1979). A method of reversible inactivation of small regions of brain tissue. *J. Neurosc. Meth.* 1:143.

McIlwain, J. T. (1975). Visual receptive fields and their images in superior colliculus of the cat. *J. Neurophysiol.* 38:219.

Robinson, D. A. (1972). Eye movements evoked by collicular stimulation in the alert monkey. *Vision Res.* 12:1795.

Roherer, W. H., White, J., and Sparks, D. L. (1987). Saccade-related burst cells in the superior colliculus: relationship of activity with saccadic velocity. *Soc. Neurosc. Abstr.* 13:1092.

Schiller, P. H., and Stryker, M. (1972). Single-unit recording and stimulation in superior colliculus of the alert rhesus monkey. *J. Neurophysiol.* 35:915.

Schiller, P. H., True, S. D., and Conway, J. L. (1980). Deficits in eye movements following frontal eye-field and superior colliculus ablations. *J. Neurophysiol.* 44:1175.

Sparks, D. L. (1978). Functional properties of neurons in the monkey superior colliculus: coupling of neuronal activity and saccade onset. *Brain Res.* 156:1.

Sparks, D. L. (1986). Translation of sensory signals into commands for control of saccadic eye movements. role of primate superior colliculus. *Physiol. Rev.* 66:118.

Sparks, D. L., and Mays, L. E. (1980). Movement fields of saccade-related burst neurons in the monkey superior colliculus. *Brain Res.* 190:39.

Sparks, D. L., Holland, R., and Guthrie, B. L. (1976). Size and distribution of movement fields in the monkey superior colliculus. *Brain Res.* 113:21.

Weber, J. T., and Harting, J. K. (1978). Parallel pathways connecting the primate superior colliculus with the posterior vermis. An experimental study using autoradiographic and horseradish peroxidase tracing methods. In: C. Nobuck (ed.). *Sensory Systems of Primates* pp. 135–149. New York: Plenum Press.

Wurtz, R. H., and Goldberg, M. E. (1972). Activity of superior colliculus in behaving monkey. III. Cells discharging before eye movements. *J. Neurophysiol.* 35:575.

V

DEVELOPMENT AND REPAIR OF SENSORY SYSTEMS

An enormous literature now exists on the experimental analysis of neural development, much of it dealing with sensory systems, in particular the visual system. Chapter 15 introduces a recently developed approach to investigations of sensory pathway development, that of neural transplantation. It was a surprise to many of us that the mammalian central nervous system (CNS) relatively readily accepts neural grafts and tolerates grafting. We now know that grafted tissue in the CNS— even between species—can survive and establish functional connections between the host and graft. The thought that grafts were protected from surveillance by the immune system, however, had to be discarded when it became clear that any graft between genetically defined strains is susceptible to rejection. This in itself, though presenting obstacles to the use of grafts to restore function, is an interesting area for future study.

The following chapter shows how neural transplants can be used to learn a great deal about factors controlling normal development of neural connections, here demonstrated in the visual system, and how such grafts eventually may become useful in repairing damage to sensory relays in human brains. But the chapter makes clear how much careful experimental work is needed before such grafts can be regarded as reliable and ethical treatment for human diseases.

Though technically difficult, the grafting operation is only the first step in the experimental procedures needed to carry these investigations forward. Survival of the graft cells and the degree of connectivity between the grafted tissue and the host brain is not always easy to ascertain. Chapter 15 describes the use of antibody label to test the connections of the graft (the antibody here labels specifically the mouse neural graft tissues, as opposed to the host rat brain, which remains unlabeled). This is an extremely useful approach because grafts are often not readily identifiable in the living host brain, and the usual anatomical methods of injecting labeling substances into the graft to trace its connections are not feasible. The antibody also reveals the limits of the graft, even if it is thoroughly fused or dispersed into the rat host neuropil. The antigenic differences of graft and host also allow for later experimental elimination of the graft to test for functional changes without having to physically locate the graft to lesion or remove it. A simple skin graft to the host from the animal that acted as the donor is sufficient to spur an immune reaction in the brain that then causes the death of the graft tissue (Lund et al., 1987).

A good understanding of normal development of the sensory system pathways and their reaction to lesions and other interference is essential to the interpretation of grafting experiments. The value of the transplantation paradigm in studying the substrates for sensory processing is potentially great, and it holds much promise as an area in which molecular biology and systems neuroscience may come together. There is wide scope for manipulating the graft tissues in tissue culture in a variety of ways before implantation is done and for using this approach to examine biochemical and molecular elements essential to neural development.

REFERENCES

Lund, R. D., Rao, K., Hankin, M. H., Kunz, H. W., and Gill, T. J. (1987). Transplantation of retina and visual cortex to rat brains of different ages: maturation, connection patterns and immunological consequences. *Ann. N.Y. Acad. Sci.* 495:227–241.

15

Neural Transplantation as an Approach to Studying Sensory Systems

RAYMOND D. LUND

The transfer of sensory information from peripheral receptors to those regions of the brain in which it is finally analyzed to allow an appropriate behavioral response is a complex process. It entails the encoding of various parameters of the particular sensation, often through parallel channels. This is followed by a relay through a chain of neurons, during which the signal is progressively modified.

The level of functional specificity encountered at each stage of analysis of the sensory signal depends, in large part, on the existence of precise interconnections among the neurons concerned. How such circuits form during development, how they are maintained, once established, and how they may be recovered or substituted after injury are questions to which a considerable amount of attention has been devoted.

NEURAL CIRCUIT FORMATION

It is apparent that the development of neural connections is a process involving a series of quite discrete stages. This has been well documented in the mammalian visual system. The first stage of development of the primary visual pathways is an outgrowth of axons from the eye to the brain. In the rat, this process begins on embryonic day (E) 14 (Weidman and Kuwabara, 1968) and is probably completed during the first postnatal week (birth occurs on E22). Two points to note about this stage are that (1) the growing optic axons seem to favor substrates available on the surface of the brain, and (2) not all axons reach the appropriate target (Bunt et al., 1983): some end in the opposite eye (Bunt and Lund, 1981), some on the inappropriate side of the brain (Martin et al., 1983), and others in the wrong nucleus (Frost, 1986). The second stage of development involves a change in growth cone configuration as individual axons reach terminal regions (Bovolenta and Mason, 1987). This stage is associated with a terminal ramification of the axon's growing tip and a relatively modest synaptogenesis involving mainly simple synaptic patterns (Lund and Lund, 1972; Lund and Bunt, 1975). During the first postnatal week in rodents, there is a period of cell death in the retina when more than half the ganglion cells are lost (see Sefton, 1986). The loss of cells is probably the result of several factors involving failure to find appropriate target cells (Bunt and Lund, 1981), axonal competition (Land and Lund, 1979), and competition among ganglion cell dendrites (Perry and Linden, 1982) and, possibly, scaling factors between retina and central target structures.

Once the stage of ganglion cell death and associated axon loss is complete, the system begins to function, as evidenced by a recordable electroretinogram (ERG) (Weidman and Kuwabara, 1968), presence of synapses in the retina (Horsburgh and Sefton, 1987), and subsequently opening of the eyelids on about postnatal day 12. This is accompanied in regions such as the superior colliculus by a massive increase in synapses formed by optic and other terminals (Lund and Lund, 1972), and the development of complex synaptic arrays including the axo-dendro-dendritic synapses in which the optic terminals play an important role (Lund, 1969).

Although considerable attention has been given to the early stages of development of the primary visual system, little is known about this late synaptic proliferation. However, it is during this period that the synaptic patterns that form the substrates for complex sensory processing in the visual system are likely to be generated. Interestingly, the geniculocortical pathway has a similar two-step synapse formation—first, the formation of synapses in a "waiting zone" deep to the cortical plate, and second, a gradual synaptic differentiation of the cortical plate (Lund, 1976; Lund and Mustari, 1977). This is seen even more clearly in larger animals (e.g., monkey—Rakic, 1977; cat—Shatz and Luskin, 1986).

Once the synapses are established, the further question arises as to whether they are substantially modifiable either in effectiveness or in location. Various theories (e.g., Marr, 1971; Changeaux and Danchin, 1976) derived from a model proposed by Hebb (1949) offer the possibility that synaptic efficacy can be improved by repeated stimulation of a circuit, and phenomena such as long-term potentiation or enhancement (see McNaughton and Morris, 1987) provide interesting models with which to examine the substrates of the effect. It is much harder to document the possibility of regular day-to-day synaptic changes not accompanying a specific entrainment procedure within the brain, although some progress has been made to this end in peripheral ganglia (Purves et al., 1986).

In recent studies, we have attempted to use transplantation techniques to develop a number of preparations in which it is possible to define direct structure—function correlates, so that we may be able to identify what circuits provide the minimum necessary substrates of particular components of vision, how these patterns of connections are assembled, and how they can be modified in maturity. This should provide us with some new approaches to the problems of circuit assembly and stability.

In this review, attention is restricted to the ability of retinas transplanted to the mammalian brain to establish functional circuits through which visual information can be relayed. There is a wealth of literature to show that optic axons in amphibians and fish can regenerate after injury and restore connections that subserve appropriate visuomotor responses (see, e.g., Sperry, 1963; Harris, 1984). In mammals, regeneration of mature optic axons after injury is normally minimal, at best (e.g., Stevenson, 1987), although it is possible to support regenerative outgrowth of some axons back to target regions such as the midbrain, by using a sciatic nerve segment (So and Aguayo, 1985; Vidal Sanz et al., 1987). Furthermore, embryonic retinas transplanted into the eye (Turner and Blair, 1986; Del Cerro et al., 1987), although they survive, appear to be unable to grow axons down to the optic nerve. However, if they are placed in appropriate locations in the brain, not only will embryonic retinas survive, but they will also make connections with the host brain.

RETINAL TRANSPLANTS

The first successful study of retinal transplantation in mammals (Tansley, 1946) showed that embryonic rat eyes could survive and differentiate when placed in rat brains. It was proposed that immediately after transplantation, the retinas degenerated and then regrew, but this has not been borne out by more recent work. One important issue not explored in the first studies was whether the transplanted retinas connected with the host brain. This has been examined in work done over the past 10 years, in which the differentiation of retinal transplants placed in the brain, their functional capacity, and their interrelation with the host brain have been studied in detail (see Lund and McLoon, 1983; McLoon et al., 1985; Lund and Simons, 1985).

Retinas taken from early embryonic mice or rats (E12–15) and placed in a variety of brain locations show many of the features that typify a normal retina. Although the transplants are often folded, and even in some instances configured into rosettes, the characteristic laminae are early identified. The specific cell types seen in normal retinas defined by soma location (McLoon and Lund, 1980b), dendritic morphology (Perry et al., 1985), and chemical markers (McLoon and McLoon, 1984) are clearly recognizable in the transplants. Both rods and cones are seen among the receptor layer (Lund et al., 1987b), and stacks of outer segment membrane are associated with the receptor cells (Lund and McLoon, 1983).

When retinas are placed in reasonable proximity to appropriate target regions, they emit bundles of axons that innervate many of the nuclei in the brainstem to which optic axons normally project (McLoon and Lund, 1980a). Such innervation is most substantial if the host optic innervation is first removed. There is no evidence of innervation by the transplants of nuclei that normally do not receive optic input.

These observations raise two questions: can transplanted retinas respond to light, and can they transmit significant information to the host brain?

The first question has been addressed by transplanting retinas over the cortex (Freed and Wyatt, 1980) or superior colliculus (Simons and Lund, 1985) and recording, within the retinal graft, gross potential responses to light stimulation. In both cases, potential changes were recorded, and these approximated the normal electroretinogram in waveform. Most important, it was found that the magnitude of the response varied with the intensity of the light stimulus (Simons and Lund, 1985). The possibility that individual classes of ganglion cell exhibit their normal responses to particular visual parameters has not yet been explored.

The question of relaying information to the host brain has been examined in animals in which the retinas were transplanted over the midbrain of neonatal rats from which one or both eyes were removed at the time of transplantation (Simons and Lund, 1985; Lund et al., 1986). Once the host animals had reached maturity, the transplants were exposed, and stimulated either photically or electrically. In the first set of studies, single- and multiunit recordings were made from the superior colliculus after the retinal transplants were stimulated with light. Under these circumstances, it was possible to record unit activity to specific parameters of the stimulus. In some cases, transient responses to light-on or to light-off were recorded; other cells responded to ambient light levels with changes in spontaneous firing rate; and in yet others, spontaneous firing was inhibited by visual stimulation (Figure 15-1B to D). These results suggest that retinal transplants are able to relay specific information about changes in light levels to cells in the superior colliculus and these cells are capable of responding quite normally to the stimuli.

We next asked whether transplant-mediated events could drive an appropriate motor response. To answer this question, we examined whether a pupillary response could be elicited in a host eye by photic stimulation of a retinal transplant (Klassen and Lund, 1987). The center for pupilloconstriction in normal animals lies in a division of the pretectal complex, the olivary pretectal nucleus. This nucleus receives a direct input from the eye, and its cells send axons to the Edinger-Westphal nucleus of the oculomotor complex. From here, axons project by way of the third cranial nerve to the ciliary ganglion, and thence to the iris sphincter muscle (Figure 15-2A).

We transplanted embryonic retinas over the midbrain of newborn rats and at the same time removed one host eye to enhance host innervation by the transplant. At maturity, the remaining optic nerve was cut intracranially (to prevent light transmitted by the host eye from entering the brain, but yet preserve the efferent pathway running in the third nerve). Two days later the transplant was exposed. Illumination of the transplant caused pupilloconstriction (Figure 15-2B and C) of the host eye; placing a screen between the light source and transplant resulted in pupillodilation. The degree of constriction depended on the intensity of illumination. Suitable controls have ruled out the possibility that incidental stimulation of the host brain was responsible for the reflex response. Damage to either the transplant or the olivary pretectal nucleus abolished the response, indicating that it was being mediated by means of the transplant through normal host efferent pathways.

The pupillary response studies show that appropriate specific connections develop between the transplant and cells in the pretectum, relaying information necessary for pupilloconstriction. Whether the reflex is mediated by a specific class of retinal cell that has made appropriate connections in this region is not known. This is of interest because evidence in birds suggests that the nucleus may be innervated normally by a particular cell class (Gamlin et al., 1984). It also remains to be seen whether other behaviors can be elicited by transplant stimulation.

Although these two experiments show that retinal transplants can relay information about light intensity changes to the host brain, they do not address another aspect of the visual signal—that of spatial localization. There is a map of visual space that is projected on most brain regions subserving visual functions. We

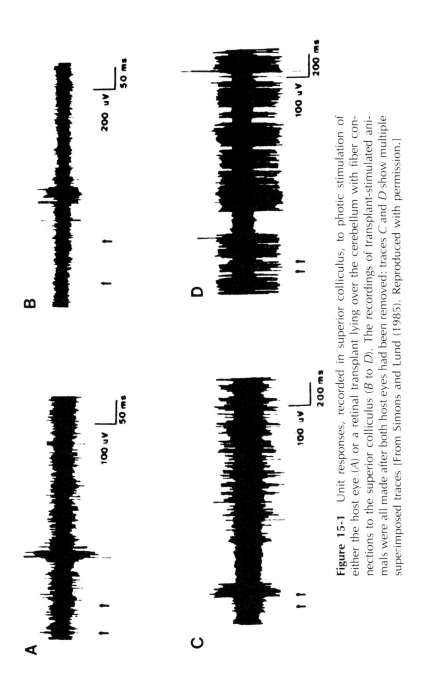

Figure 15-1 Unit responses, recorded in superior colliculus, to photic stimulation of either the host eye (A) or a retinal transplant lying over the cerebellum with fiber connections to the superior colliculus (B to D). The recordings of transplant-stimulated animals were all made after both host eyes had been removed: traces C and D show multiple superimposed traces [From Simons and Lund (1985). Reproduced with permission.]

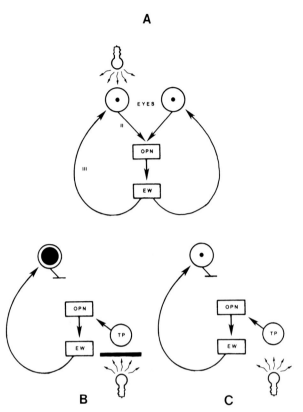

Figure 15-2 Schematic of pathway for pupilloconstriction. (*A*) In normal rats light activates neurons that pass from the eye by way of the optic nerve (II) to the olivary pretectal nuclei (OPN) on each side. From there, axons project to the Edinger-Westphal (EW) nuclei. These project, by way of the oculomotor (III) nerve through the ciliary ganglion, to the pupilloconstrictor muscles. (*B*) Experimental preparation described here with the transplant (TP) screened from light and the pupil dilated. (*C*) Illumination of transplant causes constriction of the pupil in the host eye. [From Klassen and Lund (1987). Reproduced with permission.]

wondered whether our retinal transplants would also be able to distribute a topographically ordered projection on the host brain. This has been studied anatomically (Galli et al., 1987) by placing retinas over the midbrain and in the cerebral aqueduct of newborn rats from which one or both eyes were removed at the time of transplantation. The retinas develop substantial connections with the host superior colliculus, which normally receives a topographically ordered projection from the eye. Evidence of topography can be shown by injecting two different retrogradely transported labels into separate points on the collicular surface (Figure 15-3A). If there is a topographic representation of the retina on the tectum, there would be two separated areas of the ganglion cell layer of the retina labeled, one with one dye and the other containing cells labeled with the second (Figure 15-3B). If ganglion cells of the retina had axons projecting broadly in the colliculus, then double-labeled cells might be encountered (Figure 15-3C). Finally, if cells projected to discrete but topographically inappropriate positions on the tectum, cells labeled with one or the other dye would be intermixed, but there would be no double-labeled cells (Figure 15-3D). It was the last option that was found: double-labeled cells were seen only when the two injections in the colliculus themselves overlapped. There may be several reasons why the retinal cells fail to project to appropriate parts of the tectal map. It is possible that removal of the pigment epithelium at the time of transplantation alters intercellular communication in the retina and prevents dissemination of positional information. It is also likely that the disruption of the optic nerve to form a series of bundles of axons connecting transplant and tectum contributes to the lack of topography. Certainly axons do not simply relate to a series of polar coordinate cues on the tectal surface, as appears to be the case in studies on developing chick (Thanos and Dütting, 1987).

Together these results indicate that, although transplants are able to mediate certain light-activated functions in the host brain, which are driven by intensity changes, they are unlikely to transmit visual information in which spatial parameters are important without further manipulation of the transplantation circumstances.

Multiple Transplants

The preceding section has shown that a retina can grow axons into a host brain and assemble circuits in the host neuropil of sufficient precision to elicit specific

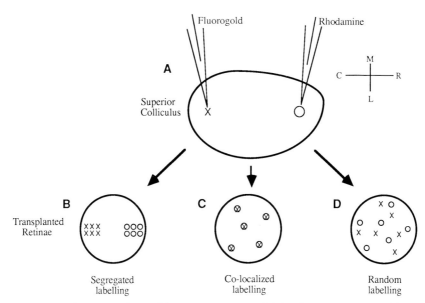

Figure 15-3 Schematic of possible consequences of injecting two retrogradely transporting dyes, one marked by an x and the other by a circle, at anterior and posterior poles of the superior colliculus, as shown in A. (B to D) Labeling patterns of ganglion cells that might be encountered in retinal transplants, (B) with a topographically ordered projection, (C) with a diffuse projection, and (D) with a disordered projection. D shows the pattern encountered in the retinal transplants. See the text for further description.

and appropriate responses to photic stimulation. The question may be raised as to whether a circuit can be assembled with more than one graft. This was examined by placing embryonic retinas together with other brain regions in the cerebral cortex of newborn rats (Sefton et al., 1987; Sefton and Lund, 1988). Retinas placed alone in the cortex failed to grow axons into the host tissue, and ultimately the ganglion cells degenerated (McLoon and Lund, 1984; Sefton et al., 1987). A similar fate attended retinas cotransplanted with embryonic cortex. However, if the retina was cotransplanted with embryonic tissue obtained from either thalamus or tectum, a different pattern was encountered. Ganglion cells survived for more than 3 months, there was substantial outgrowth of their axons from the retinal graft, and these axons distributed within local areas of the cotransplants. The projections have been shown most effectively by cotransplanting mouse retina with rat brain tissue and using mouse-specific antibodies to show not only the mouse graft, but also the terminal distribution of axons arising from it in the rat-derived tissue (Figure 15-4C). Within the tectal grafts, the axons ramified within regions, which were identified in previous studies (Lund and Harvey, 1981) as ones likely to receive input from the retina. The retinal transplant innervation of thalamic grafts was also quite discrete, providing heavy input to some cytoarchitectonically defined regions and light or no input to others. Whether the innervated regions corresponded to those thalamic regions permanently or transiently innervated by the eye under normal circumstances is not known. It is significant that the cells in the thalamic grafts appeared healthy. Since the survival of thalamic neurons normally depends on their ability to innervate the cortex, this suggests that the thalamic grafts made connections with the host cortex. Normal fiber stains and specific antibody staining (after using mouse thalamic grafts) (Zhou and Lund, unpublished) suggest this is the case, and other work examining the

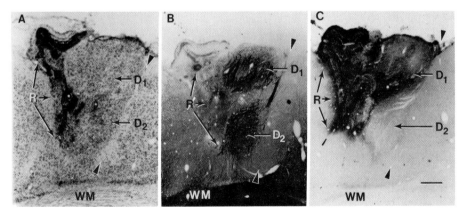

Figure 15-4 Photomicrographs of adjacent sections of a host rat cortex into which an embryonic mouse retina, R, and rat thalamus, D, had been transplanted. A is stained with cresyl violet, B for normal fibers, and C with a mouse-specific antibody to show the retinal transplant and its projections. The thalamus is divided into two lobes and, as can be seen in C, one of these is heavily innervated by the retina. The host brain is not innervated by the retinal graft. The scale bar signifies 250 μm [From Sefton et al. (1987). Reproduced with permission.]

projections of thalamic grafts placed in the cortex (Matthews, 1985) further supports this expectation.

This work raises the possibility that a circuit can be established involving a retinal transplant, a thalamic transplant, and the host cortex. Whether such a circuit might be capable of transmitting visual information to the cortex remains to be studied. In view of the positive findings of the previous section, such a proprosal does not seem unrealistic.

Plasticity of Transplant Connections

The ability to manipulate retinal transplants without direct intervention of the host nervous system, and vice versa, provides an opportunity to examine the plasticity of connections in the mature brain. To this end we studied the interactions of the retinal inputs from host and transplanted eyes in the superior colliculus. If an embryonic retinal transplant was placed over the midbrain of a neonatal rat, it provided an extensive innervation of the retinorecipient layer of the colliculus only if the host eye innervating that colliculus was removed at the time of transplantion. If the host eye was intact, axons from the transplant entered the colliculus and distributed both superficial and deep to the retinorecipient stratum griseum superficiale (McLoon and Lund, 1980a). Occasional synapses of transplant origin were seen along the surface of the striatum griseum superficiale, but never within it. However, if the host eye was removed at maturity, substantial sprouting of the transplant axons into the synaptic layers now deprived of optic input occurred (Lund et al., 1987a; Radel et al., 1987). Similar sprouting was also seen in the olivary pretectal nucleus, providing the opportunity for direct behavioral monitoring of an axonal sprouting phenomenon in the mature nervous system, since the sprouted pathway is an essential step in the relay by which the reflex is driven and, at the time the sprouting is prompted, both host and graft neuropils may be considered mature.

DISCUSSION

The most important aspects of these studies are, first, that retinas transplanted to the brains of neonatal rats make highly specific connections with cotransplants or with appropriate host brain regions and, second, that the connections made with the host brain are capable of mediating the correct light-activated response in the host.

Previous studies of transplant function have centered on two main areas, those concerned with neuromodulatory influences and those concerned with neuroendocrine controls (Azmitia and Björklund, 1987; Björklund and Stenevi, 1985). In each case the effect of the transplant appears relatively nonspecific, involving changes in levels of functional activity of the target cells involved, rather than relay of specific input–output functions. Transplants that appear to function as neuromodulators replace systems in which relatively small numbers of cells project broadly in the brain and functional recovery after injury of those cells can be instituted, in part, by local or systemic administration of the deficient transmitter or its analog. The best studied of these systems is the nigrostriatal pathway that arises from a small group of dopamine cells in the pars compacta of the substantia nigra. Damage to these cells in adults, by either lesions or administration of the specific neurotoxins 6-hydroxydopamine or MPTP, results in a characteristic set of motor disorders that are clearly quantifiable in experimental animals and are recognized as Parkinson's syndrome in humans. The loss of the dopamine pathway results in changes in excitability levels of striatal cells and, as a result, disturbs the efficacy of relay from the cortex through the striatum. This deficit can be corrected by administering dopamine (Hargraves and Freed, 1987) or transplanting nigral cells into the striatum (Björklund and Stenevi, 1979). It should be noted that these transplanted cells can reverse functional deficits even though they may not have access to their normal afferents and some of the connections made with striatal cells are abnormal (Freund et al., 1985). This is in marked contrast with the present studies, in which the appropriate response properties appear to depend on the development of discrete synaptic relays in both the transplant and its connections with the host brain.

However, a recent set of studies involving the basal ganglia parallel the relay function shown here for retina transplants. In these experiments, the striatum was destroyed by injections of ibotenic acid, and embryonic striatal cells were placed in the lesion site (Isacson et al., 1984; Pritzel et al., 1986). These cells connect with the host, and some level of functional recovery is achieved. Not only does the nigrostriatal pathway establish connections with the graft, but the corticostriatal relay functions are also redeveloped.

Besides having an important role in restoring lost neuromodulatory functions, neural grafts may therefore be capable of replacing damaged relay functions both in the visual system and in a variety of other pathways in the central nervous system. Given this possibility, it is important to define simple experimental situations in which the optimal conditions for information transfer by way of neural grafts can be established. The primary optic system has proved to be one such system, with the advantages of accessibility, ease of delivery and definition of visual signals and the opportunity to record, both physiologically and behaviorally, the responses to stimulation. Using these various measures, we can study how variations in synaptic patterns occurring under different experimental circumstances are directly reflected in function. In this way we will be able to define not only the conditions under which a transplant can best function to relay sensory information, but also the minimum substrates necessary for relay of a specific sensory signal under normal circumstances.

ACKNOWLEDGMENTS

For the personal research reported here, I am indebted to a group of colleagues, both technical and academic, whose contribution to its progress has been enormous. I thank Sharon Wesolowski for secretarial assistance and NIH for funding the research through Grant EY 05283.

REFERENCES

Azmitiza, E. C., and Björklund, A. (1987). Cell and tissue transplantation into the adult brain. *Ann N.Y. Acad. Sci.* 495.

Björklund, A., and Stenevi, U. (1979). Reconstruction of the nigrostriatal dopamine pathway by intracerebral nigral transplants. *Brain Res.* 177:555–560.

Björklund, A., and Stenevi, U. (1985). *Neural Grafting in the Mammalian CNS.* Amsterdam: Elsevier.

Bovolenta, P., and Mason, C. (1987) Growth cone morphology varies with position in the developing mouse visual pathway from retina to first targets. *J. Neurosci.* 7:1447–1460.

Bunt, S. M., and Lund, R. D. (1981). Development of a transient retino-retinal pathway in hooded and albino rats. *Brain Res.* 211:399–404.

Bunt, S. M., Lund, R. D., and Land, P. W. (1983). Prenatal development of the optic projection in albino and hooded rats. *Dev. Brain Res.* 6:149–168.

Changeaux, J. P., and Danchin, A. (1976). Selective stabilization of developing synapses as a mechanism for the specification of neural networks. *Nature* 264:705–712.

Del Cerro, M., Gash, G. M., Rae, G. N., Notter, M. F., Wiegand, S. J., Sath, S., and Del Cerro, C. (1987). Retinal transplants into the anterior chamber of the rat eye. *Neurosci.* 21:707–723.

Freed, W. J., and Wyatt, R. J. (1980). Transplantation of eyes to the adult rat brain: histological findings and light-evoked potential response. *Life Sci.* 27:503–510.

Freund, T. F., Bolam, J. P., Björklund, A., Stenevi, U., Dunnett, S. B., Powell, J. F., and Smith, A. D. (1985). Efferent synaptic connections of grafted dopaminergic neurons reinnervating the host neostriatum: a tyrosine hydroxylase immunocytochemical study. *J. Neurosci.* 5:603–616.

Frost, D. O. (1986). Development of anomalous retinal projections to nonvisual thalamic nuclei in Syrian hamsters: a quantitative study. *J. Comp. Neurol.* 252:95–105.

Galli, L., Rao, K., and Lund, R. D. (1987). The establishment of topographical order in mammalian retino-tectal projections: insights from transplants studies. *Soc. Neurosci. Abstr.* 13:513.

Gamlin, P. D. R., Reiner, A., Erichsen, J. T., Karten, H. J., and Cohen, D. H. (1984). The neural substrate for the pupillary light reflex in the pigeon (Columba livia). *J. Comp. Neurol.* 226:523–543.

Hargraves, R., and Freed, W. J. (1987). Chronic intrastriatal dopamine infusions in rats with unilateral lesions of the substantia nigra. *Life Sci.* 40:959–966.

Harris, W. A. (1984). Neural transplants in lower vertebrates. In: J. R. Sladek, and D. M. Gash (eds.). *Neural transplants: Development and Function*, pp. 43–98. New York: Plenum Press.

Hebb, D. O. (1949). *The Organisation of Behavior.* New York: John Wiley & Sons.

Horsburgh, G. M., and Sefton, A. J. (1987). Cellular degeneration and synaptogenesis in the developing retina of the rat. *J. Comp. Neurol.* 263:553–566.

Isacson, O., Brundin, P., Kelly, P. A. T., Gage, F. H., and Björklund, A. (1984). Functional neuronal replacement by grafted striatal neurones in the ibotenic acid-lesioned rat striatum. *Nature* 311:458–460.

Klassen, H., and Lund, R. D. (1987). Retinal transplants can drive a pupillary reflex in host rat brains. *Proc. Natl. Acad. Sci.* 84:6958–6960.

Land, P. W., and Lund, R. D. (1979). Development of the rat's uncrossed retinotectal pathway and its relation to plasticity studies. *Science* 205:698–700.

Lund, R. D. (1969). Synaptic patterns of the superficial layers of the superior colliculus. *J. Comp. Neurol.* 135:179–208.

Lund, R. D. (1976). The development of laminar connections in the mammalian visual cortex. *Exp. Brain Res.* (Suppl.) 1:255–258.

Lund, R. D., and Bunt, A. H. (1975). Prenatal development of central optic pathways in albino rats. *J. Comp. Neurol.* 165:247–264.

Lund, R. D., and Harvey, A. R. (1981). Transplantation of tectal tissue in rats. I. Organization of transplants and pattern of distribution of host afferents within them. *J. Comp. Neurol.* 201:191–209.

Lund, R. D., and Lund, J. S. (1972). Development of synaptic patterns in the superior colliculus of the rat. *Brain Res.* 42:1–20.

Lund, R. D., and McLoon, S. C. (1983). *Transplantation of Brain Tissue: Retinal Transplants.* Wallace, R. B. and Das, G. D. (eds.), New York: Plenum Press.

Lund, R. D., and Mustari, M. (1977). Development of the geniculocortical pathway in rats. *J. Comp. Neurol.* 173:289–306.

Lund, R. D., Hankin, M. H., Perry, V. H., Rao, K., and Simons, D. J. (1986). Retinae transplanted to rat brains. In: E. Agardh and B. Ehinger (eds.), *Retinal Signal Systems, Degenerations and Transplants,* Amsterdam: Elsevier Science Publishers.

Lund, R. D., Hankin, M. H., Rao, K., Radel, J. D., and Galli, L. (1987). Retinal transplants and plasticity of the primary optic projections in rodents. In: G. O. Ivy and T. L. Petit (eds.), *Synaptic Plasticity: A Lifespan Approach.* New York: Alan R. Liss.

Lund, R. D., and Simons, D. J. (1985). Retinal transplants: structural and functional interrelations with the host brain. In: A. Björklund and U. Stenevi (eds.) *Neural Grafting in the Mammalian CNS,* pp. 345–354. Amsterdam: Elsevier.

Lund, R. D., Rao, K., Hankin, M. H., Kunz, H. W., and Gill, T. J. (1987b). Transplantation of retina and visual cortex to rat brains of different ages: maturation, connection patterns and immunological consequences. *Ann. N.Y. Acad. Sci.* 495:227–241.

Marr, D. (1971). Simple memory: a theory for archicortex. *Phil. Trans. R. Soc. (Lond.)* B262:23–81.

Martin, P. R., Sefton, A. J., and Dreher, B. (1983). The retinal location and fate of ganglion cells which project to the ipsilateral superior colliculus in neonatal albino and hooded rats. *Neurosci. Lett.* 41:219–226.

Matthews, M. A. (1985). Transplantation of fetal lateral geniculate nucleus to the occipital cortex: connectivity with host's area 17. *Exp. Brain Res.* 58:473–489.

McLoon, L. K., McLoon, S. C., Chang, F.-L. F., Steedman, J. G. and Lund, R. D. (1985). Visual system transplanted to the brain of rats. In: A. Björklund and U. Stenevi (eds.), *Neural Grafting in the Mammalian CNS;* pp. 267–283. Amsterdam: Elsevier.

McLoon, S. C., and Lund, R. D. (1980a). Specific projections of retina transplanted to rat brain. *Exp. Brain Res.* 40:273–282.

McLoon, S. C., and Lund, R. D. (1980b). Identification of cells in retinal transplants which project to host visual centers: horseradish peroxidase study in rats. *Brain Res.* 197:491–495.

McLoon, S. C., and Lund, R. D. (1984). Loss of ganglion cells in fetal retina transplanted to rat cortex. *Dev. Brain Res.* 12:131–135.

McLoon, S. C., and McLoon, L. K. (1984). Transplantation of the developing mammalian visual system. In: J. R. Sladek and D. M. Gash (eds.). *Neural Transplants: Development Function,* pp. 99–124. New York: Plenum Press.

McNaughton, B. L., and Morris, R. G. M. (1987). Hippocampal synaptic enhancement and information storage within a distributed memory system. *TINS* 10:408–415.

Perry, V. H., and Linden, R. (1982). Evidence for dendritic competition in the developing retina of the rat. *Nature (Lond).* 297:683–685.

Perry, V. H., Lund, R. D., and McLoon, S. C. (1985). Ganglion cells in retinae transplanted to newborn rats. *J. Comp. Neurol.* 231:353–365.

Pritzel, M., Isacson, O., Brundin, P., Wiklund, L., and Björklund, A. (1986). Afferent and efferent connections of striatal grafts implanted into the ibotenic acid lesioned neostriatum in adult rats. *Exp. Brain Res.* 65:112–126.

Purves, D., Hadley, R. D., and Voyvodic, J. (1986). Dynamic changes in dendritic geometry of individual neurons visualized over periods of up to three months in the superior cervical ganglion of living mice. *J. Neurosci.* 6:1051–1060.

Radel, J. D., Galli, L., and Lund, R. D. (1987). Plasticity of innervation of superior colliculus by retinal transplants. *Soc. Neurosci. Abstr.* 13:1597.

Rakic, P. (1977). Prenatal development of the visual system in rhesus monkey. *Phil. Trans. R. Soc. Lond. B.* 278:245–260.

Sefton, A. J., (1986). The regulation of cell numbers in the developing visual system. In: J. D. Pettigrew, K. J. Sanderson, and W. R. Levick (eds.). *Visual Neuroscience*, pp. 145–156. Cambridge University Press.

Sefton, A. J., and Lund, R. D. (1988). Co-transplantation of embryonic mouse retina with tectum, diencephalon or cortex to neonatal rat cortex. *J. Comp. Neurol.* 269:548–564.

Sefton, A. J., Lund, R. D., and Perry, V. H. (1987). Target regions enhance the outgrowth and survival of ganglion cells in embryonic retina transplanted to cerebral cortex in neonatal rats. *Dev. Brain Res.* 33:145–149.

Shatz, C. J., and Luskin, M. B. (1986). The relationship between the geniculocortical afferents and their cortical target cells during development of the cat's primary visual cortex. *J. Neurosci.* 6:3655–3668.

Simons, D. J., and Lund, R. D. (1985). Fetal retinae transplanted over tecta of neonatal rats respond to light and evoke patterned neuronal discharges in the host superior colliculus. *Dev. Brain Res.* 21:156–159.

So, K.-F., and Aguayo, A. J. (1985). Lengthy regrowth of cut axons from ganglion cells after peripheral nerve transplantation into the retina of adult rats. *Brain Res.* 328:168–172.

Sperry, R. W. (1963). Chemoaffinity in the orderly growth of nerve fiber patterns of connections. *Proc. Natl. Acad. Sci.* 50:703–710.

Stevenson, J. A. (1987). Growth of retinal ganglion cell axons following optic nerve crush in adult hamsters. *Exp. Neurol.* 97:77–89.

Tansley, K. (1946). The development of the rat eye in graft. *J. Exp. Biol.* 22:221–223.

Thanos, S., and Dütting, D. (1987). Outgrowth and directional specificity of fibers from embryonic retinal transplants in the chick optic tectum. *Dev. Brain Res.* 32:161–179.

Turner, J. E., and Blair, J. R. (1986). Newborn rat retinal cells transplanted into a retinal lesion site in adult host eyes. *Dev. Brain Res.* 26:91–104.

Vidal Sanz, M., Bray, G. M., Villegas-Perez, M. P., Thanos, S., and Aguayo, A. J. (1987). Axonal regeneration and synapse formation in the superior colliculus by retinal ganglion cells in the adult rat. *J. Neurosci.* 7:2894–2909.

Weidman, T. A., and Kuwabara, T. (1968). Postnatal development of the rat retina. *Arch. Ophthalmol.* 79:470–484.

VI

MODELS OF NEURAL SENSORY PROCESSING

The next three chapters take up the question of how to define the activities of the nervous system as it deals with sensory input and how to use these definitions to develop models to explain the mechanisms of complex neuropils. Reitboeck (Chapter 16) focuses on two aspects of neural sensory processing—feature extraction and associative links. Using the visual and auditory systems as models, he discusses which elements of the sensory input provide the most information and how generalizations can be achieved. How is it, for instance, that a familiar object can be recognized visually, independent of its position in the visual field, its size, or its orientation in space? The importance of filtering mechanisms, temporal coincidence of activity, and recognition of feature against background are brought out, and possible roles for feedback and reciprocal connections are discussed.

Lehky and Sejnowski (Chapter 17) discuss the design and properties of a network model they have devised that computes, from the gray-level image of a curved surface, parameters that quantify the shape of the surface. This process is carried out independently of the illumination of the surface and the position of the surface in a local region. The receptive field properties of the model neurons in the network are similar to those of some simple and complex cells that have been found in visual cortex (Hubel and Wiesel, 1962). Although the authors clearly state that their system is not intended literally to model neural events, necessary features of the model such as feedback and changes in synaptic efficacy or weight are recognized features of the real nervous system. Important points emerge, such as the economy of discriminating systems that use relative activity of a few broadly tuned receptors versus systems in which the receptors are very narrowly tuned with great specificity and must occur in large numbers to achieve the same functional capacity as the broadly tuned system. Another point of considerable biological relevance is that the system itself "decides" how many processing units are needed for optimal function—the superfluous elements lose or do not acquire synaptic connections. Such modeling efforts are of great importance in suggesting information-processing features for neurobiologists to look for in the real nervous system.

In Chapter 18 Werner reviews the various trains of thought that psychologists and theorists have followed in describing sensory processing and behavioral phenomena since the 1940s. The student of biology may be surprised by the diversity of the philosophies and disciplines that have argued over the workings of the mind;

bench scientists may not have the patience for such arguments. However, from these battles over logic among the theorists has come a much more rigorous approach to models of neural function. In addition, these arguments have served to sharpen the neurobiologists' definitions of the phenomena they find in the nervous system and to question more closely what place their experimental findings have in explaining the overall activity of the brain.

REFERENCE

Hubel, D. H., and Wiesel, T. N. (1962). Receptive fields, binocular interactions, and functional architecture in the cat's visual cortex. *J. Physiol.* 169:106–154.

16

Neural Mechanisms
of Pattern Recognition

HERBERT J. REITBOECK

Behind all models of pattern recognition in the CNS lurks the hidden assumption that the phenomenon could be explained if the model were to do what an ideal computer program is expected to achieve: operate on the pattern until it can be represented in some standardized form that matches a template in memory. In his paper "The Many Faces of Neuro-Reductionism," Werner (1985) warns of the "homunculus watching some display in the brain," and of the specter of infinite regress hidden in that approach. Although the homunculus is hardly evident in the preceding pattern recognition scheme, he is still there, noticing the click that indicates the identity of the images.

How, then, can the brain (or a program) "understand" anything? Understanding essentially means to discover an isomorphism. A student thinks he understands what electromagnetic waves are because differential equations similar to those describing the waves describe other phenomena that can be related to everyday experiences, such as waves in water or vibrating bodies. He has discovered, or was led to see, an isomorphism between the various wave phenomena and, in a more abstract sense, between their mathematical descriptions. Now he can make predictions.

All the nervous system can achieve is the discovery of isomorphisms between various internal representations and new sensory information.

Werner (1987) has emphasized a related point: "One basic fact—self-evident though often not receiving adequate attention—is that all that is accessible to the nervous system are the states of activity of its neurons, in turn giving rise to other states of activity. In this sense, the nervous system is a self-referring system."

In the following sections, "features" are considered predominantly as efficient means of pattern coding and description. However, they can also be seen as interchangeable building blocks used in larger structures, and the isomorphism between the structures might often be more important than the blocks they are made of (a coat is a coat, independent of features such as the texture and color of its fabric). But do we not still need the homunculus to notice the click when the isomorphism is discovered? Maybe, but a machine that can efficiently search for isomorphisms has to have internal data representations that can be linked associatively. Such a machine could play with these data structures, discovering new links and analogies, and might not even care whether anyone listens to the clicking.

BIOLOGICAL VERSUS MACHINE PATTERN RECOGNITION

The Search for Relevant Information

At every instant, a large amount of sensory information is transmitted to the central nervous system of an awake animal. Only a minute fraction of this is relevant in a given behavioral context; the rest is redundant, irrelevant, or noise. A major task of the CNS is to detect *relevant information*.

Relevance in biological pattern recognition is strongly linked to *temporal change*. The underlying strategy is to analyze changes in the environment with the highest priority, to decide whether they require the organism to act. *Temporal change*, and particularly *motion*, is a powerful object feature that is easy to extract; it permits the important dichotomization into moving and nonmoving objects. This strategy greatly reduces the amount of information that has to be processed, and can considerably simplify the demands for shape analysis in lower animals. For the detection of relevance at higher levels, (features, structure, context) as it is performed by the visual and auditory system of humans and higher animals, considerably more sophisticated processing is needed. To understand the basic principles underlying this analysis, comparisons with concepts in computer pattern recognition can be helpful.

Concepts in Machine Pattern Recognition

Although structure and function of biological sensory systems are very different from present hardware and software techniques in computer pattern recognition, many of the basic problems are the same. It is legitimate, therefore, to consider whether approaches that have been successful in machine pattern recognition might be related to some functional aspects of pattern recognition in the CNS. One common link is feature detection (Werner, 1973, 1975). A wide variety of neurons have been found in the CNS that selectively respond to specific stimulus features. Feature extraction is also a fundamental principle in machine pattern recognition, and the reason for this common concept is probably the same: if the classification of complex patterns were done by means of template matching, the task would be impossible because of the astronomical numbers of templates required.[1] One means to achieve a more general pattern description, and therefore a more efficient pattern classification, is to subdivide a pattern into its elements and attributes, and to describe it through these building blocks.[2] Two main approaches in machine pattern recognition use this concept: decision theory (see, e.g., Nilsson, 1965) and structural or syntactic methods (Fu, 1974).

In the decision theory approach, the primary operation is to extract characteristic features of the pattern. These features can be image elements (object domain features), such as geometrical elements, or attributes such as texture, color, and speed, or they can be characteristic parameters in the domain of a suitable transform, such as Fourier spectra. Each pattern is then represented by a "pattern vector," whose components are the (appropriately scaled) features.

In the syntactic approach, patterns are described in terms of structural relations of their components (subpatterns and pattern primitives; see Fu, 1974; Fu and Rosenfeld, 1976). For overviews of pattern recognition methods, see also Watanabe, 1985; Pavlidis, 1977; Duda and Hart, 1973; Bongard, 1970.

The two approaches partially overlap; what are called pattern primitives in one approach might be called features in the other, and the structural relationships of features generally cannot be ignored in the decision theory approach, that is, "higher-order features" must be defined that take into account structural relations. A combination of methods from both approaches has been proposed by Fu (1983).

Decision theory offers an attractive approach to models of pattern recognition in the nervous system since the required operations—up to the classification stage—are well suited for implementation through special-purpose parallel computing networks. Syntactic methods are better suited for software implementation on a general-purpose computer. Higher stages of pattern analysis in the CNS may well be modeled by syntactic descriptions, although until now research has emphasized simpler mechanisms in lower cortical areas. Both methods use the concept of pattern decomposition into elementary descriptors or features.

NEURAL MECHANISMS OF FEATURE EXTRACTION

Feature Extraction in the Image (Object) Domain

The discovery of neurons in the striate cortex that preferentially respond to oriented bars (Hubel and Wiesel, 1959) suggested that early visual image processing might proceed in terms of line elements. Oriented bars can be taken as elementary features. The subsequent discovery of complex and hypercomplex cells (Hubel and Wiesel, 1962, 1965, 1968) seemed to support this model of initial image processing based on bars and edges.

Edge detection and line extraction have been used as image-processing operations in computer vision (Winston, 1975; for literature, see Burns et al., 1986). However, there are also arguments against the assumption that image processing at early stages in the visual system is based on oriented line elements.

Object recognition based on line analysis at early processing stages is not well suited to dealing with scenes in which the objects are embedded in a patterned background, as is usually the case in a natural environment. If object boundaries are not defined by some other means, a line-analyzing algorithm easily gets into nonterminal loops by combining object and background lines. Image analysis, in terms of structural relations of oriented line elements, therefore, would not be an efficient operating principle at the early stages of image processing in the visual system until other objectives are achieved, particularly *region definition and figure–ground separation.*[3]

An important processing operation for region definition is texture analysis; indeed, there are indications that cells with preferential response to oriented bars and edges function primarily as *texture analyzers*. "That analysis of aggregates could precede the analysis of their components" in image processing in the visual system has been emphasized by Werner (1985). A model for initial aggregate analysis (texture filtering) in the visual system is described later. Since V1 and V2 are connected to many other visual areas (for a review, see, e.g., Desimone et al., 1985), cells with orientation-specific responses may act in different functional aspects (texture analysis, line analysis, etc.) in interaction with different areas (see also Chapter 10 for interactions within area V1).

Texture filtering and region definition are very likely performed at early processing levels in the visual system. This hypothesis is supported by receptive field properties and by the arguments derived from computer vision, mentioned previously. At the same time a large body of neurophysiological evidence suggests that this is not the only function of areas V1, V2, and V3. Neurons in these regions do, indeed, respond to a wide variety of specific stimulus features, such as bars or edges of specific size, orientation, speed, direction of movement, and color (Barlow et al., 1967; Hubel and Wiesel, 1959, 1962, 1965; Dow, 1974; Poggio et al., 1975; Schiller et al., 1976; Orban et al., 1981a,b; Zeki, 1980a; DeValois et al., 1982a), stereo disparity (see Chapter 11), and fixation effects (Smith and Marg, 1974). This capacity is far beyond what would be required for texture or line analysis alone. The hypothesis that primary visual areas are involved in texture processing, however, is strengthened by the Fourier domain characteristics of receptive fields of neurons in these areas.

Fourier Transform Processing in the Visual System

Signal encoding and processing efficiency can often be improved, and signal analysis facilitated, if the signal is transformed into a different domain. A wide variety of transforms is available, such as Fourier, Haar, Walsh-Hadamard, and Mellin transforms, to name but a few. Whether or not processing in the domain of a particular transform brings any advantages over processing in the original (object) domain depends on the signals and the type of analysis to be performed. Since the late 1960s the question whether signal processing in some parts of the visual system is in the spatial frequency domain has attracted considerable interest (for a review, see Maffei, 1978). This interest was enforced by the successful application of optical and digital holography and digital filtering techniques to image processing and certain pattern recognition tasks.

Spatial Frequency Channels

Beginning in 1964, Campbell and Robson popularized and greatly extended the use of spatial Fourier techniques as a tool for studying the psychophysics of vision. The large body of evidence about functional principles of the visual system discovered since suggests that certain aspects of image processing in the visual system are based on operations in the spatial frequency domain. In their paper on contrast threshold, Campbell and Robson (1968) reported results on the perception of square-wave patterns that are in agreement with a model based on linear superposition of Fourier components. Blakemore and Campbell (1969) found that adaptation to a square-wave grating raises the threshold for the detection of the fundamental frequency and its third harmonic, as would be predicted by a Fourier model. The McCollough effect (McCollough, 1965), a much longer-lasting threshold aftereffect, is also spatial-frequency-specific (Stromeyer, 1972). Additional evidence comes from a modified Craik-Cornsweet illusion experiment performed by Campbell et al. (1971): a square-wave grating (of limited contrast and spatial frequency lower than 1 c/deg) whose fundamental frequency has been filtered out still appears as if it had square-wave luminance distribution. The visual system seems to have restituted the fundamental frequency. This effect finds an interesting analogy in an acoustic illusion involving a mixture of two or more sinusoidal sounds, with frequencies that are multiples of frequency f ($2f$, $3f$, $4f$ ...), which results in a subject hearing the fundamental frequency f although it is not present in the spectrum (see the section entitled "A Model of Acoustical Pattern Separation").

Neurons with Specific Response to Stimuli in the Spatial Frequency Domain

The Fourier hypothesis was further supported by results from electrophysiological experiments with grating stimuli, which indicate that neurons at various processing stages in the visual system respond selectively to specific ranges of spatial frequencies (Enroth-Cugell and Robson, 1966; Campbell et al., 1969; Maffei and Fiorentini, 1973; Schiller et al., 1976; DeValois et al., 1982b). Initial image-processing operations can thus be described in terms of a local Fourier analysis or, better, of local filtering operations in the spatial frequency domain. Such operations are well suited to the characterization of textures, suggesting that texture and aggregate stimulus analysis occur at early stages in the visual system.

The receptive field properties in the spatial frequency domain are determined by the spatial structure of receptive fields. This structure can be characterized, for example, by DOG (difference of Gaussian) or $\nabla^2 G$ (Gaussian smoothed second derivative) functions (Marr et al., 1979; Marr, 1982). Marcelja (1980) suggested an alternative description in terms of Gabor functions. This hypothesis is interesting, because Gabor filters constitute the optimum in the compromise between spatial resolution and spatial frequency (or time and frequency) resolution (Kulikowski et al., 1982; Daugman, 1983). The Fourier domain transfer characteristics of receptive fields have been implemented in the "difference of low-pass" (DOLP) transform (Crowley and Parker, 1984). The DOLP transform converts an image into a set of bandpass filtered images.

Object Domain versus Fourier Domain Processing

There have been arguments over whether image analysis in the visual system is in the object domain or in the spatial frequency domain (Macleod and Rosenfeld, 1974; DeValois et al., 1978, 1979; Albrecht et al., 1981). The two descriptions, of course, are not mutually exclusive but complementary (Kulikowski and Bishop, 1981; MacKay, 1981; Daugman, 1985; Kulikowski and Kranda, 1986), and the question is rather which domain gives a more transparent description of specific effects. In characterizing the receptive field properties of visual neurons at primary processing stages, both descriptions have their merits.

There remain, however, psychophysical results that are not explainable by the known transfer functions of receptive fields (e.g., the restitution of the fundamental frequency in the modified Craik-Cornsweet experiment; see Campbell et al., 1971). These results are more adequately described in terms of Fourier parameters. This can be seen as an indication that specific processing operations in the Fourier domain are taking place at higher stages in the visual system.

Fourier Domain Mechanisms in the Visual System

Signal analysis in the domain of a given transform can be particularly efficient if the kernels of the transform are eigenfunctions of the differential equations that describe the process by which the signal was generated.[4] Fourier transform processing, therefore, is well suited for the analysis of signals generated by resonant structures, as often occurs with acoustical signals. Neurons that respond to stimuli within a relatively narrow audiofrequency band and neurons that selectively respond to higher-order features in the Fourier domain, such as a specific change in frequency, have been well documented in the auditory cortex (Whitfield and Evans, 1965; Kelly and Whitfield, 1971).

In contrast to acoustical signals, images are generally not produced by harmonic generators. One might argue, therefore, that there is no reason to analyse them in terms of harmonic functions. But images may contain periodic or semiperiodic structures (e.g., texture) and such structures can be described efficiently by their Fourier spectra,[5] although no physical process is associated with the individual harmonic components. Fourier spectra, among other methods, have been used for texture description in machine pattern recognition. As far as the visual system, particularly the cortex, is concerned, the assumption of *periodicity analyzers* is compatible with anatomical and neurophysiological findings; the receptor mosaic, in combination with lateral inhibition or the columnar structure of the cortex, could provide a suitable basis for such a mechanism. The assumption of structures in the visual system that analyze signals in terms of functions that vary *sinusoidally* in space, however, does not seem plausible at first. It is surprising, therefore, that a body of psychophysical evidence supports the assumption that some type of harmonic analysis takes place at higher stages in the visual system. Since the transfer function of receptive fields of collicular and cortical cells (a broad bandpass characteristic) does not explain the spatial frequency specifity of some of the psychophysical results mentioned in the section on Spatial Frequency Channels, what other neural mechanism could make the Fourier domain so well suited to the description of these results?

Reitboeck and Wacker (1986) have shown that structures resembling digital filters, including comb filters, can easily be realized by neural circuits. Such filters require only the summation of signals that are shifted in space or time, and they may contain feedback loops (in which case the filter is said to be recursive); they are thus realizable through neural networks. A model for acoustical pattern separation based on this concept is described later. Similar circuits in the visual system could account for phenomena that lend themselves to descriptions in the Fourier domain.[6]

HIGHER MECHANISM OF PATTERN RECOGNITION IN THE VISUAL SYSTEM

The Generation of Invariances

Position Invariance

Objects must be recognized independent of their exact position in the visual field. In computer vision, a variety of techniques is available to achieve a shift-invariant pattern description. Shift-variant transforms are of particular interest for models of such operations in the CNS, because they are not object-specific and can be implemented by "hard-wired" networks of elementary computing elements. Shift-invariant transforms offer the additional advantage that other invariances can be derived from them.[7] Position-invariant pattern descriptions can be realized by such values as the absolute value of the Fourier transform, the R-transform and its derivatives (Reitboeck and Brody, 1969; Wagh and Kanetkar, 1977; Burkhardt and Mueller, 1980), moments (Hu, 1962; Teague, 1980), autocorrelation functions[8] (Bracewell, 1965), Walsh power spectra (Harmuth, 1972). Models based on such transforms have been proposed for generating shift-invariance in the visual system. The Neocognitron, a model of a self-organizing neural network (Fukushima, 1980), also generates shift-invariance.

Size and Rotation Invariance

Operations that generate size and rotation invariance (even if limited) can substantially reduce the coding and computing efforts required for pattern classification. In computer pattern recognition, scale invariance can be achieved by means of the Mellin transform (Casasent and Psaltis, 1976; West and Reitboeck, 1979). The Mellin transform can be realized through a logarithmic coordinate transform in combination with a Fourier transform[9] (see, e.g., Bracewell, 1965). Within a certain angular range, the retinal projection to the striate cortex of some mammals can be described by a cortical magnification factor (CM) that is inversely proportional to retinal eccentricity.[10] The mapping of visual space to area V1 (Daniel and Whitteridge, 1961; Cowey, 1964; Allman and Kaas, 1971) can thus be approximated by a logarithmic polar coordinate transform. In combination with a shift-invariant transform (localized, e.g., in nonvisuotopic cortical areas), this logarithmic representation of visual space could provide a mechanism for a scale-invariant object description.

Models of size invariance in the visual system based on this concept have been proposed by Cavanagh and Schwartz (for literature, see Cavanagh, 1985; Schwartz, 1985). Various problems associated with these models have been discussed by Cavanagh (1981, 1982), Schwartz (1981, 1983), and Reitboeck and Altmann (1984).

A strict Mellin correlation model requires a global Fourier transform, which is an unlikely operation to be performed in the CNS. Cavanagh (1985) suggested a model for realizing a global transform by summing local log polar frequency transforms. Because of the coarse spatial frequency resolution of cortical cells, this transform would probably not have high shape discrimination. Altmann and Reitboeck (1984) have defined a scale-invariant correlation that does not involve a global Fourier transform and proposed a model for size invariance in the visual system based on a transform that does not require neurons to perform exact arithmetic operations (Reitboeck and Altmann, 1984).

An interesting aspect of Mellin-type invariance models is that a corresponding model for auditory pattern recognition exists (Altes, 1978). The logarithmic mapping of frequencies along the basilar membrane converts frequency scalings into shifts; a subsequent shift-invariant transform results in a pitch-invariant description of auditory signals. A further coincidence in the mapping functions of V1 and along the basilar membrane is the deviation from the logarithmic map to an approximately linear representation. This holds for the foveal region (high spatial frequencies) in the visual system, and for the basilar membrane region close to the helicotrema (frequencies below 500 Hz for humans) in the auditory system. Although these deviations do not support the Mellin models, the coincidence might point to a common functional principle.

A major problem for models of size-invariant processing in primary visual areas is that a scale-invariant correlation gives a poor signal-to-noise ratio, as long as the object has not been separated from background structures. This problem is aggravated if the object is three-dimensional and changes appearance in different projections or if the object can assume a variety of shapes. This, of course, is the usual situation in the real world. Although figure/ground separation could be realized by region marking in the time domain (Reitboeck, 1983a, Reitboeck et al., 1986; see also a Model of Preattentive Texture Recognition and Region Separation section), there is still the problem that in a realistic situation the scale-invariant correlation function would have to be calculated for a very large number of templates. Other arguments remain against the hypothesis that the quasi-logarithmic mapping of visual space onto the striate cortex is functionally significant for size-invariant filtering. Ablation of V1 in the cat does not seriously affect shape recognition (Sprague et al., 1977; Hughes and Sprague, 1986). Although the cat has parallel inputs to V1, V2, and V3, one would expect shape recognition to be affected by ablation of V1 if an essential part of the size-invariance mechanism were located there. Lesions in area V1 severely impair pattern vision in monkey (Mishkin and Ungerleider, 1982). However, since the retinogeniculate inputs pass serially through V1 to higher regions in monkey, this is not an argument that the resulting loss in pattern vision is due to a damage to the size-invariance mechanism.

Pattern Processing in Higher Cortical Regions

Processing in primary cortical areas embodies aspects of both feature extraction and aggregate stimulus analysis. Higher pattern recognition functions, such as the generation of shift and size-invariant object codes, cannot be realized in areas that maintain a topological representation of the visual field. Interest has heightened in research on those extrastriate areas where higher visual functions, including the essential pattern-recognition operations, are likely to take place. The excitability of neurons in the posterior parietal cortex, for example, is influenced by attentive fixation (Mountcastle et al., 1981). Neurons of inferior temporal (IT) cortex have been found to be highly sensitive to stimulus shape (Desimone et al., 1984). IT is considered to be the last area in the cortical system for object recognition (for a review, see Desimone et al., 1985).

SPATIOTEMPORAL ASPECTS OF PATTERN RECOGNITION

Neural Assemblies

Mountcastle has emphasized that the neural correlates of the perceptive process cannot be understood at the level of the function of single cells (Mountcastle, 1966a,b). Higher mechanisms of information processing and perhaps even coding of higher-order features might only be explainable in terms of more global activity patterns within larger groups of functionally coupled neurons (Hebb, 1949; for references, see Reitboeck, 1983c).

Of particular interest are temporal signal correlations, synchronization, and phase locking within such assemblies (Malsburg, 1981; Reitboeck, 1983a; Othmer, 1985; Johannesma et al., 1986; Reitboeck et al., 1986). To study such phenomena, as well as other group activities transcending the receptive field (Nelson, 1985), experimental techniques for multiunit recordings are required.

Multiunit Recording Techniques

Early work on multiunit recording and signal analysis was done by Gerstein (1970). Since that time a number of innovative techniques for multichannel recordings of action potentials as well as field potentials have been reported (for a review, see Reitboeck, 1983c). The electrodes in these arrays, however, are fixed, which left it to chance how many electrodes carry signals suitable for processing. This disadvantage was overcome by the development of arrays of individually movable fiber microelectrodes (Reitboeck and Werner, 1983; Reitboeck, 1983b,c). A major problem in multiunit recordings is the data analysis. Improved techniques for multiunit signal analysis have been developed by Schneider et al. (1983), Eckhorn et al. (1986), and Aertsen et al. (1986).

Temporal Correlations in Neural Pattern Processing

Barlow (1985) noted that "cortical cells are thought to detect, and subsequently signal, 'suspicious coincidences' amongst the sensory and other messages they receive." Correlated activity constitutes such suspicious coincidences. Because of the cohesiveness of objects, the retinal image of a moving object induces correlated activity in groups of visual neurons. *Object movement* is a primary indicator of relevance. Correlated activity could thus be a marker for three important attributes: to define *object regions*, to indicate *movement*, and to signal *relevance*. In addition, it could serve for the linking of different object attributes represented in different cortical areas. Correlated activities are found in neurons that have a high degree of overlap in their coding properties, such as receptive field position, orientation, movement direction, and ocularity (Eckhorn and Reitboeck, 1987; Eckhorn et al., 1987). This correlation is not exclusively a result of common stimulation; it is also enhanced by neural interconnections.

A Model of Preattentive Texture Recognition and Region Separation

Humans and animals can detect objects against a patterned background. Even in situations in which object and background are similar, we can usually perceive the object as a visual entity, distinguished from the background or other overlapping objects. This *object separation* is a prerequisite for subsequent object-related processing and for generating class-invariant object descriptions. The generation of invariances (e.g., shift, size, and rotation invariance) by means of nonlinear transforms, for example, requires that object separation be achieved before the transform can be applied (Reitboeck and Brody, 1969; Reitboeck and Altmann, 1984).

A region is greatly accentuated if it moves in relation to the background. Even borders between regions of identical texture (e.g., random-dot patterns) become visible if one region is shifted against the other (Julesz, 1971; Reichardt and Poggio, 1979; Baker and Braddick, 1982). For the visual system of the fly, Reichardt and Poggio (1979) and Reichardt et al. (1983) have developed a model for figure–ground separation based on such relative movements. Human motion detectors have been analyzed by Grind et al. (1986).

A moving texture region produces correlated or even quasi-synchronous activity in those visual neurons it stimulates. Correlated activity itself could thus be a mechanism in the CNS to highlight something, to make it stand out. Even with no relative movement between the regions, the retinal image is always shifting in relation to the retina because of eye and body movements. This results in temporal variations of the activity of the receptors and of subsequent visual neurons. Spatial texture intervals are thus converted into temporal intervals within spike trains. The discharge pattern of neurons activated by regions of different textures should, therefore, be texture specific.

Based on this concept we have developed a model of preattentive texture discrimination in the visual system (Reitboeck et al., 1986). It is a model of the *preattentive* processes, and no structural texture analysis involving higher cortical mechanisms has been incorporated. Of the various types of eye movements, eye flicks, or slow saccades (because of their speed and angular range), might be the most likely candidate for the transformation of spatial texture intervals into a temporal texture code. In primates, the superior colliculus is considered to be important in generating saccadic eye movements (see Chapter 14). This is consistent with the preattentive nature of the process. For some of the texture pairs we have tested, discrimination is possible for a specific saccade direction only; this would have to be selected through a cortical mechanism. Saccades can be initiated from various visual cortical areas (Robinson and Fuchs, 1969; Schiller, 1985). This model's performance in texture discrimination tasks is in good agreement with psychophysical results and with the results of texture discrimination based on textons (Julesz, 1971, 1981a,b, 1986). Examples of preattentively indiscriminable (Figure 16-1a,c) and discriminable texture pairs (Figure 16-2a,c) are shown in

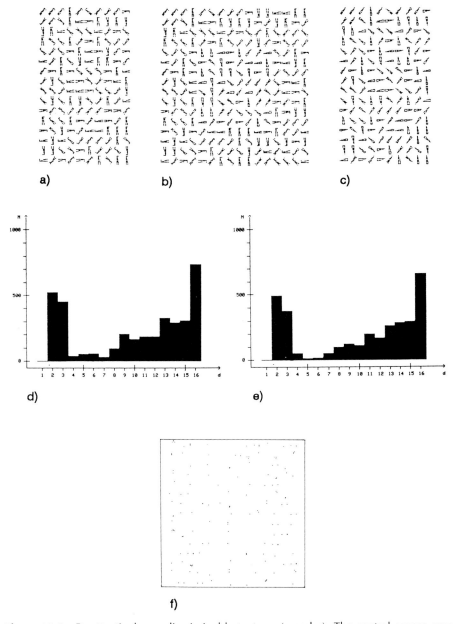

a)

b)

c)

d)

e)

f)

Figure 16-1 Preattentively nondiscriminable textures (a and c). The central square area in b contains texture c, surrounded by texture a. According to our model, the two texture regions cannot be separated, because textures a and b have very similar statistical interval distributions for any scan direction. Interval histograms for a horizontal scan are shown in d and e. The total number, N, of intervals of a given interval length is plotted against the interval length, d. No scan direction would give a characteristic difference in the probability of a particular interval on which region separation could be based. Part f shows the distribution of neural activity in layer N_3 of Figure 16-3 when layer N_2 is driven by a gating sequence of interval length $d = 6$. For textures a and b, no gating sequence would result in a texture-specific difference of activities in layer N_3. Texture region separation is therefore not possible in composites containing regions of textures a and c.

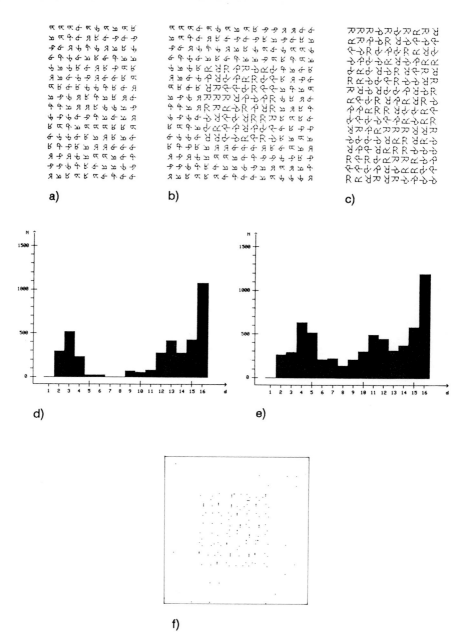

Figure 16-2 Preattentively discriminable textures (*a* and *c*). The central square area in *b* contains texture *c*, surrounded by texture *a*. According to our model, the two texture regions can be separated, because textures *a* and *b* have different statistical interval distributions for at least one scan direction. Interval histograms for a horizontal scan are shown in *d* and *e*. The total number, *N*, of intervals of a given interval length is plotted against the interval length, *d*. Many scan directions give different interval histograms for textures *a* and *c*. Part *f* shows the distribution of neural activity in layer N_3 of Figure 16-3 when layer N_2 is driven by a gating sequence of interval length $d = 6$. Many gating sequences yield distinct, texture-specific differences of activities in layer N_3, and excellent texture region separation can be achieved.

Figures 16-1 and 16-2. The histograms of the indiscriminable textures (Figure 16-1d,e) have a more similar interval distribution than those of the discriminable textures (Figure 16-2d,e).

Discriminable texture pairs contain at least one interval that has significantly different probability in the two textures. The model does not require specific neurons acting as higher-order feature detectors (texton detectors). Texture filtering and region separation can be implemented by simple neural circuits. Another interesting aspect of the model is that it provides a possible mechanism for *attention*. Figure 16-3 shows three neural layers, N_1, N_2, and N_3. N_1 is assumed to be activated by a shifting image containing different texture regions. The neurons in N_2 function essentially like gates (simplified by an array of logic AND gates). If this layer is driven by a pulse sequence that is characteristic for a particular texture, regions containing that texture will preferentially project through to layer N_3. *Directing the attention* at a specific region would mean that the layer N_2 is gated with a characteristic interval sequence of that region. The texture region separation in layer N_3 for the composite patterns in Figures 16-1b and 16-2b is shown in Figures 16-1f and 16-2f. In agreement with psychophysical results, the model can discriminate the patterns in Figure 16-2b, but it cannot discriminate the texture regions of the composite pattern shown in Figure 16-1b. No gating interval (corresponding to a specific interval length in the textures) would generate perceptible regional differences in the point densities of Figure 16-1f.

To perceive an object with different texture regions as an entity, the object

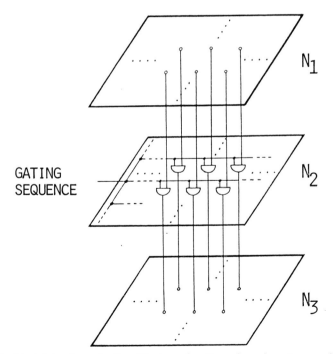

GATING
SEQUENCE

N_1

N_2

N_3

Figure 16-3 Model for the selective filtering of patterns based on temporal signal correlations. For details see the text.

regions are linked in the model, so that they share coherent events. The region linking can be achieved with a hierarchical buildup of correlated activity that modifies and partially overrides the initial texture-induced correlation. It could be induced by feedback from higher processing stages. Reciprocal interconnections between cortical areas and feedback to earlier stages (including the LGN) are well documented in the visual system (Cragg and Ainsworth, 1969; Zeki, 1969, 1980b; Tigges et al., 1973, 1974; Wong-Riley, 1978; Rockland and Pandya, 1979; Lund et al., 1981; Maunsell and van Essen, 1983; Weller and Kaas, 1983; Ungerleider et al., 1983). Correlated activity could also provide a link between aspects of pattern processing in different cortical areas. For details of the model, see Reitboeck et al. (1986).

In the first processing stage of the model, eye movements have been assumed to convert texture intervals into temporal spike intervals. Despite psychophysical evidence that eye movements play an important role in visual pattern recognition, there is also evidence that preattentive texture region separation can be achieved if the textures are shown for only a very short instant. We are presently extending the model to account for this.

A Model of Acoustical Pattern Separation

Related problems exist in acoustical pattern recognition: sounds from a specific source are generally heard against an acoustical background of other sounds. Humans are able to concentrate on a particular sound in such listening situation (selective channel attention), and animals also react to acoustical signals embedded in an acoustical background, even if the signals are well below the background level. It seems that the auditory system can generate a highly specific "matched" filter that can be tuned to a particular signal and follow its variations. What neural mechanism could provide such filter?

Most sounds that are used in acoustical communication are generated by vibrating bodies (including the vocal cords). Because of the physical boundary conditions, such sounds are composed of discrete frequencies that are multiples of a fundamental frequency. A filter with a periodic transfer function (with transmission peaks at f, $2f$, $3f$ ...) with its fundamental frequency tuned to that of a particular sound, would thus act as an acoustical pattern filter for that sound. Such "comb filters" are of particular interest as models of neural filters because they can be realized by summing time-delayed signals. This is the operating principle of *digital filters*. There are two basic types of digital filters: recursive (with feedback loops) and nonrecursive.

We have developed a model of an acoustical pattern filter based on this principle (Reitboeck and Wacker, 1986; Wacker, 1986). The separation of individual components from a mixture of signals is shown in Figure 16-4. Four signals of different fundamental frequency and (harmonic) spectral composition have been added; the contributions of the individual signals are not recognizable in the sum. Filtering almost perfectly separates the individual signals. A nonrecursive digital filter with linear-phase response has been used, to avoid shape distortions caused by phase shifts in the harmonics, so that the original signals are visually recognizable. For the separation of sound sources, a linear-phase response would not

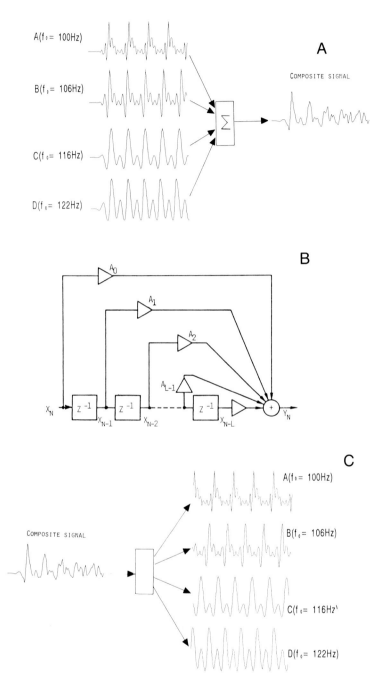

Figure 16-4   To demonstrate signal separation based on selective channel attention, four signals of different fundamental frequency and harmonic content have been added, as shown in *a*. With the digital filter in *b*, the individual signals can be restituted from the composite signal, as shown in *c*. Each stage z^{-1} in the digital filter *b* symbolizes a certain time delay. The time-delayed signals are summed with synaptic coupling strength A_0 to A_{L-1}.

be necessary; the ear is not very sensitive to phase shifts in continuous sounds. The filtered signal, of course, deviates from the original if different signals have harmonics at the same frequency.[11]

Sounds used in communication do change in their fundamental frequency. To track these signals, the filters can be programmed to dynamically follow the changes. Such comb filters can model only the first processing steps of acoustical pattern separation; subsequent processing stages must use higher-order features to keep track of complex, interrupted sound patterns, such as voices. Neural networks that have comb filter characteristics are easily realizable. Such structures act as pattern filters, without necessarily separating individual harmonic components. Digital filters can be a model for basic pattern recognition operations in both the visual and auditory system.

SUMMARY AND OUTLOOK

In this paper I have concentrated mainly on feature extraction. As important as feature extraction is, it does not explain pattern recognition in the nervous system; there is even the danger that we might carry this concept to increasingly higher processing stages until we arrive at neurons that respond to very specific stimuli only, which we might regard as the long sought-after templates. Pattern recognition in the CNS cannot be explained in terms of template matching—even if the word is used now in the broader sense of templates made more and more specific for one pattern class and tolerant of variations within that class.

What sequence of transforms is needed to create the template "dog"—invariant for all breeds of dogs, independent of the many positions a dog can assume, and independent of all the distances and angles from which the dog can be seen, and even of the hat and jacket somebody might put on him? There is no such sequence of transforms, and even the features are misleading.

The power of the nervous system lies in its ability to form associative links. Associative linking enables the CNS to detect isomorphisms. This capacity is developed to the highest (and most abstract) stage in humans. As we go down the ladder to lower animals, the capacity of shape abstraction disappears until only reaction to trigger "features" remains. What we must search for to understand the essential mechanisms of pattern recognition are concepts of how associative links are established and how the nervous system detects and signals isomorphisms.

Syntactic pattern recognition methods are well suited to generating such concepts, but intuitively I believe that the essential mechanisms in pattern recognition in the CNS are "hard-wired." Syntactic pattern recognition is not well suited for such an implementation. What other concepts could be used in the CNS to establish associative links and to signal isomorphism? An abstract marker par excellence for establishing degrees of "relatedness" (even between different sensory modalities) and for signaling degrees of "similarity" (even between objects whose appearances have little or nothing in common) is the degree of temporal signal correlation. In recent years, important discoveries have been made in the visual system about the W-, X-, Y-pathways, intracortical connections, and the topography of visual functions (see Chapters 10 and 12). The reciprocal connections that seem to be a general rule in the functional organization of the CNS could provide the means of establishing temporal signal correlations, as discussed in this chapter.

ACKNOWLEDGMENTS

I am grateful to my colleagues Drs. R. Eckhorn and J. I. Nelson for their valuable discussions and suggestions and for pointing out several literature citations. Thanks are due also to Ms. S. van Rennings for her patience in our struggle with the word processor.

NOTES

1. The number of possible patterns on a coarse black and white television screen is many orders of magnitude larger than the estimated number of elementary particles in the universe.
2. The efficiency of this approach for representing words is illustrated by a comparison of phonetic and ideographic scripts.
3. There is yet no evidence either for the existence of such higher-order feature detectors in the visual system that would respond selectively to combinations of lines or bars as one might expect in a hierarchical line-analysis scheme.
4. The kernels of the Fourier transform
$$\exp(-i\,\Omega t) = \cos \Omega t - i \sin \Omega t$$
are sinus functions. Sinus functions are also eigenfunctions of the differential equations of oscillating bodies, from which the solution of the differential equation for specific boundary conditions can be constructed.
5. Periodic structures are redundant. If such structures are described in the domain of a transform that has periodic kernels, the coding can be more efficient.
6. Another mechanism that supports the Fourier domain description of image-processing aspects in the visual system has been suggested by Eckhorn (personal communication): images are sampled by the receptors at discrete intervals; image processing in the visual system, therefore, is linked to Fourier domain descripttions by Shannon's sampling theorem. Although sampled signals can, of course, be analyzed in terms of many other transforms, certain effects, such as aliasing, are best described in the Fourier domain. (Aliasing occurs if a signal is sampled at intervals larger than half the period of the highest frequency in its spectrum. It produces new frequencies in the sampled signal that have not been present in the original signal. See, e.g., Brigham, 1974.)
7. Size and rotation invariance can be realized through a logarithmic polar coordinate transform, in combination with a subsequent shift-invariant transform. An object description that is independent of individual scale factors in two coordinate directions can be achieved by a logarithmic transform followed by separate shift-invariant transforms in x and y. With syntactic methods a position-, size-, and rotation-invariant pattern description can be generated too. It is unlikely, however, that a syntactic pattern description is used at the early processing stages in the visual system.
8. Autocorrelation is an important general concept in sensory information processing (Reichardt, 1961).
9. The logarithm converts size scalings (multiplication with a scale factor) into translations (addition of the log of the scale factor). Of course, these shifted images are logarithmically distorted. A subsequent shift-invariant transform thus results in a size-

invariant pattern description. This method works only if the original images are always presented in a specific position (e.g., by directing the gaze) or if an additional shift-invariant transform has been applied to the image before the logarithmic transform was done.

10. In primates, this holds for extrafoveal regions up to visual angles of about 20 degrees; the deviation at the fovea might be smaller than was thought previously (Dow et al., 1981). In the cat the mapping function is more complex (Epstein, 1984).

11. In the human auditory system, the acoustical background is not completely suppressed, either, when we concentrate on a particular sound.

REFERENCES

Aertsen, A., Gerstein, G., and Johannesma, P. (1986). From neuron to assembly: neuronal organization and stimulus representation. In: G. Palm and A. Aertsen (eds.), *Brain Theory*, pp. 8–24, Berlin: Springer-Verlag.

Albrecht, D. G., De Valois, R. L., and Thorell, L. G. (1981). Visual cortical neurons: are bars or gratings the optimal stimuli? *Science* 207:88–90.

Allman, J. M., and Kaas, J. H. (1971). Representation of the visual field in striate and adjoining cortex of the owl monkey (Aotus trivirgatus). *Brain Res.* 35:89–106.

Altes, R. A. (1978). The Fourier-Mellin transform and mammalian hearing. *J. Acoust. Soc. Am.* 63:174–183.

Altmann, J., and Reitboeck, H. J. (1984). A fast correlation method of scale- and translation-invariant pattern recognition. *IEEE Trans. PAMI* 6:46–57.

Baker, C. L., Jr., and Braddick, O. J. (1982). Does segregation of differently moving areas depend on relative or absolute displacement? *Vision Res.* 22:851–856.

Barlow, H. B. (1985). Cerebral cortex as model builder. In: D. Rose and V.G. Dobson (eds.). *Models of the Visual Cortex*, pp. 37–46, Chichester: John Wiley.

Barlow, H. B., Blakemore, C., and Pettigrew, J. D. (1967). The neural mechanism of binocular depth discrimination. *J. Physiol. (Lond.)* 193:327–342.

Blakemore, C., and Campbell, F. W. (1969). On the existence of neurons in the human visual system selectively sensitive to the orientation and size of retinal images. *J. Physiol.* 203:237–260.

Bongard, M. (1970). *Pattern Recognition*. Washington, DC: Spartan Books.

Bracewell, R. (1965). *The Fourier Transform and Its Applications*. New York: McGraw-Hill.

Brigham, E. O. (1974). *The Fast Fourier Transform*. Englewood Cliffs, NJ: Prentice-Hall.

Burkhardt, H., and Mueller, X. (1980). On invariant sets of a certain class of fast translation-invariant transforms. *IEEE Trans. ASSP* 28:517–523.

Burns, B. J., Hanson, A. R., and Riseman, E. M. (1986). Extracting straight lines. *IEEE Trans. PAMI* 8:425–455.

Campbell, F. W., and Robson, J. G. (1968). Application of Fourier analysis to the visibility of gratings. *J. Physiol.* 197:551–566.

Campbell, F. W., Cooper, F. G., and Enroth-Cugell, C. (1969). The spatial selectivity of the visual cells of the cat. *J. Physiol. (Lond.)* 203:223–235.

Campbell, F. W., Howell, E. R., and Robson, J. G. (1971). The appearance of gratings with and without the fundamental Fourier component. *J. Physiol. (Lond.)* 217:17–18.

Casasent, D., and Psaltis, D. (1976). Scale invariant optical correlation using Mellin transforms. *Opt. Commun.* 17:59–63.

Cavanagh, P. (1981). Size invariance: reply to Schwartz. *Perception* 10:469–474.

Cavanagh, P. (1982). Functional size invariance is not provided by the cortical magnification factor. *Vision Res.* 22:1409–1412.

Cavanagh, P. (1985). Local log polar frequency analysis in the striate cortex as a basis for size and orientation invariance. In: D. Rose and V. G. Dobson (eds.). *Models of the Visual Cortex,* pp. 85–95. Chichester: John Wiley.

Cowey, A. (1964). Projection of the retina onto striate and prestriate cortex in the squirrel monkey (Saimiri sciureus). *J. Neurophysiol.* 27:366–393.

Cragg, B. G., and Ainsworth, A. (1969). The topography of the afferent projections in the circumstriate visual cortex of the monkey studied by the Nauta method. *Vision Res.* 9:733–747.

Crowley, J. L., and Parker, A. C. (1984). A representation for shape based on peaks and ridges in the difference of low-pass transform. *IEEE Trans. PAMI* 6:156–170.

Daniel, P. M., and Whitteridge, D. (1961). The representation of the visual field on the cerebral cortex in monkeys. *J. Physiol.* 159:203–221.

Daugman, J. G. (1983). Six formal properties of two-dimensional anisotropic visual filters: structural principles and frequency/orientation selectivity. *IEEE Trans. SMC* 13:882–887.

Daugman, J. G. (1985). Representational issues and local filter models of two-dimensional spatial visual encoding. In: D. Rose and V. G. Dobson (eds). *Models of the Visual Cortex,* pp. 96–107. Chichester: John Wiley.

Desimone, R., Albright, T. D., Gross, C. G., and Bruce, C. (1984). Stimulus-selective properties of inferior temporal neurons in the macaque. *J. Neurosci.* 4:2051–2052.

Desimone, R., Schein, S. J., Moran, J., and Ungerleider, L. G. (1985). Contour, color and shape analysis beyond the striate cortex. *Vision Res.* 25:441–452.

DeValois, R. L., Albrecht, D. G., and Thorell, L. G. (1978). Cortical cells: bar and edge detectors, or spatial frequency filters? In: S. J. Cool and E. L. Smith (eds.). *Frontiers and Visual Science,* pp. 544–556. New York: Springer-Verlag.

DeValois, K. K., DeValois, R. L., and Yund, E. Y. (1979). Responses of striate cortex cells to grating and checkerboard patterns. *J. Physiol.* 291:483–505.

DeValois, R. L., Yund, E. W., and Hepler, N. (1982a). The orientation and direction selectivity of cells in macaque visual cortex. *Vision Res.* 22:531-544.

DeValois, R. L., Albrecht, D. G., and Thorell, L. G. (1982b). Spatial frequency selectivity of cells in macaque visual cortex. *Vision Res.* 22:545–559.

Dow, B. M. (1974). Functional classes of cells and their laminar distribution in monkey visual cortex. *J. Neurophysiol.* 37:927–946.

Dow, B. M., Snyder, A. Z., Vautin, R. G., and Bauer, R. (1981). Magnification factor and receptive field size in foveal striate cortex of monkey. *Exp. Brain Res.* 44:213–228.

Duda, R. O., and Hart, P. E. (1973). *Pattern Classification and Scene Analysis.* New York: Wiley-Interscience.

Eckhorn, R., and Reitboeck, H. J. (1987). Assessment of cooperative firing in groups of neurons: special concepts for multi-unit recordings from the visual system. In: E. Basar (ed.). *Springer Series in Brain Dynamics,* Berlin: Springer-Verlag. vol. 1, pp. 219 227.

Eckhorn, R., Schneider, J., and Keidel, R. (1986). Real-time covariance computer for cell assemblies is based on neuronal principles. *J. Neurosci. Meth.* 18:371–383.

Eckhorn, R., Bauer, R., Jordan, W., Epping, W., and Arndt, M. (1987). Correlated activity and coding properties of neuron groups within vertical columns of the visual cortex: a multi-electrode study. IBRO Second World Congress of Neuroscience, Budapest.

Enroth-Cugell, C., and Robson, J. G. (1966). The contrast sensitivity of retinal ganglion cells of the cat. *J. Physiol.* (Lond.) 187:517–552.

Epstein, L. I. (1984). An attempt to explain the differences between the upper and lower halves of the striate cortical map of the cat's field of view. *Biol. Cybern.* 49:175–177.

Fu, K. S. (1974). *Syntactic Methods in Pattern Recognition.* London: Academic Press.

Fu, K. S. (1983). A step towards unification of syntactic and statistical pattern recognition. *IEEE Trans. PAMI* 5:200–205.

Fu, K. S., and Rosenfeld, A. (1976). Pattern recognition and image processing. *IEEE Trans. Comput.* C-25:1336–1345.

Fukushima, K. (1980). Neocognitron: a self-organizing neural network model for a mechanism of pattern recognition unaffected by shift in position. *Biol. Cybern.* 36:193–202.

Gerstein, G. L. (1970). Functional association of neurons: detection and interpretation. In: F. O. Schmitt (ed.). *The Neuroscience, Second Study Program*, p. 648. New York: Rockefeller University Press.

Grind, W. A. van de, Koenderink, J. J., and van Doorn, A. J. (1986). The distribution of human motion detector properties in the monocular visual field. *Vision Res.* 26:797–810.

Harmuth, H. F. (1972). *Transmission of Information by Orthogonal Functions.* Berlin: Springer-Verlag.

Hebb, D. O. (1949). *The Organization of Behavior: A Neuropsychological Theory.* New York: John Wiley & Sons.

Hu, M. K. (1962). Visual pattern recognition by moment invarants. *IRE Trans. Inform. Theory* II-8:179–187.

Hubel, D. H., and Wiesel, T. N. (1959). Receptive fields of single neurons in the cat's striate cortex. *J. Physiol.* (Lond.) 148:574–579.

Hubel, D. H., and Wiesel, T. N. (1962). Receptive fields, binocular interaction and functional architecture in the cat's visual cortex. *J. Physiol.* 160:106–154.

Hubel, D. H., and Wiesel, T. N. (1965). Receptive fields and functional architecture in two nonstriate visual areas (18 and 19) of the cat. *J. Neurophys.* 28:229–289.

Hubel, D. H., and Wiesel, T. N. (1968). Receptive fields and functional architecture of monkey striate cortex. *J. Physiol.* (Lond.) 195:215–243.

Hughes, H. C., and Sprague, J. M. (1986). Cortical mechanisms for local and global analysis of visual space in the cat. *Exp. Brain Res.* 61:332–354.

Johannesma, P., Aertsen, A., van den Boogaard, H., Eggermont, J., and Epping, W. (1986). From synchrony to harmony: ideas on the function of neural assemblies and on the interpretation of neural synchrony. In: G. Palm and A. Aertsen (eds.). *Brain Theory*, pp. 25–47. Berlin: Springer-Verlag.

Julesz, B. (1971). *Foundations of Cyclopean Perception.* Chicago: University of Chicago Press.

Julesz, B. (1981a). Textons, the elements of texture perception, and their interactions. *Nature* 290:91–97.

Julesz, B. (1981b). A theory of preattentive texture discrimination based on first-order statistics of textons. *Biol. Cybern.* 41:131–138.

Julesz, B. (1986). Texton gradients: the texton theory revisited. *Biol. Cybern.* 54:245–251.

Kelly, J. B., and Whitfield, I. C. (1971). Effects of auditory cortical lesions on discriminations of rising and falling frequency-modulated tones. *J. Neurophysiol.* 34:802–816.

Kulikowski, J. J., and Bishop, P. O. (1981). Fourier analysis and spatial representation in the visual cortex. *Experientia* 37:160–163.

Kulikowski, J. J., and Kranda, K. (1986). Image analysis performed by the visual system: feature versus Fourier analysis and adaptable filtering. In: J. D. Pettigrew et al. (eds.). *Visual Neuroscience*, pp. 381–404. Cambridge: Cambridge University Press.

Kulikowski, J. J., Marcelja, S., and Bishop, P. O. (1982). Theory of spatial position and spatial frequency relations in the receptive fields of simple cells in the visual cortex. *Biol. Cybern.* 43:187–198.

Lund, J. S., Hendrickson, A. E., Ogren, M. P., and Tobin, E. A. (1981). Anatomical organization of primate visual cortex area V2. *J. Comp. Neurol.* 202:19–45.

Macleod, I. D. G., and Rosenfeld, A. (1974). The visibility of gratings: spatial frequency channels or bar detecting units? *Vision Res.* 14:909–915.

MacKay, D. (1981). Strife over visual cortical function. *Nature* 289:117–118.

Maffei, L. (1978). Spatial frequency channels: neural mechanisms. In: *Handbook of Sensory Physiology*, Vol. VIII, pp. 39–66. Berlin: Springer-Verlag.

Maffei, L., and Fiorentini, A. (1973). The visual cortex as a spatial frequency analyzer. *Vision Res.* 13:1255–1267.

Malsburg, C. von der (1981). The correlation theory of brain function. Internal report 81-2, Dept. of Neurobiology. Goettingen: Max-Planck-Institute for Biophysical Chemistry.

Marcelja, S. (1980). Mathematical description of the responses of simple cortical cells. *J. Opt. Soc. Am.* 70:1297–1300.

Marr, D. (1982). *Vision*. San Francisco: Freeman.

Marr, D., Poggio, T., and Ullmann, S. (1979). Bandpass channels, zero crossings, and early visual information processing. *J. Opt. Soc. Am.* 69:914–916.

Maunsell, H. R., and Van Essen, D. C. (1983). The connections of the middle temporal visual area (MT) and their relationship to a cortical hierarchy in the macaque monkey. *J. Neurosci.* 3:2563–2586.

McCollough, C. (1965). Color adaptation of edge-detectors in the human visual system. *Science* 149:1115–1116.

Mishkin, M., and Ungerleider, L. G. (1982). Contribution of striate inputs to the visuospatial functions of parietopreoccipital cortex in monkeys. *Behav. Brain Res.* 6:57–77.

Mountcastle, V. B. (1966a). The neural replication of sensory events in the somatic afferent system. In: J. C. Eccles (ed.). *Brain and Conscious Experience*, p. 85. Berlin: Springer-Verlag.

Mountcastle, V. B. (1966b). The functional meaning of specific and non-specific systems. In: J. Eccles (ed.). *Brain and Conscious Experience*, p. 548. Berlin: Springer-Verlag.

Mountcastle, V. B., Andersen, R. A., and Motter, B. C. (1981). The influence of attentive fixation upon the excitability of the light-sensitive neurons of the posterior parietal cortex. *J. Neurosci.* 1:1218–1235.

Nelson, J. I. (1985). The cellular basis of perception. In: D. Rose, and V. G. Dobson (eds.). *Models of the Visual Cortex*, pp. 108–122. Chichester: John Wiley.

Nilsson, N. J. (1965). *Learning Machines*. New York: McGraw-Hill.

Orban, G. A., Kennedy, H., and Maes, H. (1981a). Response to movement of neurons in areas 17 and 18 of the cat: velocity sensitivity. *J. Neurophysiol.* 45:1043–1058.

Orban, G. A., Kennedy, H., and Maes, H. (1981b). Response to movement of neurons in areas 17 and 18 of the cat: direction selectivity. *J. Neurophysiol.* 45:1059–1073.

Othmer, H. G. (1985). Synchronization, phase-locking and other phenomena in coupled cells. In: L. Rensing, and N. I. Jaeger (eds.). *Temporal Order,* pp. 130–143. Berlin: Springer-Verlag.

Pavlidis, T. (1977). *Structural Pattern Recognition.* New York: Springer-Verlag.

Poggio, G. F., Baker, F. H., Mansfield, R. J. W., Sillito, A., and Grigg, P. (1975). Spatial and chromatic properties of neurons subserving foveal and parafoveal vision in rhesus monkey. *Brain Res.* 100:25–59.

Reichardt, W. (1961). Autocorrelation, a principle for the evaluation of sensory information by the central nervous system. In: W.A. Rosenblith (ed.). *Sensory Communication,* pp. 303–317. Cambridge: MIT Press.

Reichardt, W., and Poggio, T. (1979). Figure-ground discrimination by relative movement in the visual system of the fly. Part I: experimental results. *Biol. Cybern.* 35:81–100.

Reichardt, W., Poggio, T., and Hausen, K. (1983). Figure-ground discrimination by relative movement in the visual system of the fly. Part II: towards the neural circuitry. *Biol. Cybern.* 46:1–30.

Reitboeck, H. J. (1983a). A multi-electrode matrix for studies of temporal signal correlations within neural assemblies. In: E. Basar, H. Flohr, H. Haken, and A.J. Mandell (eds.). *Synergetics of the Brain,* pp. 174–181. Berlin: Springer-Verlag.

Reitboeck, H. J. (1983b). Fiber microelectrodes for electrophysiological recordings. *J. Neurosci. Meth.* 8:249–262.

Reitboeck, H. J. (1983c). A 19-channel matrix drive with individually controllable fiber microelectrodes for neurophysiological applications. *IEEE Trans. SMC* 13:676–683.

Reitboeck, H. J., and Altmann, J. (1984). A model for size- and rotation-invariant pattern processing in the visual system. *Biol. Cybern.* 51:113–121.

Reitboeck, H. J., and Brody, T. P. (1969). A transformation with invariance under cyclic permutation for applications in pattern recognition. *Inf. Control* 15:130–154.

Reitboeck, H. J., and Wacker, H. (1986). A model of an acoustical pattern filter in the auditory system. III. Symposium Proceedings *"Music-Brain-Computer,"* H. Petsche and G. Gruber (eds.). Vienna, p. 30.

Reitboeck, H. J., and Werner, G. (1983). Multi-electrode recording system for the study of spatio-temporal activity patterns of neurons in the central nervous system. *Experientia* 19:339–341.

Reitboeck, H. J., Pabst, M., and Eckhorn, R. (1986). Texture description in the time domain. Presented at Computer simulation in brain science. Copenhagen. To appear in Rodney M. J. Cotterill (ed.). Computer simulation in brain science. *Phys. Scripta.*

Robinson, D. A., and Fuchs, A. F. (1969). Eye movements evoked by stimulation of frontal eye fields. *J. Neurophysiol.* 32:637–648.

Rockland, K. J., and Pandya, D. N. (1979). Laminar origins and terminations of cortical connections of the occipital lobe in the rhesus monkey. *Brain Res.* 179:3–20.

Schiller, P. H. (1985). A model for the generation of visually guided saccadic eye movements. In: D. Rose and V. G. Dobson (eds.). *Models of the Visual Cortex,* pp. 62–70. Chichester: John Wiley.

Schiller, P. H., Finlay, B. L., and Volman, S. F. (1976). Quantitative studies of single-cell properties in monkey striate cortex. *J. Neurophysiol.* 39:1288–1351.

Schneider, J., Eckhorn, R., and Reitboeck, H. (1983). Evaluation of neuronal coupling dynamics. *Biol. Cybern.* 46:129–134.

Schwartz, E. L. (1981). Cortical anatomy, size invariance, and spatial frequency analysis. *Perception* 10:455–468.

Schwartz, E. L. (1983). Cortical mapping and perceptual invariance: a reply to Cavanagh. *Vision Res.* 23:831–835.

Schwartz, E. L. (1985). Local and global functional architecture in primate striate cortex: outline of a spatial mapping doctrine for perception. In: D. Rose and V.G. Dobson (eds.). *Models of the Visual Cortex*, pp. 146–156. Chichester: John Wiley.

Smith, J. D., and Marg, E. (1974). Zoom neurons in visual cortex: receptive field enlargement with near fixation in macaques. *Experientia* 31:323–332.

Sprague, J. M., Leby, J., DiBerardino, A., Berlucchi, G. (1977). Visual cortical areas mediating form discrimination in the cat. *J. Comp. Neurol.* 172:441–488.

Stromeyer, C. F. (1972). Edge-contingent color after effects: spatial frequency specificity. *Vision Res.* 12:717–733.

Teague, M. R. (1980). Image analysis via the general theory of moments. *J. Opt. Soc. Am.* 70:920–930.

Tigges, J., Spatz, W. B., and Tigges, M. (1973). Reciprocal point-to-point connections between parastriate and striate cortex in the squirrel monkey (Saimiri). *J. Comp. Neurol.* 148:481–490.

Tigges, J., Spatz, W. B., and Tigges, M. (1974). Efferent cortico-cortical fiber connections of area 18 in the squirrel monkey (Saimiri). *J. Comp. Neurol.* 158:216–236.

Ungerleider, L. G., Gattass, R., Sousa, A. P. B., and Mishkin, M. (1983). Projections of area V2 in the macaque. *Soc. Neurosci. Abstr.* 9:152.

Wacker, H. (1986). A digital filter model for the separation of sound sources by the auditory system (in German). M. S. thesis, Marburg, West Germany.

Wagh, M. D., and Kanetkar, S. V. (1977). A class of translation invariant transforms. *IEEE Trans. ASSP* 25:203–205.

Watanabe, S. (1985). Pattern recognition: human and mechanical. New York: John Wiley & Sons.

Weller, R. E., and Kaas, J. H. (1983). Retinotopic patterns of connections of area 17 with visual areas V-II and MT in macaque monkeys. *J. Comp. Neurol.* 220:253–279.

Werner, G. (1973). Neural information processing with stimulus feature extractors. In: F. Schmitt and F. Worden (eds.). *The Neurosciences, Third Study Program*, Cambridge: MIT Press.

Werner, G. (ed.) (1975). *Feature Extraction by Neurons and Behavior*. Cambridge: MIT Press.

Werner, G. (1987). The many faces of neuro-reductionism. In: E. Basar (ed.). *Dynamics of Sensory and Cognitive Processing in the Brain*, Berlin: Springer-Verlag.

Werner, G. (1985). Conceptual relations between sensory neurophysiology and cognition. Paper, presented at the International Conference on Dynamics of Sensory and Cognitive Processing in the Brain, Berlin, 1985.

West, G., and Reitboeck, H. J. (1979). Zur ähnlichkeitsinvarianten Mustererkennung mittels der Fourier-Mellin-Transformation. Elektron. *Informationsverarb. u. Kybernetik* 15:507–512.

Whitfield, I. C., and Evans, E. F. (1965). Responses of auditory cortical neurons to stimuli of changing frequency. *J. Neurophysiol.* 28:655–672.

Winston, P. H. (ed.) (1975). *The Psychology of Computer Vision*. New York: McGraw-Hill.

Wong-Riley, M. (1978). Reciprocal connections between striate and prestriate cortex in squirrel monkey as demonstrated by combined peroxidase histochemistry and autoradiography. *Brain Res.* 147:159–164.

Zeki, S. M. (1969). Representation of central visual fields in prestriate cortex of monkey. *Brain Res.* 14:271–291.

Zeki, S. M. (1980a). The representation of colours in the cerebral cortex. *Nature* 284:412–418.

Zeki, S. M. (1980b). A direct projection from area VI to area V3A of rhesus monkey visual cortex. *Proc. R. Soc. Lond. B* 207:499–506.

17

Network Model for Computing Surface Curvature from Shaded Images

SIDNEY R. LEHKY AND TERRENCE J. SEJNOWSKI

As artists well know, the continuous gradations of shading within an image can be used to convey information about the three-dimensional shape of its surface. One of the problems with using shading information, however, is that the pattern of light and dark formed by light reflected from the surface depends on the direction of illumination as well as the surface shape (Ikeuchi and Horn, 1981). Nevertheless, our visual system is somehow able to separate these variables and extract shape information from shaded images (Todd and Mingolla, 1983; Mingolla and Todd, 1986). To investigate how visual cortex is organized to do this, we constructed a computer model of a neural network capable of extracting shape parameters, specifically the local surface curvatures, from the shaded images of simple geometrical surfaces.

Our primary interest was in the properties of the receptive fields in such a network, and how they compare with receptive fields actually found in visual cortex (Hubel and Wiesel, 1962). The general finding is that they are surprisingly similar to those observed biologically, and we conclude that neurons that can extract surface curvature can have receptive fields similar to those previously interpreted as bar or edge detectors. This highlights the difficulty in understanding what a neuron is doing simply by measuring its receptive field. The receptive field of a sensory neuron is necessary but not sufficient to determine its function within a network. We emphasize the importance in determining a neuron's function of its "projective field," that is, the output pattern of connections the neuron makes with other cells.

GOALS

As just stated, the purpose of this network was to extract surface curvatures from the shaded images of three-dimensional surfaces. In particular, the network was to extract the principal curvatures and directions. In differential geometry *curvature* is the rate of change of the surface normal as one traces an arc along the surface. In general the value of curvature depends on the *direction* one travels on the surface. In one direction curvature is maximum, and minimum in another. The maximum and minimum curvatures at a particular point on the surface are called the principal curvatures, and their directions, or orientations, along the surface are called the principal directions. A theorem states that the principal curvatures are always oriented at right angles to each other. If the principal curvatures are both positive, the surface is convex, or mound-shaped, and if they are negative, the surface is concave, or bowl-shaped. Principal curvatures were selected as the parameter of interest for the network because they are a relatively robust indicator of shape. Unlike the normal vector to the surface (Ikeuchi and Horn, 1981), the value of surface curvature is independent of rotations or translations of the surface (Pentland, 1984). It should be noted that the problems involved in extracting three-dimensional curvature from a surface are different from those extracting two-dimensional curvature from a contour, which have been studied physiologically by Dobbins et al. (1987).

Only simple geometrical surfaces were presented to the network. These were elliptical paraboloids, which are elliptical in cross section and parabolic in depth. These simple surfaces can be considered local approximations to small patches in a more complex image. The network presented here is therefore meant to correspond to a small circuit receiving input from only a very limited region of visual field, perhaps a region served by a single cortical column. To generate descriptions of complex images, this network would have to be replicated at different spatial locations to cover the entire visual field, and also at different spatial scales, which would all feed into higher-level networks that integrated local curvatures into more general shape descriptions, such as those suggested by Pentland (1986).

The images of elliptical paraboloids that were used could be characterized by the following parameters:

1. principal curvatures (two parameters);
2. principal orientation (one parameter);
3. illumination direction (two parameters);
4. position of surface within overall input field of the network (two parameters).

The task we desired the network to perform was to extract both the principal curvatures, and the orientation of the smaller of the two principal curvatures relative to horizontal (i.e., orientation of the elongated axis of the surface), independently of illumination direction and the position of the surface. (Recall that the orientation of the other principal curvature must be at right angles to the first, and is therefore not an independent parameter.) It should be kept in mind that surfaces also have other properties, such as the orientation of the surface normal to the plane of the image, that are not determined by the present network.

METHODS

The network model had three layers of processing units: an input unit layer, an output unit layer, and a hidden unit layer between them. Each unit connected to every unit in the subsequent layer, but there were no connections between units in the same layer. The overall organization of this trilayer network is illustrated in Figure 17-1a. In the network described here, there were 122 input units, 27 hidden units, and 24 output units. The activities of the units of the input layer were determined by the image and preprocessing, as described later. The activity level of each unit on the subsequent layers was determined by linearly summing all its inputs, weighted by synaptic strengths, and then passing the results through a sigmoid nonlinearity (see Figure 17-4). The activities of the units could assume any value between 0.0 and 1.0. These processing units have only a few of the properties of cortical neurons, such as synaptic integration and a nonlinear firing rate function, and do not consider any dynamic, or time-varying, properties of the nervous system. However, although not suitable for studying the dynamic aspects of cortical processing, a network of such units may provide useful insights into the static properties of visual representations, which is the interest here.

We used the back-propagation learning algorithm (Appendix 1) as a design technique for constructing a network with the desired characteristics (Rumelhart et al., 1986). Briefly, the network was presented with many sample images. For each presentation, responses were propagated up through the three layers to the output units. The actual responses of the output units were then compared with what the output for that image should have been. Based on this difference, synaptic weights throughout the network were slightly modified to reduce error, start-

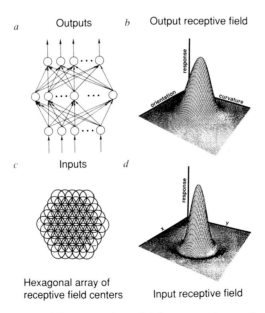

Figure 17-1 Organization of the network model for computing surface curvature from shaded images. (a) Schematic feedforward network with three layers of units. (b) Example of an output receptive field. The response of an output unit is maximum for some combination of values of the principal orientation and a principal curvature, which was different for each output unit. (c) The receptive field centers of the input units overlapped, forming a hexagonal array (Wassle et al., 1981). (d) Example of the receptive field of an on-center input unit. The responses of the units on the input layer were determined by convolution of the Laplacian receptive field with the image.

ing with synapses at the output layers themselves and then moving back down through the network (hence the term back-propagation). After thousands of image presentations, the initially random synaptic weights organized themselves into a set of receptive fields that provided the correct input/output transfer function. The back-propagation algorithm was used in this study purely as a formal technique for constructing a network with a particular set of properties. No claims are made about the biological significance of the process by which the network was created; the focus of interest is rather on the properties of the mature network.

The response properties of both the input units and output units were predefined for the network, based both on what operations we wanted the network to perform and on constraints involving biological plausibility. It was the task of the learning algorithm to organize the properties of the hidden units so as to provide the correct transfer function between the input and output units.

The input layer consisted of two hexagonal spatial arrays of units, called on-center units and off-center units, by analogy with similar terminology used for retinal ganglion cells. One such array is shown in Figure 17-1c. The two input arrays (off-center and on-center) were superimposed on each other, so that each point of the image was sampled by both types. Each of these arrays consisted of 61 units, for a total of 122 units in the input layer. The receptive field of each input unit was the Laplacian of a two-dimensional Gaussian, or in other words, the classic circularly symmetric center-surround receptive field found in the retina and lateral geniculate nucleus (Figure 17-1d). The receptive fields were extensively overlapped within the array, as shown in Figure 17-1c. The responses of each input unit to an image was determined simply by convolving the image with the unit's receptive field.

The output layer consisted of units that coded both the curvature and the orientation. The tuning curves for these units have the form of a two-dimensional Gaussian curve, as shown in Figure 17-1b; that is, they have receptive fields tuned to a local region in a two-dimensional curvature-orientation parameter space (in contrast with the input units, which are tuned to a local region of geometrical space). Such a multidimensional response is typical of those found in cells of the cerebral cortex, although cells responding specifically to curvature have not been demonstrated.

Unfortunately, a problem with this sort of nonmonotonic, multidimensional response is that the signal from a single unit is ambiguous. An infinite number of combinations of curvature and orientation give an identical response. The way to overcome this ambiguity is to have the desired value represented in a distributed fashion, by the joint activity of a population of broadly tuned units in which the units have overlapping receptive fields in the relevant parameter space (in this case curvature and orientation). The most familiar example of this kind of distributed representation is found in color vision. The responses of any one of the three broadly tuned color receptors is ambiguous, but the relative activities of all three allow one precisely to discriminate a very large number of colors. Note the economy of this form of encoding: it is possible to form fine discriminations with only a small number of coarsely tuned units, rather than a large number of narrowly tuned, nonoverlapping units (Hinton et al., 1986). The output representation of parameters in the model under consideration here follows this coarse-tuning approach.

As the training corpus for the network, we generated 2000 images of elliptic paraboloid surfaces. Each paraboloid differed in the various parameters that were listed earlier. Reflection from the surface was Lambertian (diffusely reflecting, or matte). Illumination was also diffuse, by which we mean that although light came predominantly from a particular direction, components from all directions arose from light reflected and scattered about by the general environment of the surface. Diffuse illumination served to exclude sharp shadow edges from the training set (and no edges arose from the paraboloids of the training set, which did not have any occluding boundaries), so the network had to determine curvatures solely from cues provided by shading, and not by edges. We assumed that the illumination was always from above (light tilt between 0 and 180 degrees), and that the signs of both principal curvatures were the same (i.e., the surface was convex or concave). More details about images and preprocessing are given elsewhere (Lehky and Sejnowski, 1988).

PROPERTIES OF THE NETWORK

A network containing 122 input units, 27 hidden units, and 24 output units was trained in the manner described earlier. The 2000 input images were presented to the network one at a time, and after each presentation the connection strengths were changed slightly to make the activities of the output units match more closely the desired output activities. As the learning algorithm proceeded, the network developed organized connections between the input units and the hidden units, and between the hidden units and the output units. The hidden units therefore essentially act as a map, or transform, that converts the inputs to the desired outputs. Around 40,000 trials were required before performance of the network reached a plateau at a correlation of 0.88 between actual outputs and the correct outputs. The characteristics of the network that formed are discussed later.

Figure 17-2 shows the response of the fully developed network to a typical input image. In all cases, areas of the black squares are proportional to the activity of a unit. The two hexagonal regions at the bottom represent responses of the 122 input units, 61 on-center units, and 61 off-center units, found by convolving center-surround units with the image. Responses of the 27 units in the hidden layer are represented by the 3×9 rectangular array above the hexagons. Above that, responses of the 24 output units are shown in a 4×6 array. Finally, at the top, and separated from the rest, is another 4×6 array showing the ideal outputs to that image.

The coding of parameters for the 4×6 output array are as follows. Orientation tuning of the units changes as one moves horizontally along a row, while curvature tuning changes as one moves vertically along a column, with each of these units having the two-dimensional Gaussian curvature-orientation tuning curves described earlier. Traveling along a row, the peaks of the orientation tuning curves go from 0 to 150 degrees in 30-degree increments, and there is a a sizable

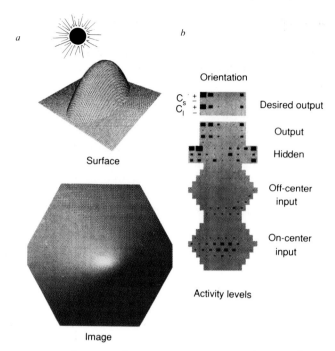

Figure 17-2   Response of a network to an image. (*a*) Typical image from a set of 2000 elliptic paraboloid surfaces used to train the network. The ratios of the two principal curvatures in this case are roughly 2:1, with the elongated axis of the surface oriented at about 5 degrees relative to horizontal. The center of the surface is shifted down and to the right, relative to the center of the overall input field of the network defined by the octagon. The illumination is coming from the upper right. (*b*) The shaded image was convolved with the Laplacian receptive field shown in Figure 17-1d at hexagonal sample points, with 61 on-centers and 61 off-centers. The area of each black square is proportional to the activity level of the unit. The converging synaptic inputs from the input layer produced a pattern of activation in the 27 hidden units, arranged in three rows of 9 units each. The hidden units, in turn projected to the output layer of 24 output units whose activity levels should be compared with the correct pattern shown above (the desired output). The 4 × 6 array of output units is arranged so that each column gives the principal orientation that produces the best output for each unit (corresponding to 0, 30, 60, 90, 120, and 150 degrees, respectively), and the row gives the sign and magnitude of each principal curvature, with C_s = small curvature and C_l = large curvature.

amount of overlap between adjacent tuning curves. The value of orientation is coded in the relative activities of all six orientations in a row, analogous to the way color is encoded. Different curvatures are represented in each of the four rows. The top two rows indicate positive or negative values for the smaller of the two principal curvatures, whereas the bottom two rows indicate the same for the other principal curvature.

However, unlike the case for orientation, the coding for curvature is ambiguous. For example, if a unit gives a small response, we do not know if the response is small because curvature is far above the peak of the curvature tuning curve for that unit or far below the peak. The ultimate cause of this ambiguity is that, in this small network, we have chosen to sample the image at only a single spatial scale. By having units in the model tuned to different, but overlapping, ranges of curvature (which would require sampling a various spatial scales), the ambiguity in curvature representation could be removed.

For the image in Figure 17-2, the output coding in the 4 × 6 array indicates that the elliptical paraboloid has principal curvatures that are both close to the peak of the curvature tuning curve (shown by the large size of some black squares) and positively valued (shown by convex surface rather than concave). The pattern of activity among the six orientation tunings indicates that the smaller principal curvature is oriented to horizontal at about 5 degrees (i.e., the elongated axis of the elliptical paraboloid is oriented at 5 degrees).

The high degree of correlation between actual and desired outputs (0.88) indicates that the network is successfully extracting the desired parameters (curvature and orientation), independent of illumination direction and the position of the surface. The next step is to examine thte patterns of synaptic weights that have developed to give these results.

Figure 17-3, called a Hinton diagram, shows all of the connections strengths for the 27 hidden units in the network. Each hidden unit is represented by one of the gray hourglass-shaped figures. Connection strengths are represented as black and white squares of varying size. The white squares are excitatory weights, the black squares are inhibitory weights, and the area of the square is proportional to the magnitude of the weight. Within the hourglass figure, two sets of connections are shown: first, the connection strengths from all input units to that particular hidden unit, and second, the connection strengths from that hidden unit to all the output units. The two hexagonal arrays on the bottom of each figure shows the connections from the on-center and off-center input arrays to the hidden unit (hidden unit receptive field), and the 4 × 6 rectangular arrays at the top are the connections from the hidden unit to units in the output layer (hidden unit projective field). Finally, the value of the bias, or negative threshold, is shown in the isolated square at the upper left corner of each hidden unit figure.

The pattern of excitatory and inhibitory connections in the two hexagonal input arrays can be interpreted as receptive fields of the hidden units. Most of the hidden units appear to be orientation-tuned to a variety of directions. These oriented fields have several excitatory and inhibitory lobes, which may occur in various phases. This is the pattern found in simple cells in cat and monkey visual cortex, which are often fit with Gabor functions (Mullikin et al., 1984). In addition to clearly orientation-selective units, a few units have receptive fields that were more or less circularly symmetric.

It is also possible to classify the hidden units on the basis of their projective fields (the 4 × 6 rectangular array at the top of each hourglass figure is Figure 17-3). Three classes of hidden units can be distinguished in this way. First, there are units that have a vertical organization in the 4 × 6 array (e.g., the one in the bottom left-hand corner of Figure 17-3). These discriminate for the orientation of the principal curvatures. Second, there are those units in which adjacent rows of the 4 × 6 array resemble each other (e.g., the unit in the top left-hand corner of Figure 17-3.) These appear to be providing information about the relative magnitudes of the two principal curvatures, but show little orientation pref-

Synaptic weights between the hidden units and the input and output layers

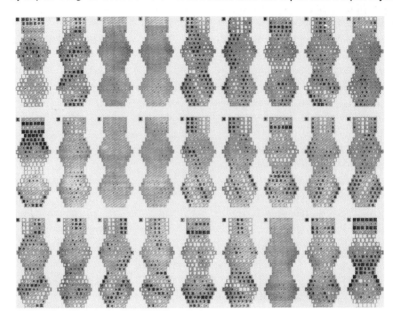

Figure 17-3 Hinton diagram showing the synaptic strengths in a network trained to extract three-dimensional curvature from images of shaded surfaces. The network contained 122 input units, 27 hidden units, and 24 output units. each of the 27 hidden units is represented by one of the gray hourglass-shaped figures, which shows the synaptic weights of all connections associated with that hidden unit, both from all the input units and to all the output units. Excitatory weights are shown as white squares and inhibitory weights as black squares; the size of the box is proportional to the magnitude of the synaptic strength. The weights from the 122 input units are given by the pair of hexagonal arrays on the bottom of each diagram and can be considered the receptive field of the hidden unit. The weights to the 24 output units are given in the 4 × 6 rectangle at the top of each figure and represent what we call the projective field of that hidden unit. The value of the bias, or negative threshold, is shown in the isolated square at the upper left corner of each hidden unit figure.

erence. Finally, there are those units in which alternative rows of the 4 × 6 array resemble each other (e.g., the first unit in the second row). These distinguish between the signs of the principal curvatures, whether they are positive or negative. In other words, they appear to be distinguishing between concave and convex surfaces.

Histograms of the activity levels in various individual hidden units in response to many different input images were collected. The hidden units of the first two classes tended to be at various intermediate levels of activity, whereas units of the third class tend to have either a very high or very low level of activity. On the basis of this difference, we would suggest that the first two classes are acting primarily as parameter filters, while the third class is acting as a feature detector for convexity/concavity.

The results of the learning procedure shown in Figure 17-3 are representative of many replications, each started with a different set of random weights. Similar patterns of receptive fields were always found, and the same three types of units could be found by examining the connections to the output units. However, there was variation in the details of the receptive fields, and in the number of units that developed no pattern of connections. It appears that only about 20 hidden units are needed to achieve the maximum performance, which was always the same average correlation of 0.88. The extra hidden units always underwent "cell death," since they served no useful role in the network and could be eliminated without changing the performance.

We have explored the responses of units in the network to bars of light, in effect conducting simulated neurophysiology. The bars were varied in position, orientation, width, and length. These bar stimuli were chosen because an extensive experimental literature uses them to study visual cortex, so that responses of real neurons are known. Although for shaded images the responses of the hidden units were generally less than half maximum, an optimally oriented and positioned bar of the same peak "luminance" provided a saturating stimulus. It was therefore necessary to use bars having low luminances to prevent responses of the network from saturating.

The responses of hidden units to bars were easily predictable from the pattern of excitatory and inhibitory connections they received from the input units. Tuning curves were measured for all stimulus parameters (bar position, orientation, width, and length). Just by looking at the pattern of connection strengths in the weights diagram (Figure 17-3), it was possible to form a good estimate of what the optimal bar stimulus would be. The ease in understanding hidden unit responses is not surprising, since the only intermediary between the stimulus and the hidden units is the array of center-surround receptive fields of the input units.

The situation was quite different when responses of output units to bar stimuli were measured. Finding an optimal stimulus required extensive trial and error. Again, this is not surprising. The responses of each output unit is determined by a weighted combination of the inputs from all 27 hidden units (although in practice a smaller number of hidden units tend to predominate). It is therefore difficult to grasp intuitively what responses to a particular stimulus will be.

Despite the complicated organization of receptive fields for output units, it was possible to obtain smooth tuning curves for the various bar parameters. The tunings were generally broader than those in hidden units. One feature of some output unit responses was the presence of strong "end-stopped inhibition." The optimal bar was often short and broad, and located off to a corner of the input field of the network. Responses dropped off precipitously when the length of this optimal bar was extended. Also, as the bar was swept across the network in a direction transverse to the optimal orientation, the output units responded over a broader region than the hidden units did (about 1.5 to 2.0 times as wide). This is not unexpected because output units receive convergent input from a number of hidden units.

The responses of units in the hidden and output layers are reminiscent of the behavior of some units that have been found in primary visual cortex (Hubel and Wiesel, 1962, 1965). The hidden units appear to have some of the properties of simple cells, and the output units tend to behave more like some types of complex cells, or perhaps hypercomplex cells. However, in drawing this analogy one should keep in mind that the relationship between hidden units and output units is strictly hierarchical (in the sense that responses of output units are entirely synthesized from inputs from the preceding hidden units), while the extent to which such a hierarchical relationship holds for simple and complex cells remains controversial in the biological literature (Ferster and Koch, 1987).

Although the response properties of the processing units in our model were similar in some respects to the properties of neurons in visual cortex, how these response properties arise could be quite different. For example, projections from the lateral geniculate are all excitatory, and within striate cortex feedforward projections are either excitatory or inhibitory (Lund, 1987; see also Chapter 10). In this model, however, we have allowed feedforward connections from a single unit to be both excitatory and inhibitory, a situation not observed in visual pathways. Also, the oriented responses of cells in visual cortex may arise in part from inhibitory interneurons (Sillito, 1975), whereas in this model orientation specificity arises entirely from the pattern of convergence from a previous stage. However, our network is not meant to be a literal model of actual cortical circuitry, which is certain to be much more complex, but rather is a model of the representation in visual cortex of information about surfaces in the visual field. The same representations could be constructed differently in different systems, as indeed the orientation tuning of neurons in different species might also have different origins although they serve the same function.

Nonetheless, we believe our model can evolve toward a more detailed account of real cortical circuitry as more information is found about detailed patterns of connectivity between different cell types in cortex. Toward this end, we have also constructed a network similar to the one presented here, except that we required connections arising from the input units (both on-center and off-center) to be purely excitatory, instead of allowing these units to form both excitatory and inhibitory connections, as was done here. This is meant to mimic the observation that principal cells in the lateral geniculate nucleus form only excitatory connection with the cortex. The average performance of this restricted network was nearly identical to that of the previous network, and the same response properties were found for the hidden units as before. Evidently, sampling the image with both on-center and off-center units provides a degree of redundancy, so that excitatory off-center units in this network can substitute for the inhibitory inputs from on-center units in the previous one, and vice versa.

DISCUSSION

If one simply looked at the various receptive fields developed by this neural network model (see Figure 17-3), one might be tempted to classify them into categories such as bar detectors or edge detectors. However, having constructed this network, we know that they are engaged in something entirely different. They are extracting information about surface curvatures from the continuous gradations of shading in an image. The network demonstrates that detecting bounding contours is not the only possible function of cells with receptive field properties such as those found in visual cortex, and it makes the more general point that it is difficult to tell what a cell is doing just be examining its receptive field.

The function of a single unit in the hidden layer of the network was revealed only when its projective field to the output units was examined. The projective field provides the missing information needed to interpret the computational role of the unit in the network, and this can be inferred only indirectly by examining the next stage of processing.

The network model provides an alternative interpretation of cortical receptive field properties, namely, that they can be used to detect shape parameters from the gradual shading rather than sharp edges. The information contained in the shaded portions of objects in images can partially activate the simple and complex cells in visual cortex, and these responses can be used to extract curvature parameters from the image. It might prove interesting to test the response properties of cortical cells with curved surfaces similar to those used here. In particular, it should be possible to record the responses of endstopped complex cells in visual cortex to shaded images. One expectation is that different cells will respond preferentially to convex or concave images.

It is worth repeating what is perhaps the most important lesson to be drawn from this entire modeling study: knowledge of the receptive field of a unit does not appear sufficient to deduce the function of that unit within a network. This study therefore raises questions about the standard interpretations of receptive fields of real neurons, in both visual areas and other sensory systems.

ACKNOWLEDGMENT

The work described here was supported by a Presidential Young Investigator Award and a grant from the Sloan Foundation to T. J. Sejnowski and G. F. Poggio.

APPENDIX 1: Back-Propagation Learning Algorithm

The properties of the nonlinear processing units used in the model network for the curvature problem include (1) the integration of diverse low-accuracy excitatory and inhibitory signals arriving from other units, (2) an output signal that is a nonlinear transformation of the total integrated input, including a threshold, and (3) a complex pattern of interconnectivity. The output of a neuron is a nonlinear function of the weighted sum of its inputs, and this can be approximated by the output function shown in Figure 17-4. This function has a sigmoid shape: it monotonically increases with input, it is 0 if the input is very negative, and it asymptotically approaches 1 as the input becomes large. This roughly describes the firing rate of a neuron as a function of its integrated input: if the input is below threshold there is no input, the firing rate increases with the input, and it saturates at a maximum firing rate. The behavior of the network does not depend critically on the details of the sigmoid function, but the one we used is given by

$$s_i = P(E_i) = \frac{1}{1 + e^{-E_i}} \qquad (17\text{-}1)$$

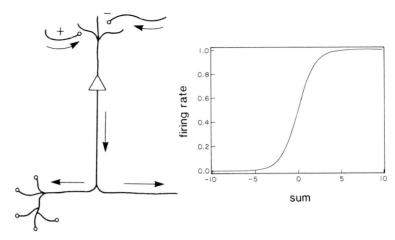

Figure 17-4   Firing rate of a model neuron as a function of its integrated synaptic input. The excitatory and inhibitory influences were summed to determine the output (normalized to lie between 0 and 1).

where s_i is the output of the ith unit and the total input E_i is

$$E_i = \sum_j w_{ij} s_j \qquad (17\text{-}2)$$

where w_{ij} is the weight from the jth to the ith unit. The weights can have positive or negative real values, representing an excitatory or inhibitory influence.

In addition to the weights connecting them, each unit also has a threshold. In some learning algorithms the thresholds can also vary. To make the notation uniform, the threshold was implemented as an ordinary weight from a special unit, called the true unit, that always had an output value of 1. This fixed bias acts as a threshold whose value is the negative of the weight.

Back-propagation is an error-correcting learning procedure that was introduced by Rumelhart et al. (1986). It works on networks with multilayered feedforward architectures. There may be direct connections between the input layer and the output layer, as well as through the hidden units. A superscript is used to denote the layer for each unit, so that $s_i^{(n)}$ is the ith unit on the nth layer. The final output layer is designated the Nth layer.

The first step is to compute the output of the network for a given input using the procedure described on successive layers. The goal of the learning procedure is to minimize the average squared error between the computed values of the output units and the correct pattern, s_i^*, provided by a teacher:

$$\text{Error} = \sum_{i=1}^{J} (s_i^* - s_i^{(N)})^2 \qquad (17\text{-}3)$$

where J is the number of units in the output layer. This is accomplished by first computing the error gradient on the output layer:

$$\delta_i^{(N)} = (s_i^* - s_i^{(N)})P'(E_i^{(N)}) \tag{17-4}$$

and then propagating it backwards through the network, layer by layer.

$$\delta_i^{(n)} = \sum_j \delta_j^{n+1} w_{ji}^{(n)} P'(E_i^{(n)}) \tag{17-5}$$

where $P'(E_i)$ is the first derivative of the function ($P(E_i)$ in Figure 17-4).

These gradients are the directions in which each weight should be altered to reduce the error for a particular item. To reduce the average error for all the input patterns, these gradients must be averaged over all the training patterns before the weights are updated. In practice, it is sufficient to average over several inputs before the weights are updated. Another method is to compute a running average of the gradient with an exponentially decaying filter:

$$\Delta w_{ij}^{(n)}(u + 1) = \alpha \Delta w_{ij}^{(n)}(u) + (1 - \alpha)\delta_i^{(n + 1)} s_j^{(n)} \tag{17-6}$$

where α is a smoothing parameter (typically 0.9) and u is the number of input patterns presented. The smoothed weight gradients $\Delta w_{ij}^{(n)}(u)$ can then be used to update the weights:

$$w_{ij}^{(n)}(t + 1) = w_{ij}^{(n)}(t) + \epsilon \Delta w_{ij}^{(n)} \tag{17-7}$$

where the t is the number of weight updates and ϵ is the learning rate (typically 1.0). The error signal was back-propagated only when the difference between the actual and desired values of the outputs was greater than a margin (typically 0.1). This ensures that the network does not overlearn on inputs that it is already getting correct. This learning algorithm can be generalized to networks with feedback connections (Rumelhart et al., 1986), but this extension will not be discussed further.

The definitions of the learning parameters here are somewhat different from those used in Rumelhart et al. (1986). In the original algorithm, ϵ is used rather than $(1 - \alpha)$ in Eq. 17-6. Our parameter α is used to smooth the gradient in a way that is independent of the learning rate, ϵ, which appears only in the weight update (Eq. 17-7). Our averaging procedure also makes it unnecessary to scale the learning rate by the number of presentations per weight update.

REFERENCES

Dobbins, A., Zucker, S. W., and Cynader, M. S. (1987). Endstopped neurons in the visual cortex as a substrate for calculating curvature. *Nature* 329:438–441.

Ferster, D., and Koch, C. (1987). Neuronal connection underlying orientation selectivity in cat visual cortex. *Trends Neurosci.* 10:487–491.

Hinton, G. E., McClelland, J. L., and Rumelhart, D. E. (1986). Distributed representations. In: D. E. Rumelhart and J. L. McClelland (eds.). *Parallel Distributed Processing: Explorations in the Microstructure of Cognition. Vol. 1: Foundations.* Cambridge: MIT Press.

Hubel, D. H., and Wiesel, T. N. (1962). Receptive fields, binocular interactions, and functional architecture in the cat's visual cortex. *J. Physiol.* 160:106–154.

Hubel, D. H., and Wiesel, T. N. (1965). Receptive fields and functional architecture in two non-striate visual areas (18 and 19) of the cat. *J. Neurophys.* 28:229–289.

Ikeuchi, K., and Horn, B. K. P. (1981). Numerical shape from shading and occluding boundaries. *Artificial Intell.* 17:141–184.

Lehky, S. R., and Sejnowski, T. J. (1988). Network model of shape-from-shading: Neural function arises from both receptive and projective fields. *Nature* 333:452–454.

Lund, J. S. (1987). Local circuit neurons of macaque monkey striate cortex: I. Neurons of laminae 4C and 5A. *J. Comp. Neurol.* 257:60–92.

Mingolla, E., and Todd, J. T. (1986). Perception of solid shape from shading. *Biol. Cybern.* 53:137–151.

Mullikin, W.H., Jones, J. P., and Palmer, L. A. (1984). Periodic simple cells in cat area 17. *J. Neurophys.* 52:372–387.

Pentland, A. P. (1984). Local shading analysis. *IEEE Trans. Pattern Anal. Mach. Intell.* 6:170–187.

Pentland, A. P. (1986). Perceptual organization and the representation of natural form. *Artificial Intell.* 28:293–331.

Rumelhart, D. E., Hinton, G. E., and Williams, R. J. (1986). Learning internal representations by error propagation, In: D. E. Rumelhart and J. L. McClelland, *Parallel Distributed Processing: Explorations in the Microstructure of Cognition. Vol. 1: Foundations.* Cambridge: MIT Press.

Sillito, A. M. (1975). The contribution of inhibitory mechanisms to the receptive field properties of neurones in cat striate cortex. *J. Physiol. (Lond.)* 250:305–329.

Todd, J. T., and Mingola, E. (1983). Perception of surface curvature and direction of illumination from patterns of shading. *J. Exp. Psych.: Hum. Percept. Perform.* 9:583–595.

Wässle, H., Boycott, B. B., and Illing, R.-B. (1981). Morphology and mosaic of on- and off-beta cells in the cat retina and some functional considerations. *Proc. R. Soc. Lond. B* 212:177–195.

18

Knowledge and Natural Selection: Five Decades on the Path to Naturalizing Epistemology

GERHARD WERNER

This volume directs attention to an aspect of neuroscience that is traditionally addressed as neuroreductionism. It seeks to account for the facts of behavior and subjective experience in terms of neural processes and events. Reductionism can be viewed in two complementary ways: as elimination of one theory by another (specifically: the mental by the neurological) or as formulation of "bridging laws" that relate properties in the reduced theory to properties in the reducing (presumably more basic) theory. The doctrine of reductionism is the neurosciences has taken many forms (Churchland, 1979; Churchland, 1986; Werner, 1988), but the various approaches are largely impeded by ambiguities in the problem statement: there is lack of agreement over what is to be reduced (i.e., whether the "folk psychology" of common sense generalization of mental states is a suitable candidate for reduction; (see Stich, 1983) and uncertainty whether reductionism in the physical sciences is a suitable model to emulate (Fodor, 1968; Clark, 1980; Cummins, 1983).

This overview is intended to sketch various ideas in and around the neurosciences that have unfolded during the past half century in their historical succession, mutual interdependence, transience, and sometimes ironic twists and turns and blind alleys. Though the course has not been straight, important baselines were secured which provide the current operational perspective for empirical inquiry and remove obstacles to a principled view of reductionism in neurobiology.

THE RISE AND DECLINE OF CYBERNETICS

The intellectual climate of the 1940s is a suitable starting point. On the philosophical scene, logical empiricism reigned supreme as a result of efforts over the preceding two decades to account for the possibilities of knowledge of the external world in a rigorous manner. This position is best exemplified by Russell's logical atomism (Russell, 1922) and Wittgenstein's *Tractatus* (Wittgenstein, 1961); knowledge of reality is based on direct acquaintance with predicates of sense qualities (e.g., sense data of the form "little patch of green color"), which can be aligned by a logically perfect language into coherent and true knowledge of reality. Observation records, and logical constructions derived from them, were considered the necessary and sufficient conditions for knowing reality. The solid foundation of logic developed by Russell and Whitehead in the *Principia Mathematica* (Russell and Whitehead, 1913) offered the recipe that seemed to obviate all obstacles to generating true knowledge or to identify pseudoproblems arising from improper use of language.

In the late 1930s the applicability of this doctrine to clarifying the role of the nervous system in generating knowledge of the external world gained specific direction from two signal contributions in seemingly unrelated fields: Allan Turning's paper "On Computable Numbers with Application to the Entscheidungsproblem" (Turing, 1937), which showed that a binary code and a few simple instructions are sufficient to carry out any conceivable computable procedure, and a publication by Shannon proving that relay and switching circuits can implement the logical calculus of Boole's *Laws of Thought* (Shannon, 1938).

Considering the all-or-none principle of neuronal discharges a binary code, McCulloch and Pitts combined the doctrine of logical empiricism, Turing's principle, and Shannon's proof to the powerful theory that neuronal nets can individuate statements in propositional logic and therefore generate true knowledge of the external world from the registration of sense data (atomic sensations, in Russell's terminology). Their publication *Logic Calculus of Ideas Immanent in Nervous Activity* (McCulloch and Pitts, 1943) and several companion papers demonstrated that "activities which we are wont to call mental are rigorously deducible from present neurophysiology," given the conformity of the all-or-none law of neural activity with the logic of propositions.

The repercussions of this and related studies by McCulloch and others were considerable: at last, neuroscience seemed to have rigorous conceptual tools to elucidate mechanisms by which the nervous systems gains knowledge of the external world. The excitement over this possibility was further fueled by a paper of Rosenblueth, Wiener, and Bigelow (Rosenblueth et al., 1943) entitled *Behavior, Purpose and Teleology*, which demonstrated definitively that the notion of teleology, previously held to be a specifically biological concept, can be formulated as the engineering concept of feedback. This pioneering paper was the foundation for Wiener's influential book *Cybernetics*, published a few years later (Wiener, 1948). Concurrently, Shannon and Weaver (Shannon and Weaver, 1949) defined information as a measurable quantity, and their paradigm of information theory provide a tool of seemingly inexhaustible applicability for quantitative studies in psychology and neurophysiology.

The numerous landmark symposia held during these years of intensive fermentation [e.g., the meetings sponsored by the Josiah Macy foundation (von Foerster, 1950), and the Hixon Symposium (Jeffress, 1951)] attest to the forcefulness of the ideas that attracted scientists of many denominations, including such luminaries as the mathematician von Neuman, to the emerging interdisciplinary study of the nervous system. Needless to say, many of us whose careers in the neurosciences were then in their formative years found the fascination of the these ideas irresistible. At the threshold of the next decade, as if to complete the agenda of the 1940s, Ashby published the monograph *The Design of a Brain* (Ashby, 1952), which proved that the concepts of information theory and feedback in dynamic systems are sufficient to design machines with adaptive behavior and, accordingly, adequate for a comprehensive account of the brain's funciton.

Despite these successes, there were strains in the conceptual building: McCulloch soon realized that logics of higher order than that of the *Principia Mathematica* are required for making behavioral decisions on the basis of information received from the environment. To supplement his own untiring efforts to develop a suitable logic, he enlisted the philosopher Gunther, who has worked persistently for the past 40 years on formalizing Hegel's dialectic into a many-valued logic that would fulfill the requirements placed on the brain as a decision organ (Gunther, 1976). Similarly, von Neuman proposed that a higher-order complex logic may have to be invented to account for the mode of function of the nervous system (von Neuman, 1951).

However, more fundamental than these strains was the declaration of bankruptcy of logical empiricism that had served as an essential pillar [in Pepper's terms, as the "root metaphor" (Pepper, 1942) of the neurosciences in the 1940s. By 1950, some 20 years after its heyday, logical empiricism was declared dead by philosophers such as Wittgenstein (in his later work), Duhem, Quine, Goodman, Carnap, and many others (for overview, see Romanos, 1984)]. Essentially, it became apparent that the relationship between observation records (sense data) and knowledge is far more complex than was envisioned by the logical empiricists: knowledge is not stabilized by reference to isolated observations, but by the coherent network of, and interrelations with, other knowledge elements within a theory. Hence, knowledge and truth do not reside in individual propositions, but in an entire web of interrelated propositions, that is, in whole theories. Accordingly, it is no longer acceptable to base the nervous system's acquisition of knowledge on propositional logic operating on sense data. Instead, the new doctrine proclaims that knowledge of what-there-is (ontology) is relative to arbitrary decisions that assign knowledge elements (i.e., words in a descriptive language) to their referents in the world.

The discomfort caused by this relativistic ontology, which shook the theoretical edifice of the 1940s, paradoxically turned to the neurosciences' advantage. The new look in epistemology prompted authors such as Campbell (1959) and Quine (1969) to reconcile relativism of knowledge with its obvious survival value by assimilating the notion of evolution. The result was a program of evolutionary epistemology—that is, the declaration of epistemology as a branch of natural science for studying how the brain conducts its epistemic business as a necessary by-product of natural selection. Thus originated the trend of naturalizing epistemology, which continues to engage numerous investigators and provides a basis for interpreting processes of perception and cognition in the context of ethology (see e.g., Lorenz, 1943; Vollmer, 1983; Shepard, 1984).

THE ASCENDANCY OF COGNITIVE SCIENCE

Concurrently with the unfolding of this direction, an entirely different line of thought with precisely opposite consequences for the neurosciences originated with the work of A. Newell, C. Shaw, and H. Simon. First formulated in a memorandum of the Rand Corporation in 1950, it was subsequently elaborated by Newell and Simon as the physical symbol theory (Newell and Simon, 1976). The basic idea, stated here with gross oversimplification, is this: persons emit symbol strings in the form of language; hence, what goes on inside their heads (as thoughts) is also symbol strings. The designation *physical symbol theory* is intended to convey the idea that the systems under consideration obey the laws of physics. Symbols, as physical patterns, occur as components of larger entities (i.e., symbol structures or expressions). The systems also contain a collection of procedures that transform expressions into other expressions. Hence, such a system is a machine that, in time, generates an evolving succession of symbol structures. If expressions designate objects in the real world, it is then possible to choose transformations on the expressions so that the evolving symbol structures correspond to a trajectory of events in the real world. Such systems, which encompass the Turing machine concept and the spirit of Boolean logic, are postulated to meet the necessary and sufficient conditions for intelligent action. Accordingly, the physical symbol theory can be subject to empirical tests in the form of artificial intelligence, that is, computer programs that mimic human intelligent behavior. The first empirical test of this theory was conducted by Newell and Simon, who demonstrated that the computer program Logical Theorist can prove theorems in the *Principia Mathematica*.

The physical symbol theory has two implications: (1) symbolic behavior arises because humans have the character of a physical symbol system, and (2) this theory offers important conceptual tools for the study of human cognition, while pursuits at the level of neuronal processes and functions were declared unprofitable for studying cognition, at least for the time being. The rapidly gaining momentum of this theory captured the imagination of many investigators as an "intuition pump" (Dennett, 1980) for inquiries in psychology and artificial intelligence, forming what was designated as cognitive science. At the same time, the neurologically focused cybernetics of the 1940s faded into the background, although remaining alive in some organized circles (e.g., the Society for General Systems Research).

By the mid-1950s, this transition was in full swing. Cognitive science molded its vocabulary after computer science and the physical symbol theory: it defined psychological states as representations of states of the world in symbol structures (e.g., as a language of thought, see Fodor, 1975) and psychological processes as computations over these representations. The idea of representations as mental states occupies a central role in cognitive science as the medium between the organism's brain and the world. Mental states can give rise to other mental states by computational transformation, and can cause behavior whereby the representation mediates between the organism and reality (indirect realism). This is the doctrine of *computational representational functionalism*, a term that reflects the functional role of mental states that are the product of computations on representations (Block, 1980).

Functionalism has implications for the reductionist program, since descriptions of psychological states can be considered neutral with respect to the medium of their instantiation, whether a brain or machine. Hence, Putnam argued that reductionism loses its relevance; it does not make sense to reduce psychological descriptions to descriptions of their mechanistic implementation because any machine can implement them (Putnam, 1960).

With the "strong representational theory of mind" (Stich, 1983) as a reference point, four explicit postulates have been developed by its principal proponents (Pylyshyn, 1984; Fodor, 1983):

1. Mental representations are languagelike structures over which semantic relations of inference, goals, beliefs, (propositional attitudes), obtain.
2. Mental processes (propositional attitudes) cannot be accounted for by any means other than the postulated representations.
3. Central processes of representation and computation are outside the purview of mechanistic (i.e., neurologic) interpretations because of their holistic, diffuse (isotropic) nature, akin to the weblike knowledge structures of Quine's epistemology.
4. The concepts and techniques of neuroscience are applicable only to the study of the transducers of information at the input stage of the central processor, to which a stereotypical, reflexlike computational organization is attributed.

The trajectory sketched in the foregoing took an intriguing turn: starting with the physical symbol theory, which excluded neurosciences *by fiat*, strong cognitive science arrived at a *principled* exclusion of neuroscience from the study of what has traditionally been referred to as the higher neural functions.

In actual practice, cognitivism exists in many varieties, some less radical than the strong representational theory, and seeks to maintain some continuity with the behavioristic tradition (Richelle, 1986). Although it does not necessarily embrace all the implications of the strong representational theory of mind, psychology has harvested rich returns by employing the concepts of representation and computation as heuristic tools. Concurrently, computer simulations of psychological theories became the arbiters of their validity. But, although cognitivism, as one of the offspring of the physical symbol theory, flourished in psychology as the computer metaphor (Massaro, 1986), artificial intelligence as its other offspring began to call into question the sufficiency of the foundations on which it was building its edifice.

THE MANY MEANINGS OF REPRESENTATION

From an operational as well as a conceptual point of view, the practitioners of artificial intelligence came to realize that problem solving by incremental steps, following each other sequentially in a Turning machine paradigm, is but one relatively restricted problem-solving format; it applies only when the steps in the solution path and the solution space are deterministically defined. However, different conditions generally obtain because multiple independent knowledge sources are involved in the nondeterministic generation of the solution path. Among the attempts to conceptualize alternative computing architectures that would be commensurate with this condition, Minsky's idea of the society of mind (Minsky, 1986) stands out, not in the least because of its inspiration by significant analogies between machine intelligence and human performance, the latter with an eye on neurological baseline facts. Essentially, systems capable of complex performance are viewed as being composed of many relatively autonomous "agents," each capable of executing merely a minor task and interacting with but a few other agents of equally limited competence, and influencing others indirectly. An organization of this kind functions partly serially and partly with many agents acting in parallel. In this organization the notion of representation assumes a functional connotation, as the way in which agents communicate with each other in generating an action or decision, distinctly different from an abstract description of reality.

At a more fundamental level, Haugeland reminded cognitivists that the formal (syntactic) operations of computers require an interpreter to become semantically and pragmatically relevent (Haugeland, 1981). Hence, the symbol structures inside the computing device exist in two domains, as it were: the syntactic and the semantic. But the connection between these two domains requires the function of an "interpretant" that assigns to the symbol structures reference in the external world. The representation of states in the world as symbol structures in the machine is the programmer's work. This realization strips the concept of representation of the almost magical properties it seemed to have acquired in the careless generalization of the computer metaphor.

Based on different considerations, Newell, one of the architects of the physical symbol theory, also demystified representations by explicitly identifying a *knowledge level* as an essential aspect of artificial intelligence programs that has been tacitly included in the computation–representation paradigm: it ensures the organization of programs for the achievement of goals, guided by certain assumptions about the structure of reality and a principle of rationality (Newell, 1982). This concept of knowledge level characterizes artificial intelligence programs as intentional systems, in Dennett's sense (Dennett, 1978), whose behavior can be ascribed to the possession of certain information, and rational operations guided by certain suppositions and goals; this is no less than Brentano's original claim for intentionality as the dividing line between the physical and mental (Brentano, 1874). Among the beliefs of the knowledge level is a commitment to an ontology in which elements of the symbol structure individuate objects and events in the world in accord with the structure of language: the "ontological relativity thesis" of Quine points to the essential arbitrariness of this assignment (Quine, 1960).

These considerations epitomize the central issue: whatever truth, meaning and reference the representation internal to the system may have, it is derivative in the sense of being supplied by a programmer or user (Searle, 1980). As far as organisms are concerned, the return to information-processing considerations is now thought by some (see following) to offer the missing link the programmer provides to the machine: the success of this endeavor would settle the perennial philosophical problem of the "mystery of the original meaning" for grounding the reference of concepts in reality (Haugeland, 1985).

One of the aspirations of the 1950s was to find some way of extending Shannon and Weaver's information measure of the capacity of communication channels (Shannon and Weaver, 1949) to characterize information content (i.e., semantic information; see MacKay, 1969). Acknowledgement of the need to establish reference for the cognitivist's representations motivated a return to these aspirations. Sayre places the emphasis on the dyadic coupling between observing agent and observed environment (Sayre, 1986). Pylyshyn referred to the same notion in considering the implications of equipping a machine with transducers and allowing it to interact freely with both natural and linguistic environments (Pylyshyn, 1978). Under these conditions, representations can be generated internal to the machine, making derivative interpretations superfluous. Dretske (Dretske, 1986) seems to allude to the same idea by suggesting, as does Sayre, that robotics may be a more suitable "intuition pump" for cognitive science than artificial intelligence.

The point of this digression is to substantiate the view that a narrow interpretation of the physical symbol theory, with its fundamentalist extrapolation to cognitive science may turn out to have been premature; although it inspired a great number of studies in psychology, at the same time, it stood in the way of a naturalistic and biological foundation of cognitivism.

Beginning in the 1970s, signs of discomfort over the state of research in psychology began to be expressed in various circles. Newell put the dilemma bluntly in a paper entitled "You Can't Play 20 Questions with Nature and Win" (Newell, 1973). His argument was forceful and clear: explanations of observed phenomena are apt to be viewed as binary decisions between opposite alternatives in a conceptual system, without adequate regard being given to the task structure and the subject's process mechanism. In 1973 he seemed to anticipate, with this criticism of contemporary psychology, what his concept of the knowledge level was intended to remedy for artificial intelligence in 1980. In any case, the advent of the computer metaphor had provided a neat conceptual framework, suited for explaining psychological phenomena as dichotomous decisions (e.g., propositional versus analog representations, serial versus parallel processing). However, the tacit assumptions of the framework pointed to weaknesses in cognitivism's edifice.

In a series of publications, Turvey and colleagues argued that computational models based on the Turing machine principle, unlike organisms, are prisoners of the representational range with which they were initially endowed (Carello et al., 1984). This is but one example of the growing literature on "uncomputability" in relation to thinking and learning (Kugel, 1986). Representations that symbolize in a languagelike medium (predicates, feature lists) have also become the target of criticism: their range of action is though to be restricted to decisions between predicates supplied in advance (Balzano and McCabe, 1986). These and related assessments are reminiscent of the facetious criticisms directed by behaviorists against Tolman's notorious rat, sitting in the corner of its cage and "thinking" while starving to death (Tolman, 1932). The call was for open systems to accommodate novelty in transactions with an ever-changing environment—shades of a much-debated issue of the cybernetic age (von Foerster, 1960).

Motivated in part by the wish to escape the fate of Tolman's rat, efforts were set in motion to recruit the ecological outlook Gibson had persistently, and in relative isolation, pursued for several decades (Gibson, 1966). U. Neisser's book *Cognitive Psychology*, published in 1967 as a landmark in the emergence of cognitive psychology, was superseded in 1976 by the same author's ecologically inspired book *Cognition and Reality*. Ecological psychology speaks of spatiotemporal events as information structures that are actively and directly obtained by the organism in reciprocity with its environment. In its original and rigorous form, ecological psychology has no place for representation or computation. The traditional duality between observer and observed is repudiated in favor of a "direct pick-up" of information by the organism, which is specifically "attuned" to its ecological niche (for an excellent overview, see Michaels and Carello, 1981). The current spectrum of discussions ranges from attempts to accommodate aspects of the ecological position as transplants into cognitivism to the absorption of the host (cognitivism) by the transplant. On the conservative side, the question is whether substitution of the languagelike representations of cognitivism by events described as "cognitive units" would remedy some flaws (Jenkins, 1980), or whether more radical redefinition of the concept of representation in cognitivism is needed (Hoffman and Nead, 1983) or perhaps a more inclusive concept of symbols (Kolers and Smythe, 1984). Are these more cases of Newell's 20-question game—this time between computational-representational cognitivism and the fundamentally anticomputation-representation stance of ecological psychology?

Practicing neuroscientists are sometimes at risk of adopting a position of isolationism in relation to these conceptual complexities. The risk lies in adhering to experimental paradigms that, in a seemingly unobtrusive way, inherit suppositions from a shared territory with other fields of study; consider, for instance, the concept of stimulus as predicate (feature) in contrast to the ecologist's events (Werner, 1988). Shared vocabulary with other disciplines is another liability: the term *representation* is one such instance. Neurophysiologists have long used representation in an operational sense—neural activity in the central nervous system represents (i.e., signals) the impact of energy on peripheral receptors; a sensory surface is topologically represented by a map in the nervous system. Palmer's incisive analysis identified a dozen ways in which the term representation is applied for different purposes (Palmer, 1979): a common denominator is the existence of two related, but functionally separate domains. In a neurologically grounded cognitive science, representations form a cascade from the external world to the neural to the mental domain. Ecologists sought to do away with this complexity by describing the physical world in psychological terms and negating representations.

What about the representation of which the neurophysiologist speaks? The habitual pattern of describing neurophysiological observations of stimulus-evoked activity in the nervous system defines the experimental situation as a tripartite system: the brain under study, the environmental event that constitutes the stimulus, and the experimenter as the observer of the environment–brain interaction. The observer can characterize this interaction as an information transfer from environment to organism. As the agent who monitors both, he can describe this relation as a representation of the external event in the brain. From this perspective, representation is an observer-specific descriptive construct. This is inherent in the grammar of the word representation, which requires two referents and an arbiter (the latter often not explicity identified): "A represents B, as judged by C." Maturana has insisted that the ascription of any other signficance to a neural representation qua organism as explanation of regularities in its behavior is false; if the observer is removed from the triade, the organism's responses to the environment are strictly accountable as states of activity in its neurons, set in motion by the external pertubation and recursively attaining a new equilibrium state, primarily determined by the organism's internal structure and organization (Maturana, 1970). In the dyadic environment–brain system, the brain is ignorant of a representandum that can be known only by an observer beholding both the environment and the organism's response (Maturana and Varela, 1980).

The consequences of this analysis of the meaning of representation in the neurophysiologist's vocabulary are significant; a specific case is the elucidation of the notion of "local sign," first raised in Lotze's *Psychophysics* (Lotze, 1884). In the neurophysiologist's terms, the question that coresponds to Lotze's problem is this: How can a place label be assigned to an external event that is signaled by impulses carried in peripheral nerves? Koenderink addressed this question in three parts: (1) environmental events are registered by the brain as changing fluxes of nerve impulses in bundles of peripheral nerve fibers, (2) neural activity intrinsic to the nervous system is the only informational currency available (this is a corollary of the foregoing comments on the observer fallacy of neural representations), and (3) the distribution of activity over a matrix of neural elements (e.g., a fiber bundle) confers geometric structure to the disposition of the peripheral receptors. Correlated activity in two adjacent neural elements implies proximity of the peripheral sites from which activity originated, and correlation of activity in one element with that of others in a group implies inclusion (Koenderink, 1984a).

The implications are significant. The "wiring diagram of the hardware" determines the possible orders of simultaneity and succession in which neurons in the receiving matrix can fire when stimuli impinge on the receptive surface; and the totality of these orders of neural activity are isomorphic with some abstract space. This is the only sense in which it is permissible to say that a matrix of neural elements carries (represents) geometric information. Local sign and geometric order are based on a logical calculus on the possible relations of simultaneity and succession of signals in afferent nerves. The pattern of relations is one of partial order, and partial orders are closely related to geometries.

Koenderink has supported this idea with rigorous mathematical arguments (Koenderink, 1984b, 1984c). In the most general sense, the implication is that the semantics of the system resides in its hardward structure (i.e., the neighborhood relations in the afferent channels). These theoretical considerations provide a context for attributing specific functional significance to the experimental findings that the nervous system shuffles and sorts afferent pathways in their ascending course, according to modality and topological criteria (Werner, 1970); these processes are the ways in which the nervous system derives its own semantics from its evolution.

For the neuroscientist, these reflections mandate a restriction of the scope of representation, relative to the form in which it has become a centerpiece in cognitivism: the meaning of representation is restricted to a figure of speech to describe the organism–environment dyad as if the organism harbored a reified model of the world. Correspondingly, Winograd considered the conceptual restructuring in artificial intelligence that results from this change of perspective, notably in reference to understanding language. An utterance in language, traditionally conceived as a description of a state in the world, is turned into a perturbation in an active cognitive system that seeks to make sense of the world. The issue is not that the utterance has meaning because it corresponds to a referent in the world, but that it acquires meaning by virtue of "a triggering process within the speaker whose cognitive structure depends on prior history and current process activity" (Winograd, 1980). This is a surprising turn toward hermeneutics, the study of the dialectic process of interpretation; the cross-references between Maturana's concept of the nervous system and hermeneutics are that the external world (like the text in hermeneutics) encounters and interacts with an organism (reader) in an initial state, and then sets a process of changes in motion, which is determined by prior history and internal structure. This process is said to be the "hermeneutic circle" (for overview of hermeneutics, see Palmer, 1969).

Once again, this digression is intended to point to the interdigitation of the neuroscientist's epistemic concerns with the totality of endeavors that seek to illuminate what it means to know of the world.

It is another ironic twist in the history of neurosciences that the emerging reservations about the cognitivist's view of representation were already circumvented in the cybernetic age, long before the ascendancy of cognitive science. Donald MacKay proposed in the mid-1950s considering the nervous system (or any part of it) in terms of state transitions. An initial state is characterized by some probability distribution function that specifies transitions into another state, given a certain external event. The probability distribution function is determined by the nervous system's internal structure and its past history. Successful behavior obtains when the state transition leads to an adaptive response (MacKay, 1956). The cyclic nature of this process is readily apparent, because any response to the environment generates a new external event to which the nervous system reacts with another change of state, and so on. Nervous system and environment form a dyadic system in constant flux, converging to mutual adaptation. Hence, the world is not represented as an analog model of sorts (as the cognitivist would have it), but rather as a "matching state of conditional readiness for action," based on the nervous system's internal structure, which evolved in part with its coupling with the world (MacKay, 1984).

WHAT KIND OF COMPUTATION?

Thus far in this overview, the plausibility of the cognitivist's view of representation has been the principal subject of critical assessment. The second fundamental notion of cognitivism—namely, computation—has also not remained imune from examination of its appropriateness to information processing by the nervous system. Restrictions on computability were traditionally based on Turing's universal machine concept, but as more ambitious artificial intelligence programs were considered, pragmatic criteria were included in the constraint of what is computable. Feasible bounds on time and storage requirements became a realistic limit (Feldman and Ballard, 1982). Likewise, the discrepancy between time requirements in the serial machine (often referred to as the "von Neuman architecture") and the speed of behavioral decision making by organisms in complex situations became an undeniable obstacle to adopting the Turing model for the brain (Posner, 1978). Concepts of computation in parallel (Fahlman, 1979; Minsky, 1986) stimulated technological developments to build massively parallel computers and increased interest in the basic properties of computation in connection machines (Hillis, 1985). Neurologically inspired information processing models (Arbib, 1981) and new models of cognitive processes (Anderson et al., 1977; Ratcliff, 1978; McClelland and Rumelhart, 1981) added strong appeal to the systematic study of systems with parallel distributed processing capability (Hinton and Anderson, 1981). Several essential premises are the reference points for fixing the ideas about this currently very active line of pursuit (Rumelhart and McClelland, 1986; McClelland and Rumelhart, 1981), whose history dates back to the 1950s (Rosenblatt, 1962) and has been enriched by Grossberg's steady pursuit of mathematical foundations (Grossberg, 1980). The basic configuration is a set of processing units with a specified pattern of connectivity that determines the response of the system to an input from the environment. The degree to which activity in one unit influences activity in other elements (strength of connection) is subject to modification by an algorithm and is also a function of the prior history of the system; in this sense, the system's memory is in the strength of the connections, and this determines the course of further processing. Unlike the propositional calculus over sentential representations, the dominant mode of computation is an iterative computation of connection strengths, with the system as a whole settling into a solution that satisfies multiple weak constraints. The entire system describes the environment as a stable (emergent) state of activity in the interconnected units; this form of representation is actively maintained in interaction with the environment by the connection strengths between processing units and is available as a readiness to respond with an input-specific pattern of activity, distributed over the system's elements.

The conceptual implications of parallel computation extend well beyond a switch in computational architecture and its consequences for the notion of representation; they concern the basic position in the philosophy of mind, which had become the bedrock of cognitivism. Functionalism, introduced in this overview in connection with the strong representational theory of mind, posits that mental states are a form of software, capable of giving rise to other mental states, irrespective of the material implementation in any particular hardware; a simplistic simile is a computer program that runs on any kind of machine of a certain type of architecture. Although subject to some criticisms, notably for neglecting the specific characteristics of conscious experience (Block, 1978; Churchland, 1984), functionalism became a prevalent doctrine, undoubtedly drawing strength from successes with the applications of the physical symbol theory in artificial intelligence; for instance, much weight was given to apparent formal similarities between thinking by humans and machines. But here is the snag: in parallel computation, hardware and software are much more interdependent than functionalism allows. The al-

gorithms devised to operate the connection machine (Hillis, 1985), the Boltzman
machine (Hinton and Sejnowski, 1986), or simulations of parallel computation
(Fahlman, 1979) are vastly different. Hence, Thagard concludes that "hardware
may well matter to the mental" (Thagard, 1986), suggesting that the specifics of
neural connectivity are an important determiner of the possible algorithms the
brain can sustain, and there may be algorithms that are specific for the brain's
hardware.

Thinking in terms of parallel distributed processing models has an intuitive
appeal and offers substantive contributions to the concept of emergent properties,
especially as an escape hatch from stalemates in the reductionism debate. Cognition can be studied as the consequence of interactions between connected elements in structures organized in the style of the nervous system. Interactions of
this type give rise to phenomena requiring their own language of description,
which is not predictable from the behavior of its component elements in isolation.
Describing all that can be known about the brain does not furnish an adequate
vocabulary for describing behavior. On the other hand, a phenomenological description of behavior remains a crude and, possibly misleading, approximation
unless it is grounded in the description of brain states. This reflection seems to
put cognitivism in a balanced perspective—it provides a macrotheory, developed
in deliberate disregard for a microtheory. The times seems ripe to dig the tunnel
from both ends.

The lesson of the last 50 years seems clear: the arguments that led the cognitivists to assign a highly restricted role to the neurosciences in the partnership
for establishing knowledge of the world is now seen to rest on inadequate premises. Dispelling the fallacies of doctrinaire cognitivism reaffirms the soundness of
the continuous thread that links the empirical study of organizing principles of the
nervous system (Edelman and Mountcastle, 1978) with Sherrington's vision in
Man on His Nature. Speaking of the "perceiving mind" and the "perceived world,"
he wrote "nature in evolving us makes them two parts of the knowledge of one
mind and that mind our own. We are the tie between them" (Sherrington, 1951).
This is the blueprint of a naturalized epistemology as an empirical science.

REFERENCES

Anderson, J. A., Silverstein, J. W., Ritz, S. A., and Jones, R. S. (1977). Distinctive
    features, categorical perception and probability learning: some applications of neural
    model. *Psych. Rev.,* 84:413–451.
Arbib, M. (1981). Perceptual structures and distributed motor control. In: B. V. Brooks,
    (ed.). *Handbook of Physiology, Vol. II,* Bethesda, MD: American Physiological
    Society.
Ashby, W. R. (1952). *Design for a Brain.* London: Chapman and Hall.
Balzano, G. J., and McCabe, V. (1986). An ecological perspective on concepts and cog-
    nition. In: V. McCabe and G. J. Balzano (eds.) *Event Cognition: An Ecological
    Perspective.* Hillsdale, NJ: Lawrence Erlbaum.
Block, N. (1978). Troubles with functionalism. In: C. W. Savage (ed.). *Perception and
    Cognition,* Minneapolis: University of Minnesota Press.

Block, N. (1980). Introduction: what is functionalism? In: N. Block (ed.). *Readings in Philosophy of Psychology, Vol. 1*, Cambridge: Harvard University Press.

Brentano, F. (1874). *Psychologie from Empirischen Standpunkt*. Leipzig: Leipzig.

Campbell, D. T. (1959). Methodological suggestions from a comparative psychology of knowledge processes. *Inquiry* 2:152–182.

Carello, C., Truvey, M. T., Kugler, P. N., and Shaw, R. E. (1984). Inadequacies of the computer metaphor. In: M. S. Gazzaniga (ed.). *Handbook of Cognitive Neuroscience*, New York: Plenum Press.

Churchland, P. M. (1979). *Scientific Realism and the Plasticity of Mind*. Cambridge: Cambridge University Press.

Churchland, P. M. (1984). *Matter and Consciousness*. Cambridge: MIT Press.

Churchland, P. S. (1986). *Neurophilosophy*. Cambridge: MIT Press.

Clark, A. (1980). *Psychological Models and Neural Mechanisms*. Oxford: Clarendon Press.

Cummins, R. (1983). *The Nature of Psychological Explanation*. Cambridge: MIT Press.

Dennett, D. C. (1978). Intentional systems. In: D. C. Dennett (ed.). *Brainstorms*. Cambridge: MIT Press.

Dennett, D. C. (1980). The milk of human intentionality. *Behav. Brain Sci.* 3:428–430.

Dretske, F. (1986). Minds, machines and meaning. In: C. U. Mitcham and A. Huning (eds.). *Boston Studies in the Philosophy of Science. Vol. 90*. and *Philosophy and Technology II*, Dordrecht: D. Reidel.

Edelman, G. M., and Mountcastle, V. B. M. (1978). *The Mindful Brain*. Cambridge: MIT Press.

Fahlman, S. E. (1979). *NETL: A System for Representing and Using Real-World Knowledge*. Cambridge: MIT Press.

Feldman, J. A., and Ballard, D. H. (1982). Connectionist models and their properties. *Cogn. Sci.* 6:205–54.

Fodor, J. A. (1968). *Psychological Explanation*. New York: Random House.

Fodor, J. A. (1975). *The Language of Thought*. New York: T. Y. Cromwell.

Fodor, J. A. (1983). *The Modularity of Mind*. Cambridge: MIT Press.

Gibson, J. J. (1966). *The Senses Considered as Perceptual Systems*. Boston: Houghton Mifflin Co.

Grossberg, S. (1980). How does the brain build a cognitive? *Psychol. Rev.* 87:1–51.

Gunther, G. (1976). *Beitraege zur Grundlegung einer operationsfaehigen Dialektik*. Hamburg: Felix Meiner.

Haugeland, J. (1981). Semantic engines. In: J. Haugeland (ed.). *Mind Design*. Cambridge: MIT Press.

Haugeland, J. (1985). *Artificial Intelligence: The Very Idea*. Cambridge: MIT Press.

Hillis, W. D. (1985). *The Connection Machine*. Cambridge: MIT Press.

Hinton, E. G., and Anderson, J. A. (1981). *Parallel Models of Associative Memory*. Hillsdale, NJ: Lawrence Erlbaum.

Hinton, E. G., and Sejnowski, T. J. (1986). Learning and relearning in Boltzman machines. In: D. E. Rumelhart and J. L. McClelland (eds.). *Parallel Distributed Processing, Vol. 1*. Cambridge: MIT Press.

Hoffman, R. R., and Nead, J. M. (1983). General contextualism, ecological science and cognitive research. *J. Mind Behav.* 4:507–559.

Jeffress, L. A. (1951). *The Hixon Symposium. Cerebral Mechanisms of Behavior*. New York: John Wiley & Sons.

Jenkins, J. J. (1980). Can we have a fruitful cognitive psychology? In: J. H. Flowers (ed.). *Nebraska Symposium on Motiviation-Cognitive Processes*. Lincoln, NE: University of Nebraska Press.

Koenderink, J. J. (1984a). The concept of local sign. In: A. J. van Doorn, W. A. van de

Grind, and J. J. Koenderink (eds.). *Limits in Perception*. Utrecht: VNU Science Press.

Koenderink, J. J. (1984b). Simultaneous order in nervous nets from a functional standpoint. *Biol. Cybern.* 50:35–41.

Koenderink, J. J. (1984c). Geometrical structures determined by the functional order in nervous nets. *Biol. Cybern.* 5:43–50.

Kolers, P. A., and Smythe, W. E. (1984). Symbol manipulation: alternatives to the computational view of mind. *J. Verb. Learn. Verb. Behav.* 23:289–314.

Kugel, P. (1986). Thinking may be more than computing. *Cognition* 22:137–198.

Lorenz, K. (1943). Die angeborenen formen moeglicher erfrung. *Ziet. Tierpsychologie* 5:235–409.

Lotze, H. (1884). *Mikrokosmos*. Leipzig: Hirzel.

MacKay, D. M. (1956). Towards an information-flow model of human behavior. *Br. J. Psychol.* 47:30–43.

MacKay, D. M. (1969). *Information, Mechanism and Meaning*. Cambridge: MIT Press.

MacKay, D. M. (1984). Mind talk and brain talk. In: M. S. Gazzaniga (ed.). *Handbook of Cognitive Neuroscience*. New York: Plenum Press.

Massaro, D. W. (1986). The computer as a metaphor for psychological inquiry: considerations and recommendations. *Behav. Res. Meth. Instr. Compu.* 18:73–92.

Maturana, H. R. (1970). Neurophysiology of cognition. In: P. Garvin (ed.). *Cognition: A Multiple View*. Washington, DC: Spartan Press.

Maturana, H. R., and Varela, F. J. (1980). Autopiesis and cognition. In: R. S. Cohen, and M. W. Wartofsky (eds.). *Boston Studies in the Philosophy of Science, vol. 42*, Boston: Reidel Pub. Co.

McClelland, J. L., and Rumelhart, D. E. (1981). An interactive activation model of the effect of context in perception. *Psychol. Rev.* 88:375–407.

McCulloch, W., and Pitts, W. (1943). A logical calculus of the ideas immanent in nervous activity. *Bull. Math. Biophys.* 5:115–133.

Michaels C. F., and Carello, C. (1981). *Direct Perception*. Englewood Cliffs, NJ: Prentice Hall.

Minsky, M. (1986). *The Society of Mind*. New York: Simon & Schuster.

Newell, A. (1973). You can't play 20 questions with nature and win. In: W. G. Chase, (ed.). *Visual Information processing*. New York: Academic Press.

Newell, a. (1982). The knowledge level. *Artificial Intell.* 18:87–127.

Newell, A., and Simon, H. A. (1976). Computer science as empirical inquiry: symbols and search. *Commun. ACM* 19:113–126.

Palmer, R. E. (1969). *Hermeneutics*. Evanston, IL: Northwestern University Press.

Palmer, S. E. (1979). Fundamental aspect of cognitive representation. In: E. Rosch and B. B. Lloyd, (eds.). *Cognition and Categorization*. Hillsdale, NJ: Lawrence Erlbaum.

Pepper, S. C. (1942). *World Hypotheses*. Berkeley: Univeristy of California Press.

Posner, M. I. (1978). *Chronometric Explorations of the Mind*. Hillsdale, NJ: Lawrence Erlbaum.

Putnam, H. (1960). Minds and machines. In: S. Hook, (ed.). *Dimensions of Mind*. New York: New York University Press.

Pylyshyn, Z. W. (1978). Computational models and empirical constraints. *Behav. Brain Sci.* 1:93–127.

Pylyshyn, Z. W. (1984). *Computation and Cognition*. Cambridge: MIT Press.

Quine, W. V. O. (1960). *World and Object*. Cambridge: MIT Press.

Quine, W. V. O. (1969). *Ontological Relativity and Other Essays*. New York: Columbia University Press.

Ratcliff, R. A. (1978). A theory of memory retrieval. *Psychol. Rev.* 85:59–108.

Richelle, M. N. (1986). Some varieties of cognitivism. In: P. Eelen, and O. Fontane, (eds.). *Beyond the Conditioning Framework*, Hillsdale, NJ: Lawrence Erlbaum.

Romanos, G. D. (1984). *Quine and Analytic Philosophy*. Cambridge: MIT Press.

Rosenblatt, F. (1962). *Principles of Neurodynamics*. New York: Spartan Press.

Rosenblueth, A., Wiener, N. and Bigelow, J. (1943). Behavior, teleology and purpose. *Phil. Sci.* 10:18–24.

Rumelhart, D. E., and McClelland, J. L. (1986). *Parallel Distributed Processing, vol. 1*, Cambridge: MIT Press.

Russell, B. (1922). *Our Knowledge of the External World*. London: Allen & Unwin.

Russell, B., and Whitehead, A. N. (1913). *Principia Mathematica*. Cambridge: Cambridge University Press.

Sayre, K. M. (1986). Intentionality and information processing: an alternative model for cognitive science. *Behav. Brain Sci.* 9:121–166.

Searle, J. R. (1980). Minds, brains and programs. *Behav. Brain Sci.* 3:417–424.

Shannon, C. E. (1938). A symbolic analysis of relay and switching curcuits. *Trans. AIEE Inst.* 57:1–11.

Shannon, C. E., and Weaver, W. (1949). *The Mathematical Theory of Communication*. Urbana, IL: University of Illinois Press.

Shepard, R. N. (1984). Ecological constraints on internal representation: resonant kinematics of perceiving, imagining, thinking, and dreaming. *Psychol. Rev.* 91:417–447.

Sherrington, C. (1951). *Man on His Nature*. Cambridge: Cambridge University Press.

Stich, S. P. (1983). *From Folk Psychology to Cognitive Science*. Cambridge: MIT Press.

Thagard, P. (1986). Parallel computation and the mind-body problem. *Cog. Sci.* 10:301–318.

Tolman, E. C. (1932). *Purposive Behavior in Animals and Man*. New York: Century.

Turing, A. M. (1937). On computable numbers, with an application to the entscheidungsproblem. *Proc. Lond. Math. Soc.* 42:230–265.

Vollmer, G. (1983). *Evolutionaere Erkenntnistheorie*. Stuttgart: S. Hirzel.

von Foerster, H., (1950). *Cybernetics: Transactions of the 7th Conference*. New York: Josiah Macy Jr. Foundation.

von Foerster, H. (1960). On self-organizing systems and their environments. In: M. C. Yovitz and S. Camerson, (eds.). *Self-organizing Systems*. New York: Pergamon Press.

von Neuman, J. (1951). The general and logical theory of automata. In: L. A. Jeffress (ed.). *Cerebral Mechanisms of Behavior*. New York: John Wiley & Sons.

Werner, G. (1970). The topology of the body representation in the somatic afferent pathway. In: F. O. Schmitt and F. G. Worden (eds.). *The Neurosciences—Second Study Program*. Cambridge: MIT Press.

Werner, G. (1988). The many faces of neuroreductionism. In: E. Basar (ed.). *Dynamics of Sensory and Cognitive Processing by the Brain*. Berlin: Springer-Verlag.

Wiener, N. (1948). *Cybernetics, or Control and Communication in the Animal and the Machine*. Cambridge: MIT Press.

Winograd, T. (1980). What does it mean to understand language? *Cogn. Sci.* 4:209–241.

Wittgenstein, L. (1961). *Tractatus Logico-Philosophicus*. London: Routledge and Kegan Paul.

Index